WADSWORTH-KTL ANA
BACTERIOLOGY MAN

SIXTH EDITION

Hannele Jousimies-Somer, Ph.D.

Paula Summanen, M.S.

Diane M. Citron, B.S.

Ellen Jo Baron, Ph.D.

Hannah M. Wexler, Ph.D.

Sydney M. Finegold, M.D.

Veterans Administration West Los Angeles Healthcare Center,
and Departments of Medicine, and Microbiology and Immunology,
UCLA School of Medicine, Los Angeles, California

National Public Health Institute (KTL)
Department of Bacteriology
Anaerobe Reference Laboratory,
Helsinki, Finland

PUBLISHING COMPANY

PUBLISHING COMPANY
P. O. Box 68
Belmont, California 94002

Managing Editor: Stuart Hoffman
Photography: Wadsworth and KTL Laboratories

Printed in Korea

Library of Congress Cataloging-in-Publication Data

Wadsworth-KTL anaerobic bacteriology manual / Hannele R. Jousimies-Somer ... [et al.] -- 6th ed.
p. cm.
Third ed. published as: Wadsworth anaerobic bacteriology manual / by Vera L. Sutter.
Includes bibliographical references and index.
ISBN 0-89863-209-9
1. Anaerobic bacteria--Identification--Handbooks, manuals, etc. I. Jousimies-Somer, Hannele. II. Sutter, Vera L., 1924- Wadsworth anaerobic bacteriology manual.
QR89.5 .W334 2002
616'.014--dc21

2001049728

The authors and the publisher do not imply an endorsement of the products mentioned in this book.

9 8 7 6 5 4 3 2 1

DEDICATION

We dedicate this sixth edition of the Wadsworth-KTL Anaerobic Bacteriology Manual (formerly the Wadsworth Anaerobic Bacteriology Manual) to the memory of a number of outstanding anaerobists of the recent and more distant past–the giant L. Pasteur; the one who first called attention to the common occurrence of a variety of anaerobes in clinical infections–M.A. Veillon; the master anaerobist, A.-R. Prévot; surgeons W.A. Altemeier and F.L. Meleney; Ivan Hall; Louis D.S. Smith, the outstanding group from Virginia Polytechnic Institute–W.E.C. (Ed) Moore, E. (Betty) Cato, and J.L. Johnson), V.R. (Bud) Dowell, Jr. of the Centers for Disease Control; T. Rosebury; and A.C. Sonnenwirth. The previous edition of this Manual was dedicated to our colleague, Vera Sutter. We also dedicate this edition to several outstanding contemporary anaerobic bacteriologists, now retired or partially retired–Henri Beerens, L.V. (Peg) Holdeman-Moore, Madeleine Sebald, and Trevor Willis. There are many others–too numerous to mention.

Contents

short
even

Preface

Anaerobic bacteria are important pathogens in many different types of infection and may be involved in a major way in certain other processes, such as the bacterial overgrowth syndrome. The most common types of infection in which anaerobes are often important are skin and soft tissue, pleuropulmonary, intra-abdominal, female genital tract, and oral and dental infections, but essentially all types of infections, at any site in the body, may involve anaerobes. Most often, anaerobes are found as part of a mixed flora but solitary infections occur as well. The indigenous flora of the body is the principal reservoir of these organisms but some clostridial infections may have an exogenous source. The spectrum of severity of anaerobic infections varies from severe and life threatening to mild. Characteristically, anaerobic infections are marked by tissue necrosis and suppuration.

On the other hand, surprisingly little is known about how our complex indigenous microflora influences our physiology and well-being; interest is picking up in this area. A great deal of research on probiotics is going on and promises new insights into body homeostasis. Unfortunately, much of the probiotic work on the past has not been scientifically rigorous and there has been much commercialism.

Although there are currently many antimicrobial agents with good activity against anaerobic bacteria, there are often certain anaerobes that display resistance (e.g., nonsporeforming anaerobic gram-positive bacilli are usually resistant to metronidazole despite the generally excellent activity of this drug). Antimicrobial resistance is an important problem with anaerobes, as with many other types of microbes. We are seeing resistance on the part of some anaerobes to virtually all of our most effective anti-anaerobic drugs.

While clinicians and microbiologists are more knowledgeable about anaerobic bacteria and their role in disease than was true a decade or two ago, there is a real tendency recently to do fewer anaerobic cultures and to minimize the workup on specimens that are cultured for these organisms. This is related to several things–availability of generally effective antimicrobials, the time it takes for useful information about the infecting flora to be generated in the laboratory, and economic considerations. Ironically, this nearsighted approach usually involves administration of one of our top-flight antibiotics, which may actually lead to increased expense; it also leads to increased resistance to these drugs. If this minimalist trend is allowed to continue, it is easy to predict that these problems may be endangering this important branch of bacteriology by means of deterioration of the skills of microbiologists, and jeopardizing the principles of sound antimicrobial usage policies in hospitals and the community.

Preface

The approach in this Manual, directed primarily at the laboratorian, emphasizes practical schemes for growing, identifying, and susceptibility testing of the most commonly encountered and clinically most important anaerobes. Classical techniques of anaerobic bacteriology are still very important. This new edition of the Manual also stresses cost considerations and rapid identification methodology. Certain very fastidious anaerobes require specialized techniques, but these delicate organisms are not often involved in significant infection. We do present some material on the more rigorous procedures that are crucial for studies of indigenous flora and the newer identification techniques. The manual deals mainly with organisms of importance in human infection but gives some practical information on identification of the key anaerobic organisms of animal origin that are implicated in bite-wound infections.

This Manual, as indicated by the new name, is now a joint project of the Wadsworth VA Anaerobic Bacteriology Research Laboratory of Los Angeles and the Anaerobe Reference Laboratory of the National Public Health Institute (KTL) of Finland, located in Helsinki. We welcome our new collaboration warmly. The two laboratories have been engaged jointly in a variety of research projects of mutual interest over two decades and a number of personnel have spent time in both laboratories.

This monograph was first published almost three decades ago. Many individuals, too numerous to name, have been directly involved in producing the sixth edition or have made valuable suggestions that have resulted in continuing improvements. We express our sincere appreciation to them. We do offer a special note of gratitude to Marja-Liisa Väisänen and to Arja Kanervo for their major contributions of providing data and testing reactions. Maureen McTeague and Jane Flynn also helped a great deal with practical suggestions for the clinical laboratory and information re: oral flora. Denise Molitoris was heavily involved in antimicrobial susceptibility testing.

Sydney M. Finegold, M.D.
Hannele Jousimies-Somer, Ph.D.

chapter 1

Introduction to Anaerobic Bacteriology

Anaerobic bacteria are intimately associated with humans as indigenous microflora, outnumbering aerobes 10:1 to 1000:1 in a number of areas. Many anaerobes behave as opportunistic pathogens when they encounter a permissive environment within the host. Very few major syndromes are due to exogenous anaerobes. Special laboratory procedures have been developed for the isolation, identification, and susceptibility testing of this diverse group of bacteria; this manual presents a rational approach to that process.

Since many anaerobes grow more slowly than do facultative or aerobic bacteria, and particularly since clinical specimens yielding anaerobic bacteria commonly contain several organisms and sometimes very complex mixtures, considerable time may elapse before the laboratory is able to provide a final report. Indeed, at times it may literally take weeks before certain specimens with a complex flora are worked up definitively. Questions then arise as to whether the bacteriologic data are really beneficial to the clinician and whether it is reasonable to do detailed analyses on very complex specimens.

Aside from the time factor, what data are useful to the clinician? Is the clinician interested in accurate species and subspecies identification, or will general identification to clinically meaningful taxa or groupings, with empiric or determined susceptibility data, suffice to permit effective treatment of the patient? Even when the clinician is interested in detailed, accurate bacteriologic data—if only for academic reasons—does cost-benefit analysis warrant providing them? Just how far should the laboratory go in processing anaerobic cultures? These are difficult questions, and the answers vary according to specific circumstances such as the patient's clinical condition and response to therapy, type and source of the specimen, type and capability of the laboratory, nature of the laboratory (routine clinical, teaching, or reference), significance (predominant growth of a clinically important species) or complexity (mixed culture of more than five species) of the growth on primary plates, and so on. Hard and fast rules cannot be easily determined. Nevertheless, all laboratories performing anaerobic cultures should be able to isolate and store strains and have access to reference laboratory services when indicated.

It is mandatory that the clinician begin therapy empirically, with the assistance of information obtained from Gram-stained preparations of appropriate specimens. Essentially all mixed aerobic-anaerobic infections are of endogenous origin; therefore, the clinician or his/her consultant microbiologist must know the usual infecting flora in various specific infections, the composition of the normal flora at the site where an infection arises, and the ways in which that flora may have been modified by various disease states, prior antimicrobial prophylaxis or therapy, and other factors.

In addition, the clinician or microbiologist must be aware of the usual antimicrobial susceptibility patterns of the organisms that might be involved in infection at various sites in the body. This information includes knowledge of the patterns of susceptibility in the particular hospital in which the clinician practices, since there will be variability from one institution to another, depending primarily on patterns of antimicrobial use.

Clearly, it is important that the laboratory provide as much information as possible as soon as it can after receipt of the specimen. This is particularly true, of course, in the case of very ill patients. It is the physician's responsibility to call such patients to the attention of the microbiologist. A series of reports on each individual specimen would provide the clinician with information in optimal fashion. The initial report, at least in the case of very sick patients, should be an interpretation of the Gram stain and any other direct examination of the specimen. The microbiologist must not hesitate to express judgment on likely possibilities, not only from direct examination, but (later) from observation of colonial characteristics and cellular morphology. This presumes, of course, that the microbiologist is well informed and applies rational judgment.

After 18 to 24 hours, examination of aerobic cultures (and certain special anaerobic cultures such as those on selective media for the *B. fragilis* group and clostridia) permits the microbiologist to provide a more reliable interim report. Similarly, examination of routine anaerobic cultures after 48 hours provides additional information, especially when selective and/or differential media are used. A number of rapid diagnostic procedures may be employed to expedite presumptive or definitive identification. In this sixth edition we provide in Chapter 5 procedures for rapid and cost-efficient identification not only for the use of routine hospital laboratories but also as the first approach to more definitive identification for more advanced laboratories.

THE INDIGENOUS FLORA

A knowledge of the presence of specific anaerobes as normal flora at various sites in the body is important in several ways. Since most anaerobic infections arise in proximity to mucosal surfaces, where anaerobes predominate as normal flora, information on which organisms make up the indigenous flora at these sites enables one to anticipate the presence of certain organisms in particular specimens and thus to assist the clinician in choosing the proper drugs for initiating therapy. This information also helps the microbiologist choose selective and other media that might be particularly useful. Knowledge of the normal flora of various regions may also allow one to judge more readily whether a given isolate is significant. For example, *Propionibacterium* in a single blood culture most often represents contamination from the patient's skin, particularly if growth does not appear until after several days. Conversely, the presence of a particular organism in blood cultures may suggest the portal of entry for the bacteremia and even the patient's underlying condition.

Table 1-1 indicates the incidence of certain anaerobes as normal flora at various sites in humans. Further details are provided below. Additional data on anaerobes as indigenous flora are found in Rosebury's classic book (277) and in other references (23;112;115;147;148;151;152;263;300;315;324;337;374). The three major sites of normal colonization of mucosal surfaces are the oral cavity, the intestinal tract, and the female genital tract. The skin is a lesser site. Infections relating to these sites of carriage typically involve organisms found normally at those locations as well as other organisms (usually nosocomial pathogens) that colonize patients in the hospital setting. Organisms introduced in the course of surgical procedures or various manipulations may also play a role, of course.

The Oral Flora

The indigenous oral flora consists of various streptococci, particularly those of the viridans group (including the anginosus group, or former "*Streptococcus milleri*" group, *S. anginosus, S. constellatus,* and *S. intermedius*) and *Abiotrophia* and *Granulicatella* species (formerly called nutritionally variant streptococci), as well as various nonsporeforming anaerobic bacteria (186;277;288;300;324;335;337). Colonization with anaerobic bacteria starts soon after birth and the species diversity steadily increases with a certain order or rank (184;186). The nonsporeforming anaerobes include various *Prevotella, Bacteroides, Porphyromonas, Fusobacterium, Campylobacter,* and *Veillonella* species (among the gram-negative species), as well as *Actinomyces, Bifidobacterium, Collinsella, Eggerthella, Eubacterium, Lactobacillus, Pseudoramibacter, Slackia, Peptostreptococcus, Micromonas,* and *Finegoldia* species. The *Bacteroides fragilis* group (most strains of which are β-lactamase producers) is a very rare finding from the oral flora but it may take part in anaerobic pulmonary infections in less than 10% of patients with such infections (73). Other gram-negative anaerobic rods may also produce β-lactamase, however, and therefore may be resistant to β-lactam agents such as penicillin. Included among these are the pigmented rods in the genus *Prevotella* (30–50%), *Porphyromonas* (<10%), *Prevotella oris, P. buccae,* and the *Prevotella oralis* group (116;250). The proportion of β-lactamase-positive anaerobic gram-negative rods steadily increases by age in children during the first year of life (up to 89%) and high frequencies correlate with prior antimicrobial exposure (250). The most common *Fusobacterium* species encountered is *Fusobacterium nucleatum,* but *Fusobacterium necrophorum* is also an important pathogen and capable of causing a life-threatening infection: post-anginal sepsis syndrome or Lemierre's disease. Other organisms associated with infections in the oral cavity include *Porphyromonas gingivalis, Bacteroides forsythus,* and *Campylobacter rectus,* all common in adult periodontitis, and *Prevotella (Mitsuokella) dentalis* (root canal infections). *Actinobacillus actinomycetemcomitans,* a capnophilic gram-negative bacillus, is important in juvenile periodontitis. Species of *Capnocytophaga* and *Leptotrichia* are important pathogens in oral mucositis or associated bacteremic infections in immunocompromised hosts.

If a patient aspirates in the hospital setting (a relatively common situation), the subsequent aspiration pneumonia may involve not only anaerobic bacteria and viridans streptococci, but also such nosocomial pathogens as *Staphylococcus aureus,* various members of the family *Enterobacteriaceae,* and *Pseudomonas* species. Organisms from the oral cavity, of course, often are involved in various oral and dental infections (abscesses and cellulitis), head and neck infections, and central nervous system (CNS) infections in addition to pulmonary and pleural infections; they may also be involved in soft tissue infections in "skin poppers" (users who inject illicit drugs subcutaneously), especially if they lubricate the needle with saliva.

The Gastrointestinal Flora

The flora of the gastrointestinal tract is of interest in connection with intraabdominal infection. Normally the stomach has a sparse flora consisting typically of only ≤ 100

table 1-1. Incidence of various anaerobes (or microaerophiles/

	GRAM-POSITIVE							
	Actinomyces	Bifidobacterium/Lactobacillus	Propionibacterium	Eubacterium and related organisms	Mobiluncus	Clostridium	C. difficile	Cocci
Skin	0	0	2	+/–	0	0	0	1
Upper respiratory tract[1]	1	+/–	1	+/–	0	0	0	2
Mouth	2	1	+/–	2	0	+/–	0	2
Intestine	+/–	2	+/–	2	+/–	2	+/–	2
Genito-urinary tract[2]	+/–	1-2	+/–	+/–	+/–	+/–	0	2

Key: 0, not found or rare; +/–, irregular; 1, usually present; 2, usually present in large numbers; GNB, gram-negative bacilli.

[1] Includes nasal passages, nasopharynx, oropharynx, and tonsils.

[2] Includes external genitalia, urethra, vagina, and endocervix.

organisms/ml of gastric juice and representing mainly swallowed oral flora (95;147). In patients with achlorhydria or a relatively high gastric pH as a result of therapy (e.g., with H_2 antagonists), counts of organisms in the stomach may reach 10^6–10^7/ml or even higher (95). Certain disease states, such as bleeding or obstructing peptic ulcer, gastric ulcer, or gastric carcinoma, lead to significant colonization of the stomach with a diverse flora that includes members of the *Enterobacteriaceae* and various anaerobes, even the *B. fragilis* group (248).

The upper small bowel also has a relatively sparse flora under normal circumstances. In the distal ileum, however, counts of bacteria reach 10^4–10^6/ml, and both coliforms and various anaerobes may be encountered (95;147).

The colonic flora is much more profuse and diverse than the small bowel flora. In the distal colon, total counts average 10^{11}–10^{12}/g of feces, with anaerobes outnumbering other organisms by a ratio of 1000:1 (118). Members of the *B. fragilis* group are dominant in the cultivable indigenous flora of the large bowel. Within the *B. fragilis* group, the species that is most prevalent in the normal flora is *Bacteroides thetaiotaomicron.* This organism is also the member of the *B. fragilis* group most resistant to antimicrobial

facultatives) as indigenous flora in humans

	GRAM-NEGATIVE										
	B. fragilis group	**Bilophila**	**Campylobacter**	**Capnocytophaga/Leptotrichia**	**Desulfomonas/Desulfovibrio**	**Fusobacterium**	**Porphyromonas**	**Prevotella**	**Sutterella**	**Other GNB**	**Cocci**
	0	0	0	0	0	0	0	0	0	0	0
	0	0	1	1	0	1	+/–	1	0	1	1
	0	+/–	2	1	0	2	1	2	0	1	1
	2	1	1	+/–	1	1	1	1	1	1	1
	+/–	+/–	+/–	+/–	0	1	+/–	1	0	1	1

agents. In addition to the *B. fragilis* group, *Fusobacterium prausnitzii,* other gram-negative anaerobes, and *Eubacterium rectale,* the colonic flora includes large numbers of clostridia and anaerobic cocci as well as nonsporeforming, gram-positive anaerobic rods, which are primarily of low pathogenicity. Recent 16S rDNA sequencing studies on the diversity of bacterial species or groups in fecal material suggest that only about 24% of the molecular species recovered correspond to described organisms with available sequences. The three main phylogenetic groups contained 95% of the clones: the *B. fragilis* group, *Clostridium coccoides* group, and *Clostridium leptum* subgroup (315;374). Among the facultative organisms, *Escherichia coli* and various streptococci and enterococci predominate (118).

In biliary tract infection, *E. coli*—and sometimes similar organisms, such as *Klebsiella* species—and enterococci are the organisms most commonly encountered. *Clostridium perfringens* is not a frequent isolate but may cause devastating disease. In patients in the older age group with repeated biliary tract infection related to carcinoma or to intermittent or unrelieved biliary tract obstruction or with repeated biliary tract surgery, the *B. fragilis* group can also be important. Data on the association of the noncultivated "molecular species" with infections are still scarce, but gradually accumulating (266).

The Flora of the Female Genital Tract

The predominant organisms encountered normally in the female genital tract are various species of facultative H_2O_2-producing *Lactobacillus* and *Peptostreptococcus. C. perfringens,* other clostridia, and various gram-negative and gram-positive nonsporeforming rods are encountered irregularly. *P. bivia* and *P. disiens* tend to dominate among the gram-negative rods, but pigmented *Porphyromonas* and *Prevotella* species, the *B. fragilis* group, and other *Prevotella* and *Bacteroides* species may be seen as well. *Actinomyces* and *Eubacterium nodatum* are important in women with actinomycotic infection related to intrauterine devices (actinomycosis). Lack of H_2O_2-producing lactobacilli and predominance of *Mobiluncus, Gardnerella vaginalis, Prevotella bivia/disiens, Prevotella corporis, Porphyromonas levii, Peptostreptococcus prevotii, Peptostreptococcus tetradius, Peptostreptococcus anaerobius,* viridans streptococci, *Ureaplasma urealyticum,* and *Mycoplasma hominis* are recognized as the bacterial vaginosis (BV) complex (151) (see Appendix A). Bacterial vaginosis has been associated with preterm labor and a wide variety of genital tract infections, including chorioamnionitis, postpartum endometritis, amnionitis, posthysterectomy vaginal cuff cellulitis, and pelvic inflammatory disease (104;150;327). Prepubertal girls and postmenopausal women who have not received estrogen replacement therapy harbor lactobacilli, yeast, and BV-associated microorganisms less often than women of child-bearing age. The exact effects of estrogen replacement therapy on the vaginal flora of postmenopausal women have not been extensively evaluated (148;152).

Among the nonanaerobes, Group B streptococci, other streptococci, and *E. coli* and related organisms predominate. In sexually active women with pelvic inflammatory disease, gonococci, chlamydia, and *Ureaplasma* may be found; *Mycoplasma hominis, Gardnerella vaginalis,* and other organisms may be pathogens in postpartum infections.

OPTIMAL ANAEROBIC BACTERIOLOGY IN THE FACE OF RESTRICTED RESOURCES

The desirability of performing optimal anaerobic bacteriology despite limited resources is a real dilemma. Because of the complexity of mixed anaerobic infections and the labor-intensive nature of anaerobic bacteriology, the complete workup of a single culture may cost between $500 and $1000 or even more. It is important that the microbiologist select approaches that are reasonable in cost yet do not compromise good patient care (114;117).

Costs may be reduced or minimized in several ways in the anaerobic bacteriology laboratory: (1) rejection of inappropriate specimens; (2) use of simple, direct tests to provide presumptive evidence that anaerobes are present; (3) use of selective and differential media for rapid isolation and presumptive identification of specific anaerobes; (4) identification of various isolates to a level appropriate for the specimen source, the type of bacteria present, and the patient's clinical status; and (5) use of simple tests for classification of anaerobes into general groups or definitive identification in the case of certain species (278).

Specimens that may be contaminated with normal flora (e.g., throat swabs, expectorated sputum, vaginal swabs) and specimens from relatively minor wounds or lesions that will respond to simple incision and drainage should not be cultured anaerobically. In many anaerobic infections the infecting flora is fairly predictable. If the patient is not very ill, empiric therapy with minimal bacteriologic study is appropriate.

Foul or putrid odor of a specimen is definitive evidence that anaerobic bacteria are part of the infecting flora. Other evidence useful in determining whether anaerobes are present and perhaps in determining which ones are present include gram-stained preparations with morphology typical of certain anaerobes, fluorescence under ultraviolet light (indicative of pigmented *Prevotella* or *Porphyromonas*), the presence of certain volatile or organic acids on gas chromatography, and (on occasion) the results of DNA probes and direct 16S rDNA sequencing (85;226;266).

Any laboratory engaged in anaerobic cultivation should be able to recover in pure culture all anaerobes present in clinical specimens, to maintain them in a viable state, and to do at least preliminary characterization. This, with identification of certain key organisms such as the *B. fragilis* group and *C. perfringens,* may provide the clinician with the data needed to successfully manage patients with anaerobic infections. The anaerobes described above, along with the pigmented *Prevotella* and *Porphyromonas, Fusobacterium nucleatum,* and the anaerobic cocci (these groups are also readily identified), are present in approximately two-thirds of all clinically significant infections involving anaerobes.

The use of selective and differential solid media facilitates the work of the anaerobic bacteriologist (114;117;118). Bacteroides bile esculin (BBE) (200) agar is an excellent medium highly selective for the *B. fragilis* group, most strains of which produce large black colonies with blackening of the surrounding area, and for *Bilophila,* which forms translucent colonies with black centers. Good growth of the *B. fragilis* group may be obtained on BBE medium within 18–24 hours. The appearance of typical colonies of appropriate size on this medium is good presumptive evidence that the *B. fragilis* group is present in a specimen. Other media that are useful include kanamycin-vancomycin laked blood agar, phenylethyl alcohol agar, and cycloserine cefoxitin fructose agar (for *Clostridium difficile* (123)). **Nonselective media should always be used along with selective media.** In research laboratories use of two different nonselective media is recommended as each may promote the growth of different groups of bacteria (296).

As indicated above, the identification of isolates should be tailored to the source of the specimen, the patient's clinical status, and the bacteria present. Close communication (in both directions) between the microbiologist and clinician is required. For very sick patients, an expanded and expedited workup may be of enormous help to the clinician and may indeed be lifesaving. For less significant infections, such a detailed workup would merely waste resources.

Whenever possible, simple tests should be used for general—and, to the extent that they permit, definitive—identification of anaerobes (114;117). Included in this category would be colonial and microscopic morphology, spot indole, catalase, lipase, lecithinase, the nitrate disk test, the bile disk test, growth stimulation tests, and tests of sensitivity to special potency antibiotic disks (vancomycin, kanamycin, and colistin). Rapid tests, based on the presence of preformed enzymes, may also be useful and produce clinically meaningful results in a timely fashion (see Chapter 5).

In general, the *B. fragilis* group should be identified since it is the most commonly encountered group of anaerobes that may be involved in serious infections and since its members are among the anaerobes most resistant to antimicrobial agents.

Although we recognize that many laboratories may find it difficult to go beyond what has been outlined in the preceding paragraphs and that cost may be a problem, we nonetheless urge full, definitive identification whenever it is possible. As an example, identification of species within the *B. fragilis* group may provide useful information. *B. fragilis* and *B. thetaiotaomicron* are commonly found as pathogens; recovery of *B. merdae* might mean contamination of a specimen with normal bowel flora.

Similarly, species identification within the *B. ureolyticus*-like group is important. *Sutterella wadsworthensis* is much more pathogenic than other *B. ureolyticus*-like organisms and is much more resistant to antimicrobial agents. If an organism initially felt to be *B. fragilis* from a patient with bacteremia of unknown source was subsequently identified as *B. splanchnicus,* this would indicate the bowel as the likely source, whereas if it were *B. fragilis,* the portal of entry might also have been the female genital tract or elsewhere. Identification of a *Clostridium* isolated from the blood as *C. septicum* provides the clinician with a valuable clue, as there is a strong association between bacteremia with this organism and malignancy or other disease of the colon, especially the cecum. Exact identification of an organism isolated from a patient with two or more episodes of infection helps distinguish between recurrence (which may imply tumor, foreign body, or an undrained abscess) and a new infection.

Quantitation of results is of particular importance in anaerobic bacteriology since anaerobic infections are frequently mixed and the relative importance of various organisms in complex mixtures may often be deduced from this type of information. Formal quantitation is not usually necessary; designation as "heavy growth," "moderate growth," "few colonies," and so on is adequate.

In academic centers it is important that definitive anaerobic bacteriologic studies be carried out. One major reason is to avoid deterioration of our present skills; also, such centers are engaged in teaching clinicians and microbiologists as well as pathologists. Moreover, academic centers engage in basic and clinical research that provides information on recovery of new taxa and changes in infection patterns and susceptibility results; these data help guide empiric therapy in other settings.

CLINICAL BACKGROUND

Incidence of Anaerobes in Infection

Anaerobes may cause any type of infection in humans. In a number of infections, anaerobic bacteria are the predominant pathogens or are commonly found; these are listed in Table 1-2. When information regarding incidence of anaerobes in these infections is available, it has been indicated.

It must be emphasized, however, that the majority of published studies on anaerobic infections have been retrospective. The bacteriologic methods, particularly the anaerobic methods, were not uniform and, in many instances, not optimal. Therefore, some of these incidence figures are undoubtedly low. Several articles emphasizing various aspects of anaerobic infections are also listed in Table 1-2 and are recommended for those wishing to read further on this subject.

Other studies, not included in Table 1-2, incorporate inaccuracies related to uncertainty about the clinical significance of isolates and to selection of certain types of specimens that would certainly contain elements of the normal flora. In this type of study, of course, the incidence of anaerobes may be falsely high.

Table 1-3 details our own recent experience with recovery of anaerobes from clinical specimens.

Clues to Anaerobic Infection

Certain hints suggest to the microbiologist that a given specimen is likely to contain anaerobic bacteria:

1. Foul odor of specimen.
2. Location of infection in proximity to a mucosal surface.
3. Infection secondary to human or animal bite.
4. Gas in specimen.
5. Previous therapy with aminoglycoside antibiotics (such as gentamicin or amikacin) or other drugs poorly active against anaerobes (older quinolones, trimethoprim/sulfamethoxazole) in the absence of concomitant effective coverage of anaerobes.
6. Black discoloration of blood-containing exudates; these exudates may fluoresce red under ultraviolet light (infections involving pigmented gram-negative anaerobes).
7. Presence of "sulfur granules" in discharges (actinomycosis).
8. Unique morphology on Gram stain.
9. Failure of organisms seen on Gram stain of original exudate to grow aerobically.
10. Growth in anaerobic zone of fluid media or of agar deeps.
11. Anaerobic growth on media containing 75 to 100 µg/ml of kanamycin, neomycin, or paromomycin (or medium also containing vancomycin, in the case of gram-negative anaerobic bacilli) or on other selective media such as BBE, CCFA, and RIF (see Appendix A).
12. Characteristic colonies on anaerobic agar plates (for example, *F. nucleatum* and *C. perfringens*).
13. Red fluorescence under ultraviolet light may indicate young colonies of pigmented gram-negative anaerobic rods (growth from anaerobic blood agar plate) whose pigment is not well developed.

Taxonomic Changes

Table 1-4 presents recent taxonomic changes. Clearly, we must adopt all approved changes in clinical microbiology laboratories. It is prudent to include the previous name in reports for a period of time (as long as one year) to *(continued on p. 13)*

table 1-2. Infections commonly involving anaerobes

	% of pos. cultures yielding anaerobes *	Proportion or % of cultures positive for anaerobes yielding only anaerobes	Reference
Bacteremia	2.8	–	(285)
	3.9	–	(356)
	4.2	–	(143)
	5.7	–	(234)
Bacteremia secondary to oral surgery	45–70	–	(145)
Ocular infections	38	10/43	(165)
Corneal ulcers	7	9/11	(259)
Central nervous system			
Brain abscess	62	24	(68)
	83	63	(42)
	73	–	(3)
	63	–	(292)
Subdural empyema	77	81	(42)
	29 (84 cases)		(326)
Epidural abscess	39 (41 cases)		(326)
Head and neck			
Chronic sinusitis	100	62	(38)
	78	62	(51)
	48	–	(164)
Acute sinusitis	7		(139)
	2	5/8	(170)
Chronic otitis media	63	20	(52)
	59	1	(267)
Acute otitis media	27	48	(43)
Cholesteatoma	92	1/11	(158)
Neck space infections	100	3/4	(21)
Wound infection following head and neck surgery	95	0	(24)
Peritonsillar abscess	87	17	(224)
	82	17	(166)
Dental, oral, facial			
Orofacial, of dental origin	95	40	(349)
	67	–	(146)
	94	4/10	(67)
Root canal infection	100	10/10	(373)
	100	18/55	(131)

table 1-2. *(continued)*

	% of pos. cultures yielding anaerobes *	Proportion or % of cultures positive for anaerobes yielding only anaerobes	Reference
Dental, oral, facial *(continued)*			
Periapical granuloma	67	51	(353)
	86	12/14	(159)
Periapical abscess	94	16/30	(49)
Periodontal abscess	100	0/9	(247)
Dental abscess, endodontic origin	100	8/12	(50)
Thoracic			
Aspiration pneumonia	85 (summary of 7 studies)	–	(20)
	93	1/2[1]	(22)
	62	1/3	(205)
Hospital-acquired pneumonia	35	–	(20)
Community-acquired pneumonia	28 (summary of 2 studies)	–	(20)
Lung abscess	78 (58–100); summary of seven studies	–	(84)
	90	–	(20)
Bronchiectasis	27		(291)
	17 (18 cases)	3/3	(27)
Empyema (nonsurgical)		19/46	(73)
	22 (summary of 7 studies)	–	(20)
	31	39	(7)
	36	36	(44)
Abdominal			
Intra-abdominal infection (general)	90	1/3	(232)
	94	1/7	(132)
	50	–	(308)
Appendicitis with peritonitis	100	–	(357)
	92	8/71	(26)
Liver abscess	52	1/3	(284)
	64	44	(46)
Other intra-abdominal infection (postsurgery)	93	1/6	(134)
Wound infection following bowel surgery	60 (33 cases)	5/20	(257)
Biliary tract	45	0	(297)
	41	2/117	(103)

table 1-2. ***(continued)***

	% of pos. cultures yielding anaerobes *	Proportion or % of cultures positive for anaerobes yielding only anaerobes	Reference
Obstetric-gynecologic			
Miscellaneous types	100	1/3	(339)
	74	1/3	(329)
	52	–	(328)
Posthysterectomy (abdominal) wound infection	43	35	(223)
Postpartum endometritis	66	35	(310)
Pelvic abscess	88	1/2	(10)
Vulvovaginal abscess	75	1/4	(254)
Vaginal cuff abscess	98	1/30	(138)
Septic abortion, sepsis	69	18/20	(66)
	67		(282)
	63		(303)
Pelvic inflammatory disease	25	1/14	(65)
	48	1/7	(105)
Soft tissue and miscellaneous			
Nonclostridial crepitant cellulitis	75	1/12	(209)
Pilonidal abscess	88 (41 cases)		(361)
	96 (75 cases)	76	(41)
Bite wound infections:			
Cat bites	67	0	(330)
Dog bites	57	4	(330)
Human bites	50–56[2]		(130)
	57	6	(195)
Diabetic foot infections	95	1/20	(124)
	40	–	(129)
Infected diabetic gangrene (deep tissue culture)	85	1/11	(287)
Skin and soft tissue abscesses	44	5/70	(322)
	38	6	(331)
Cutaneous abscesses	62	1/5	(222)
	55	49	(201)
Decubitus ulcers with bacteremia	16	7/10	(54)
	63	10/12	(64)
	50	–	(228)
Decubitus ulcers, infected	57	–	(102)
	50	24	(37)

table 1-2. *(continued)*

	% of pos. cultures yielding anaerobes *	Proportion or % of cultures positive for anaerobes yielding only anaerobes	Reference
Arthritis, purulent		89	(45)
	93	44	(137)
Osteomyelitis	40	1/10	(197)
	22	23	(136)
		67	(45)
Gas gangrene (clostridial myonecrosis)	100		(323)
	100		(11)
Soft tissue and miscellaneous			
Breast abscess, nonpuerperal	53	5/8	(194)
	83	18/34	(40)
	79	5/41	(100)
	71	10/10?	(86)
Perirectal abscess	77 (74 cases)		(361)
	90 (144 cases)	21	(47)

* % of all cultures in the case of bacteremia

[1]Aspiration pneumonia occurring in the community rather than in the hospital involves anaerobes to the exclusion of aerobic or facultative forms two-thirds of the time.

[2]Includes clenched fist injuries

allow clinicians time to adapt. The changes indicated are based on genetic analysis and also take phenotypic characteristics into consideration. Use of the newer terminology allows us to be more precise in our identification. The importance of definitive identification has been discussed in the section entitled "Optimal Anaerobic Bacteriology in the Face of Restricted Resources."

table 1-3. Incidence of specific anaerobes in various

	Blood	CNS	Head and Neck	Dental	TTA and Pleural	Lung	Perirectal abscess
No. of specimens	[b]	103	225	8	651	128	17
No. of specimens yielding anaerobes	83	16	123	8	48	28	15
Bacteroides fragilis	26				1	3	5
B. thetaiotaomicron	4						2
Other *B. fragilis* group	12	1	2		1	3	7
Bilophila wadsworthia					1	1	
Campylobacter gracilis			4		1	2	
Campylobacter/ B. ureolyticus			5		2	1	2
Fusobacterium nucleatum	3	4	23	9	10	5	1
F. necrophorum			2				
F. mortiferum/varium	1						
Other *Fusobacterium* sp.	2		2				1
Porphyromonas asaccharolytica		1	3				2
P. gingivalis			2	3			
P. levii–like organisms			2		1		
Porphyromonas spp.			1				
Prevotella melaninogenica/ denticola	1	1	7	1	3	2	
P. intermedia-nigrescens-pallens	1	1	23	5	9	3	1
P. loescheii, P. corporis			7	1	3	1	
P. bivia/disiens			5		2		1
P. oralis group			2		2	1	
P. oris/buccae		1	29		5	1	1
Prevotella spp.	1		9	5	1	2	4
Desulfomonas/Desulfovibrio							
Sutterella wadsworthensis							1
Veillonella spp.	2		15		2	7	
Other gram-negative cocci							2
Gemella morbillorum			7			1	
Peptostreptococcus anaerobius					1		
P. asaccharolyticus	1		6				2

[a] Includes foot ulcer and osteomyelitis specimens

[b] Total number of specimens was not available

infections (Wadsworth VA Medical Center experience 1991–1997)

Decubitus ulcer	Miscellaneous skin soft tissue and bone infections[a] below the waist	above the waist	Appendiceal tissue or abscess	Peritoneal fluid	Gall bladder	Abdominal, other
22	777	327	183	[b]	155	165
16	241	98	133	170	17	46
9	33	6	70	61	2	18
	9		41	23	1	12
6	31	4	177	140	3	44
	2	1	49	32		6
		3	1			
	16	4	7	2		1
3	18	26	23	24		7
	3		13	9		
1	2		8		1	5
	3	5	1	3	1	2
2	12		2	5		1
			14	10		1
	12	2		1		
2	6	2	9	3		
	12	10				
	9	9	15	8		
	7	10		1		
2	15	3	2	2		1
	3	3	3	3		
	10	7	2	6	2	2
1	22	8				3
			7	6		2
	2	1	17	19		
	11	16	1	5	1	2
	1					2
	1	2	13	2		2
1	13	3	22	12		2
6	26	4	1	2		

table 1-3. *(continued)*

	Blood	CNS	Head and Neck	Dental	TTA and Pleural	Lung	Perirectal abscess
F. magna	1		13		1		3
M. micros	1	1	16	2	8	4	3
P. prevotii/tetradius	1		8				1
Peptostreptococcus spp.	1		1		1	1	3
Anaerobic *Streptococcus* spp.		2	13		3	4	
Staphylococcus saccharolyticus			2		2	1	
Clostridium perfringens	16						
Other *Clostridium* spp.	13		2		1	1	4
Actinomyces spp.		1	19		4	4	1
Bifidobacterium spp.			7				
Eggerthella lenta	1		1		1	1	
Eubacterium spp.	2		3				
Lactobacillus spp.	1		11	2	4	2	
Propionibacterium acnes	1	8	22		16	5	
Propionibacterium spp.	2	1	8		1	1	1
Pseudoramibacter alactolyticus					3		

[a] Includes foot ulcer and osteomyelitis specimens.

[b] Total number of specimens was not available.

Decubitus ulcer	Miscellaneous skin soft tissue and bone infections[a] below the waist	Miscellaneous skin soft tissue and bone infections[a] above the waist	Appendiceal tissue or abscess	Peritoneal fluid	Gall bladder	Abdominal, other
3	64	8	2	1		1
3	13	21	37	24		7
6	41	7	1	1		
2	12	3	3	1		1
	10	8	8	4	1	4
	2	1		1		1
	5	3	1	4	1	1
	15	10	36	41	1	12
	21	15	10	4		5
	2	1	1	4		1
		2	21	12		2
	5	1	11	6		2
	9	9	19	10		6
1	15	13	5	23	9	4
	2	1				
			6	4		

table 1-4. Recent taxonomic changes and trends (since the 5th edition of *Wadsworth Anaerobic Bacteriology Manual*)

Current Nomenclature	Synonym/Taxonomic position	Reference
Gram-negative bacilli		
Anaerobiospirillum thomasii	New species	(211)
Bacteroides distasonis	Related to *Porphyromonas* cluster	(256)
*Bacteroides forsythus**	Related to *Porphyromonas* cluster	(256)
Bacteroides furcosus	Related to *Porphyromonas* cluster	(256)
Bacteroides putredinis	Possibly related to *Rikenella*	(256)
Bacteroides pyogenes[a]	*Bacteroides tectum* homology group II (some strains)	(119)
Bacteroides splanchnicus	Possibly represents a new genus	(256)
Bacteroides tectus	*Bacteroides tectum,* related to *Bacteroides fragilis*	(108)
Butyrivibrio species	Related to *Clostridium* subphylum, cluster XIVa	(365)
Campylobacter gracilis	*Bacteroides gracilis* (some strains)	(347)
Campylobacter hominis	New species	(192a)
Campylobacter showae	New species	(107)
Capnocytophaga granulosa	New species	(379)
Capnocytophaga haemolytica	New species	(379)
Catonella morbi	Related to *Clostridium* subphylum, cluster XIVa	(367)
Centipeda periodontii	Related to *Selenomonas* species	(334)
Dialister pneumosintes	Related to *Sporomusa* branch of *Clostridium* subphylum, cluster IX	(369)
Fusobacterium varium	*Fusobacterium pseudonecrophorum*	(13)
Johnsonella ignava	Related to *Clostridium* subphylum, cluster XIVa	(367)
Leptotrichia sanguinegens	New species	(140)
Mitsuokella multiacida	Related to *Sporomusa* branch of *Clostridium* subphylum, cluster IX	(334)
Porphyromonas cangingivalis[a]	New species	(78)
Porphyromonas canoris[a]	New species	(207)
Porphyromonas cansulci[a]	New species	(78)
Porphyromonas catoniae	*Oribaculum catoniae*	(370)
Porphyromonas crevioricanis[a]	New species	(153)
Porphyromonas gingivicanis[a]	New species	(153)
Porphyromonas gulae[a]	New species	(119a)
Porphyromonas levii[a]	*Bacteroides levii*	(293)
Porphyromonas macacae[a]	*Bacteroides macacae, Porphyromonas salivosa*	(206)

[a]Of animal origin

table 1-4. continued

Current Nomenclature	Synonym/Taxonomic position	Reference
Prevotella albensis[a]	*Bacteroides ruminicola* subsp. *ruminicola* biovar 7, *Prevotella ruminicola* (some strains)	(12)
Prevotella brevis[a]	*Bacteroides ruminicola* subsp. *brevis* biovars 1,2, *Prevotella ruminicola* (some strains)	(12)
Prevotella bryantii[a]	*B. ruminicola* subsp. *brevis* biovar 3, *Prevotella ruminicola* (some strains)	(12)
Prevotella dentalis	*Mitsuokella dentalis*	
	Hallella seregens	(366)
Prevotella enoeca	New species	(229)
Prevotella heparinolytica	Related to *Bacteroides fragilis* group	(256)
Prevotella nigrescens	New species; *P. intermedia* (some strains)	(294)
Prevotella pallens	New species	(185)
Prevotella ruminicola[a]	*B. ruminicola* subsp. *ruminicola* biovar 1	(12)
Prevotella tannerae	New species	(229)
Prevotella zoogleoformans	Related to *Bacteroides fragilis* group	(256)
Selenomonas species	Related to *Sporomusa* branch of *Clostridium* subphylum, cluster IX	(334)
Sutterella wadsworthensis	New genus and species, *Campylobacter* (*Bacteroides*) *gracilis* (some strains)	(359)
Tissierella praeacuta	Related to *Clostridium* subphylum, cluster XII	(110)
Gram-negative cocci		
Acidaminococcus fermentans	Related to *Sporomusa* branch of *Clostridium* subphylum, cluster IX	(108)
Megasphaera elsdenii	Related to *Sporomusa* branch of *Clostridium* subphylum, cluster IX	(334)
Veillonella species	Related to *Sporomusa* branch of *Clostridium* subphylum, cluster IX	(334)
Nonspore forming gram-positive bacilli		
Actinobaculum schalii	New genus and species; *Actinomyces*-like	(193)
Actinomyces europaeus	New species	(121)
Actinomyces funkei	New species	(193a)
Actinomyces graevenitzii	New species	(265)
Actinomyces naeslundii	Possibly homologous with *Actinomyces viscosus*	(161)
Actinomyces neuii subspecies *neuii*	New subspecies	(122)
Actinomyces neuii subspecies *anitratus*	New subspecies	(122)
Actinomyces radicidentis	New species	(79a)
Actinomyces radingae	CDC coryneform group E; APL1	(377)
Actinomyces turicensis	CDC coryneform group E; APL10	(377)
Actinomyces urogenitalis	New species	(248a)
Arcanobacterium bernardiae	*Actinomyces bernardiae*	(255)

[a]Of animal origin

table 1-4. continued

Current Nomenclature	Synonym/Taxonomic position	Reference
Arcanobacterium pyogenes	*Actinomyces pyogenes*	(255)
Atopobium minutum	New genus, *Lactobacillus minutus*	(79)
Atopobium parvulum	*Streptococcus parvulus* *Peptostreptococcus parvulus*	(79)
Atopobium rimae	*Lactobacillus rimae*	(79)
Atopobium vaginae	New species	(276)
Bulleidia extructa	New genus and species	(91)
Bifidobacterium denticolens	*Bifiobacterium eriksonii,* *Bifidobacterium dentium* (some strains)	(81)
Bifidobacterium inopinatum	*Bifidobacterium dentium* (some strains)	(81)
Collinsella aerofaciens	New genus, *Eubacterium aerofaciens*	(175)
Cryptobacterium curtum	New genus and species	(243)
Eggerthella lenta	*Eubacterium lentum* (some strains)	(350)
Eubacterium infirmum	New species	(63)
Eubacterium minutum	*E. tardum* (some strains)	(351)
Eubacterium saphenum	New species	(346)
Eubacterium sulci	*Fusobacterium sulci*	(160)
Eubacterium tenue	Related to *Clostridium* species	(77)
Holdemania filiformis	New genus and species	(372)
Lactobacillus paraplantarum	New species	(83)
Lactobacillus uli[a]	New species	(252)
Mobiluncus curtisii subspecies *curtisii*	Phylogeneally represent one species	(340a)
Mobiluncus curtisii subspecies *holmesii*		
Mogibacterium pumilum	New genus and species	(244)
Mogibacterium timidum	*Eubacterium timidum*	(244)
Mogibacterium vescum	New genus and species	(244)
Pseudoramibacter alactolyticus	New genus, *Eubacterium alactolyticum*	(371)
Slackia exigua	New genus; *Eubacterium exiguum*	(350)
Slackia heliontrinireducens	*Peptostreptococcus heliontrinireducens*	(350)
Endospore-forming gram-positive bacilli		
Clostridium argentinense	*Clostridium botulinum*, group G	
	Clostridium subterminale (some strains)	
	Clostridium hastiforme (some strains)	(316)
Clostridium piliforme	*Bacillus piliformis*	
	Tyzzer's bacillus	(98)
Filifactor alocis (no spores)	*Fusobacterium alocis*	(160)
Filifactor villosus	*Clostridium villosum*	(77)

[a]Of animal origin

table 1-4. continued

Current Nomenclature	Synonym/Taxonomic position	Reference
Gram-Positive Cocci		
Abiotrophia defectiva	New genus, *Streptococcus defectivus*	(178)
Granulicatella adiacens	New genus, *Streptococcus adjacens, Abiotrophia adiacens*	(76)
Granulicatella elegans	New species, *Abiotrophia elegans*	(76)
Finegoldia magna	New genus, *Peptostreptococcus magnus*	(240)
Micromonas micros	New genus, *Peptostreptococcus micros*	(240)
Peptostreptococcus harei[b, c]	New species	(239)
Peptostreptococcus ivorii[b]	New species	(239)
Peptostreptococcus lacrimalis[b, c]	New species	(198)
Peptostreptococcus lactolyticus[d]	New species	(198)
Peptostreptococcus octavius[d]	New species	(239)
Peptostreptococcus trisimilis	Provisional new species	(238)
Peptostreptococcus vaginalis[d]	New species	(198)
Ruminococcus hansenii	*Streptococcus hansenii*	(109)
Ruminococcus productus	*Peptostreptococcus productus*	(109)
Streptococcus pleomorphus	Related to *Clostridium* subphylum, cluster XVI	(77)

[a]Of animal origin.

[b]Newly proposed genus by Ezaki et al. is *Peptoniphilus*. Ezaki T, Kawamura Y, Li N, Li Z-Y, Zhao L, Shu S-E. Proposal of the genera *Anaerococcus* gen. nov., *Pentoniphilus* gen. nov. and *Gallicola* gen. nov. for members of the genus *Peptostreptococcus*. Int. J. Syst. Evol. Microbiol. 2001; 51: 1521–1528.

[c]Newly proposed genus by Rajendram et al. is *Shleiferella*. Rajendram D, Shah HN, Gharbia SE, Murdoch DA. Reclassification of *Peptostreptococcus asaccharolyticus* (Distaso 1912) Ezaki, Yamamoto, Ninomiya, Suzuki and Yabuuchi, 1983 as *Schleiferella asaccharolytica* comb. nov.; *Peptostreptococcus indolicus* (Christiansen 1934) Ezaki, Yamamoto, Niomiya, Suzuki and Yabuuchi 1983 as *Schleiferella indolica* comb. nov., *Peptostreptococcus lacrimalis* Li, Hashimoto, Adnan, Miura, Yamamoto and Ezaki 1992 as *Schleiferella lacrimalis* comb. nov. and *Peptostreptococcus harei* (Murdoch, Collins, Willems, Hardie, Young and Magee 1997) as *Schleiferella harei* comb. nov. Anaerobe 2001; 07: 93–101.

[d]Newly proposed genus by Ezaki et al. is *Anaerococcus*. See reference above.

[*]*Lactobacillus uli* was reclassified as *Olsonella uli* (Dewhirst et al., IJSEM. 2001; 51; 1797–1804). *Bacteroides forsythus* was reclassified as *Tannerella forsythensis*. (M. Sakamoto et al., IJSEM, In press, published online 29 November 2001) *Desulfomonas pigra* was reclassified as *Desulfovibrio piger*. (J. Loubinoux et al., IJSEM, In press.)

chapter 2

Specimen Collection and Anaerobic Culture Techniques

Specimen Collection

Principle

All appropriate specimens collected so as to exclude normal flora and those from normally sterile body sites should be cultured anaerobically. Anaerobic flora are particularly prevalent on mucosal surfaces of the gastrointestinal and genital tracts; specimens collected from these sites should not ordinarily be cultured for anaerobic bacteria. The oral cavity also contains large numbers of anaerobes in saliva and gingival secretions; unless special oral microbiology protocols are being used, specimens not excluding these materials should not be cultured for anaerobes. The following specimens are among those likely to be contaminated with normal flora and should not be cultured anaerobically under routine circumstances:

Throat swabs

Nasopharyngeal swabs

Gingival or any other intraoral surface swabs

Expectorated sputum

Sputum obtained by nasotracheal or endotracheal suction

Bronchial washings or other specimens obtained via a bronchoscope (except those obtained via a protected double-lumen catheter or properly done bronchoalveolar lavage)

Gastric and small bowel contents (except in the case of "blind loop" or bacterial overgrowth syndrome)

Large bowel contents (except for *C. difficile, C. botulinum, Anaerobiospirillum* species, and other specific etiologic agents)

Ileostomy, colostomy effluents

Feces (except large bowel contents)

Voided or catheterized urine

Vaginal or cervical swabs

Female genital tract cultures collected via the vagina (except suction curettings or other specimen collected via a double-lumen catheter)

Surface swabs from decubitus ulcers, perirectal abscesses, foot ulcers, exposed wounds, eschars, pilonidal sinuses, and other sinus tracts

Any material adjacent to a mucous membrane that has not been adequately decontaminated

table 2-1. Specimen collection for anaerobic microbiology

Note: aspirates should be performed after decontamination of intact skin surface with alcohol followed by povidone-iodine.

Site of infectious process	Appropriate specimen	Collection method
Head and neck	Aspirate	Percutaneous needle aspiration
	Tissue biopsy	Surgically obtained
Periodontal	Subgingival pocket plaque	Paper points or scaler into anaerobic VMGA III transport medium
	Aspirate	Needle aspiration
Pulmonary	Lung aspirate	Percutaneous needle aspiration of lung
	Tissue biopsy	Surgically obtained
	Deep bronchial secretions	Transtracheal aspirate or protected bronchial brush
Joint	Joint fluid	Needle aspirate obtained at surgery
Abdominal	Peritoneal fluid	Needle aspirate obtained at surgery
	Abscess contents	Aspirate obtained at surgery or under CT or ultrasound guidance (avoiding contamination with bowel contents)
	Bile	Bile obtained at surgery
	Tissue biopsy	Surgically obtained
Female genital tract	Peritoneal fluid	Culdocentesis
	Endometrial material	Endometrial suction or protected collector
	Tissue biopsy (Others)	Surgically obtained
Bone	Biopsy	Curettings or scrapings obtained surgically
	Aspirate	Aspirate of deep tissue via uninvolved skin surface
Other soft tissue	Tissue biopsy	Surgically obtained
	Aspirate	Percutaneous needle aspiration
	Tissue	Curettings
Urine	Bladder urine	Suprapubic aspirate

Swabs from most sites (except those obtained during surgery such as neurosurgery, where the volume of material or site of infection precludes obtaining fluid or tissue without harming the patient)

Specimen Collection Methods

The importance of collecting specimens so as to avoid contamination with normal flora cannot be overemphasized. Indigenous anaerobes are often present in such large numbers (10^9–10^{12}) that even minimal contamination with normal flora will yield very misleading results and lead to much wasted effort by the laboratory. Recommended collection methods to avoid indigenous flora are listed in Table 2-1.

Methods for collection of appropriate specimens from selected infectious processes, particularly for those sites for which special techniques are often necessary, are described in the following section. In general, a swab is not an ideal collection method; only when no other options are available and resulting information is judged to be potentially important clinically should a swab be accepted for anaerobic culture. An example of such infection is an animal bite wound involving a small puncture and abscess.

Any aspirated material from a normally sterile body site is acceptable; skin preparation should be the same as that used for venipuncture for blood culture: vigorous scrubbing with 70% alcohol from the center of the site outward in widening concentric circles to remove surface dirt and oils, followed by an iodine preparation (applied in widening concentric circles from the center outward and allowed to remain wet for 30 s (tincture of iodine) or 1 min (aqueous iodophor)). The iodine preparation should be removed with additional alcohol after the procedure, as many people are sensitive to iodine.

Abscesses

As for all loculated fluid collections, material should be aspirated via needle and syringe through disinfected, uninvolved, and intact tissue. If the abscess fluid is exposed at surgery and a needle is contraindicated due to the small volume of material (< 0.2 ml) or proximity to critical tissues (such as may occur with brain abscess), fluid may be aspirated through a flexible plastic catheter or directly into a sterile syringe without a needle. A fresh, sterile needle should be used to inject the material from the syringe into an anaerobic transport vial, so as not to carry over contamination from the needle used to collect the specimen. The septa or stoppers on some anaerobic transport vials are thin enough that a Vacutainer adapter system can be used for injection to avoid needles.

Sinus Tracts or Deep, Draining Wounds (Figure 2-1)

The skin surface surrounding the sinus tract should be cleaned thoroughly, first with 70% alcohol and then with an iodine preparation, which is allowed to remain wet on the skin for a minimum of 30 s (tincture) or 1 min (aqueous iodophor). Both disinfectants should be applied in a circular motion, moving outward in concentric circles to a radius of approximately 2 cm beyond the sinus tract. Iodine preparation should also be instilled into the opening of the sinus tract at the same time to disinfect the proximal area. The iodine is removed by aspirating, and small curettings of material from deep within the tract are obtained. Surface curettings should be discarded and deeper curettings should then be obtained for culture.

Alternatively, if pus is draining, aspiration of material from within the tract via a flexible plastic catheter and syringe usually provides adequate material for culture. Cultures prepared from a swab inserted into the exterior opening are not as likely to predict correctly the etiologic agent of the infection; secondary colonizers contaminating the wound will be recovered in greater numbers.

Oral or Gingival Abscesses (49;189;272;352)

The flora present in oral secretions are commonly found in perioral abscesses; therefore it is essential to avoid contamination during the collection process. **Swabs from oral sites are never acceptable for anaerobic culture.** Cultures of noninfected oral sites, such as subgingival plaque, should be attempted for research purposes only and only by laboratories experienced in the specialized techniques required for such studies (see Appendix A).

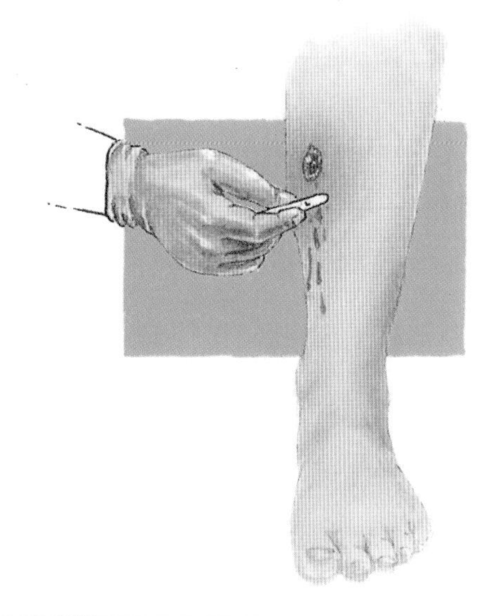

STEP 1: Thoroughly cleanse skin surface surrounding wound with povidone-iodine. Allow to remain wet for 1 min. Wash with ethanol to remove iodine. Allow ethanol to dry.

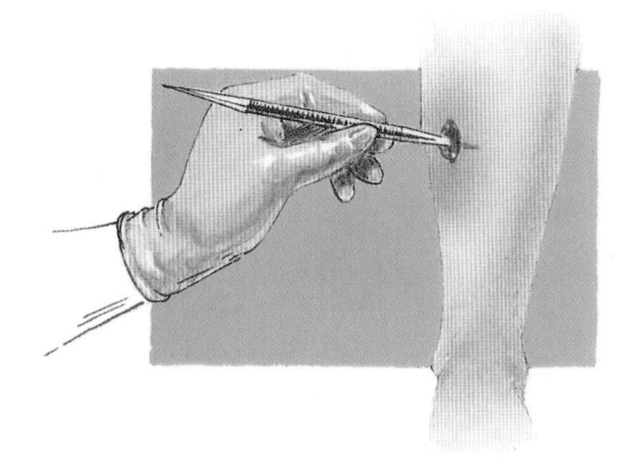

STEP 2: Obtain curetting from deep interior of wound or sinus tract.

NOTE: Swabs are not recommended because they are unlikely to yield clinically significant anaerobes.

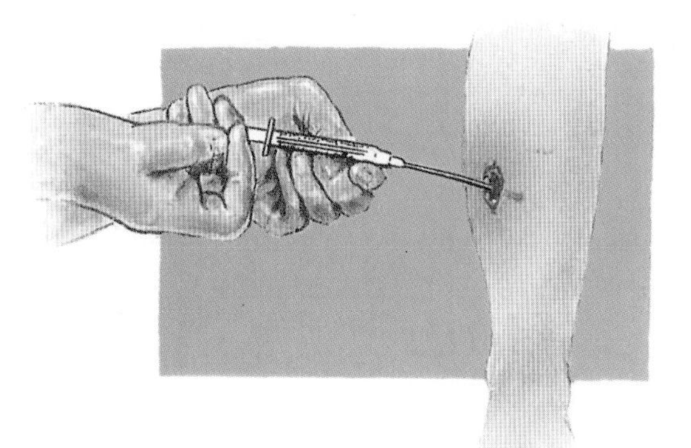

ALTERNATIVE METHOD: Obtain exudate from deep interior of wound using a plastic catheter and syringe.

Figure 2-1 Aspiration of draining sinus tract or deep wound. (From Finegold, Baron, and Wexler, *A Clinical Cuide to Anaerobic Infections,* 1992. Star Publishing Co.)

Aspirated abscess material is the most desirable specimen. Aspiration and drainage via the external skin surface (extraoral route) provides the preferred specimen. If the oral cavity mucous membranes must be entered, the site should be isolated with cotton rolls, dried, and swabbed vigorously with povidone-iodine or chlorhexidine, which is allowed to remain on the site for 1 min before the needle is inserted. Abscess contents are then aspirated. The needle is replaced with a fresh sterile needle, and after pushing the air out of the new needle with the specimen, the remainder of the specimen is expelled into an anaerobic transport vial.

If periodontal areas are being sampled directly, a suitable reduced transport medium should be available for direct inoculation. The area to be sampled should be isolated with cotton rolls and dried with sterile cotton swabs; 70% ethanol may be used to further disinfect and dry the surface. Supragingival plaque must be removed by a sterile scaler. To obtain the sample, either a sterile scaler (nickel-plated Morse 00 is recommended) or sterile Gracey curette can be used to gently scrape material against the root surface from the depth of the sulcus.

Sterile paper points are an alternative; insert three fine paper points to the depth of the pocket and leave in place for 10 s. Whatever the method of obtaining the specimen, it must be placed immediately into an anaerobic transport medium such as VMGA III or other prereduced liquid medium (chopped meat broth, thioglycolate, Ringer's, or yeast extract). Because of the fastidious and oxygen-sensitive nature of many oral anaerobes, prompt processing (preferably in an anaerobic chamber) is essential to obtaining clinically relevant results.

Paranasal Sinus Secretions (170)

Material collected via the nares or upper respiratory tract secretions submitted on swabs are not appropriate for anaerobic culture. Maxillary sinus secretions are obtained by first spraying the nasal cavity below the anterior portion of the inferior turbinate with 10% xylocaine and then swabbing the area to be punctured with a cotton swab soaked in 2% tetracaine to anesthetize and disinfect the area. After waiting 15–20 min, material from the maxillary antrum is aspirated from the sinus by a properly trained physician using a sterile 2 mm diameter puncture needle attached to a 20 ml syringe. If no material is obtained initially even after tilting the patient's head, the needle can be left in place and the syringe filled with 1–2 ml sterile, (reduced) physiologic saline, which is injected into the sinus and re-aspirated.

After the specimen is obtained, a fresh needle is substituted, air is pushed out of the new needle with sample, and the rest of the sample is injected into an anaerobic transport vial. Material from other sinuses is collected only during the course of surgical procedures. Endoscopically obtained secretions are less satisfactory; they should be collected with a double-lumen catheter and injected immediately into an anaerobic transport vial (48).

Superficial Ulcers (Figure 2-2)

Although these types of infections, such as decubitus ulcers and foot ulcers in diabetics, commonly involve anaerobes, they are also predisposed to contamination with fecal, or other, organisms. Thus, collection of material from below the surface is essential. Microorganisms recovered on surface swab specimens represent those found in deeper tissue in less than half of all cases (295). Several strategies for removing contaminated surface tissue are available, including surface debridement with pulsed or

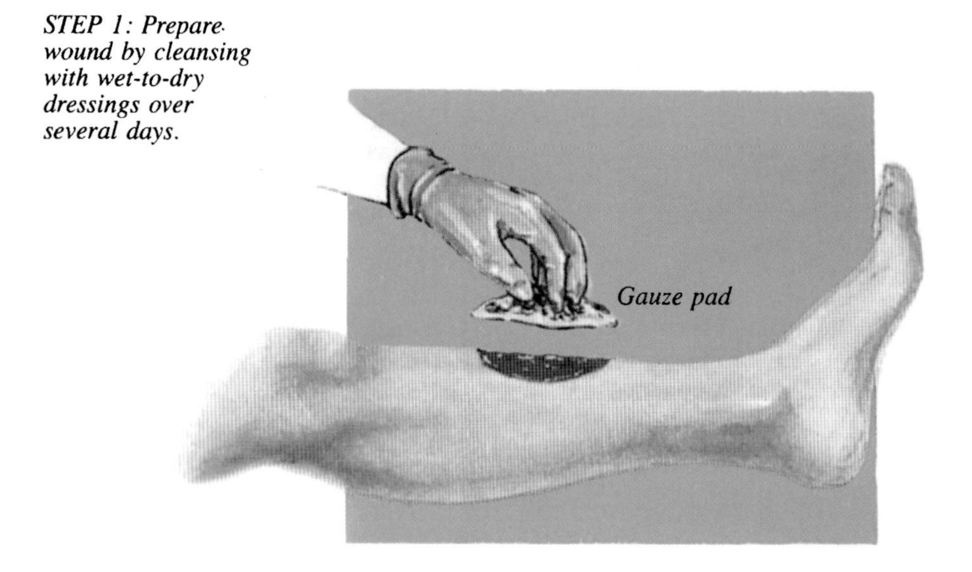

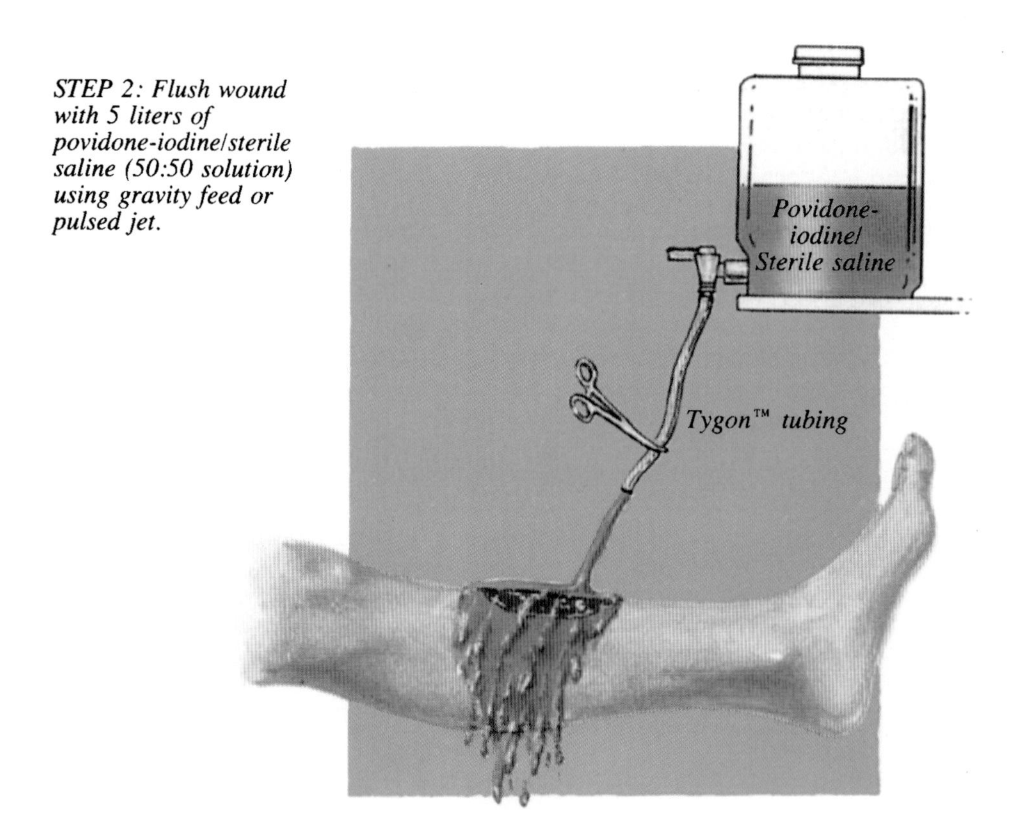

Figure 2-2 Obtaining specimen from contaminated surface wound or ulcer when aspiration is not possible. (From Finegold, Baron, and Wexler, *A Clinical Guide to Anaerobic Infections,* 1992. Star Publishing Co.)

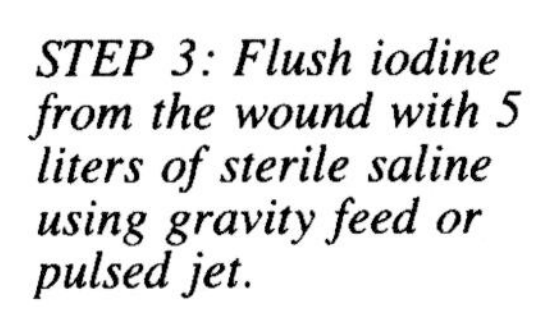

STEP 3: Flush iodine from the wound with 5 liters of sterile saline using gravity feed or pulsed jet.

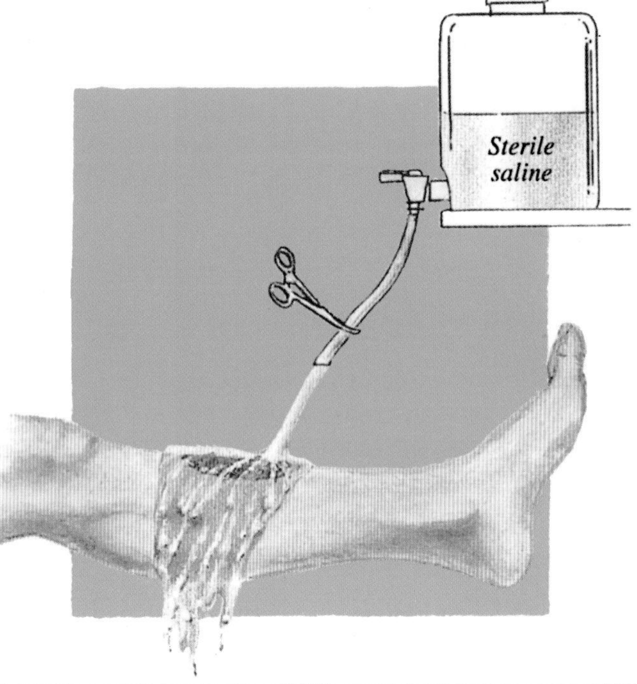

STEP 4:
Use curette to obtain tissue from base of decontaminated ulcer.

LESS RECOMMENDED ALTERNATIVE: Use swab to obtain specimen from base of decontaminated ulcer, avoiding contact with skin.

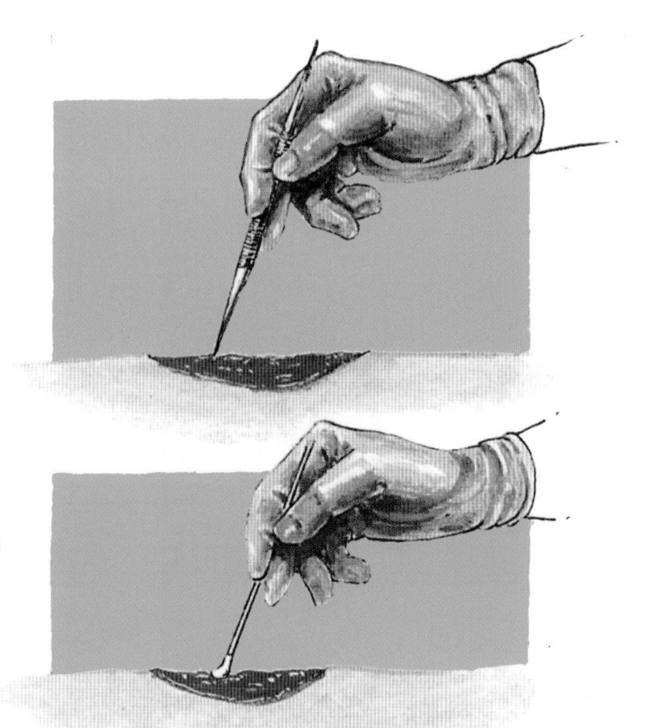

gravity-feed iodophor followed by sterile saline, surface debridement by application of wet-to-dry gauze dressing changes over several days, or direct manual debridement with sterile gauze. Following each of these procedures, superficial materials or curettings are discarded, and material from the base of the ulcer is collected by curette or vigorous swabbing; the specimen should be placed immediately into an anaerobic transport tube. The diabetic foot usually has reduced sensation so that this procedure is not painful to the patient.

Alternatively, collection of purulent material from under skin flaps or from deep pockets using a needle and syringe inserted through disinfected, uninvolved skin may be used. Each of these methods carries some risk of spreading the infection, either into contiguous tissue or into the bloodstream.

Submarginal irrigation-aspiration has been suggested as a less traumatic alternative method for obtaining anaerobic specimens from draining decubitus ulcers and infected diabetic foot ulcers in cases where aspiration fails to yield secretions (102); however, this method tends to underestimate the isolates present compared to carefully collected swab specimens (163).

Respiratory Tract Secretions

Most respiratory tract specimens collected via the oral cavity are not suitable for anaerobic culture because of the substantial normal flora contamination encountered in specimen collection. Specimens in this category include expectorated sputum, aspirated sputum, induced sputum, bronchial washings, standard bronchoalveolar lavages, and bronchial brushings obtained without a protected double-lumen catheter. Respiratory specimens acceptable for anaerobic culture include lung tissue, percutaneous lung or transtracheal aspirates, bronchial brushings collected via a double-lumen protected catheter, protected bronchoalveolar lavage, and thoracentesis fluid (218).

Protected bronchial brush specimens must be placed immediately into reduced anaerobic transport fluid (obtained from the laboratory before beginning the procedure) to serve as the initial diluent and to maintain anaerobic conditions (see Appendix A). Other specimens obtained with a needle and syringe should be injected into an anaerobic transport vial after a fresh needle has been substituted.

Female Genital Tract Specimens

Specimens obtained via the vagina are contaminated with large numbers of the same organisms likely to be involved in the pathologic process; therefore, specialized collection methods for anaerobic culture are required. Although culdocentesis yields excellent specimens, this procedure is used only occasionally at present. Instructions for collection of specimens and diagnosis of bacterial vaginosis are presented in Appendix A.

After the cervical os has been adequately disinfected by wiping off excess mucus and then swabbing with povidone-iodine, a protected sampling device may be inserted for collection of intrauterine material. The endometrial suction curette (Pipelle) adequately recovers organisms from the upper genital tract. The AccuCulShure sintered plastic collection device may also prove to be useful. Both devices include a double-lumen collector and a self-contained transport system. The Pipelle uses suction to obtain cellular material from the uterine wall (see Appendix A); this is the most desirable specimen. Pathogens other than bacteria, including *Chlamydia,* mycoplasmas, and viruses, should also be sought in these specimens.

Specimens from upper female genital tract infections, such as tuboovarian abscess material, should be collected at the time of surgery in the same manner described for abscesses or tissue.

Swabs of vaginal discharge are not appropriate for anaerobic culture, although a Gram stain can be used to diagnose bacterial vaginosis, a process whose etiology involves mixed anaerobic bacteria (188).

Specimens Collected at Time of Surgery

Any tissue obtained at surgery should be appropriate for anaerobic culture. The surgeon should place a small portion of tissue (at least 5 mm^3 in size) representing the infected site (necrotic or gangrenous tissue, abscess wall, etc.) directly into an anaerobic transport tube with a screwcap and an agar plug in its base for maintaining moisture and reducing oxygen. A large piece of tissue (> 1 cm^3) may be transported in a sterile specimen container (or urine cup) if laboratory processing occurs within 1–2 h.

Other Specimens

Certain other types of specimens may be suitable for anaerobic culture if they are collected to exclude air and contamination, such as urine obtained via a suprapubic bladder tap. Other specimens, although not suitable for routine anaerobic culture, may be processed to determine the presence of a particular species or group (e.g., stool culture for *Clostridium difficile* only, vaginal discharge material for *Bacteroides fragilis*–group organisms). A Pap smear is far more sensitive than culture for demonstration of *Actinomyces* species or *E. nodatum* from a woman with an intrauterine device (97). For culture of such specimens, selective media must be used.

SPECIMEN TRANSPORT

Specimens must be protected from the deleterious effects of oxygen until they can be cultured. In a proper anaerobic transport medium, anaerobic bacteria may survive for up to several days, depending on the nature of the specimen. Purulent specimens contain numerous reducing compounds and are more protective of anaerobic viability than are clear fluids. Anaerobes survive particularly well in pieces of tissue, especially larger ones. Certain specimens, such as stool and urine, contain enzymes that degrade bacterial components; these specimens must be plated as soon as possible. Specimens should be transported and held at room temperature; incubator temperatures will cause differential bacterial overgrowth or loss of some strains and cold temperatures will allow increased oxygen diffusion.

Aspirated material must not be transported in the syringe. Not only is the transport of an unsheathed (even corked) needle extremely unsafe, but even without a needle attached, the syringe itself may easily be jostled or pushed, which will eject the material and create a hazard. Aerosols may be created, nearby people and the environment may become contaminated, or the specimen may be lost. Several companies produce vials containing reducing substances in an anaerobic environment, capped with a Hungate-style cap or with a rubber septum, through which the specimen can be injected (Figure 2-3). These systems maintain relative numbers and viability of anaerobes very well, and can even be used to transport exudates and fluid specimens overnight (18;39).

Body fluids may be transported in sterile tubes if anaerobic transport vials are not available. Purulent samples >1.0 ml in volume will maintain viability of anaerobes adequately for several hours. Clear fluids should be placed into a small container (tube, vial) with as little airspace above the fluid level as possible. The tubes should be kept upright to avoid mixing with air and should be maintained at room temperature.

Anaerobic transport vials usually contain modified Cary-Blair or other medium containing agar, reducing substances to scavenge excess oxygen, and an oxygen tension indicator (usually resazurin). Tissue pieces and curettings should be placed into either an anaerobic transport vial with a screwtop Hungate cap (Anaerobe Systems) or into a small, sterile tube that is immediately placed under anaerobic conditions. Sealable plastic bags and components to generate an anaerobic atmosphere inside the bag are available commercially (BBL, Becton-Dickinson, Hardy Diagnostics, Merck); these are very useful for transporting tubes or small containers of specimens (with caps loose until the atmosphere is anaerobic) or tissues in Petri dishes (Figure 2-4). Very large pieces of tissue may be transported in a sterile urine cup if they are to be processed within a reasonable time after collection. The interior of such specimens will probably remain reduced over some time (up to several hours at least) if they remain moist.

For swabs (the least desirable specimen), anaerobic transport medium in soft agar deeps is available from BBL and Anaerobe Systems (Figure 2-5). The swab (plastic

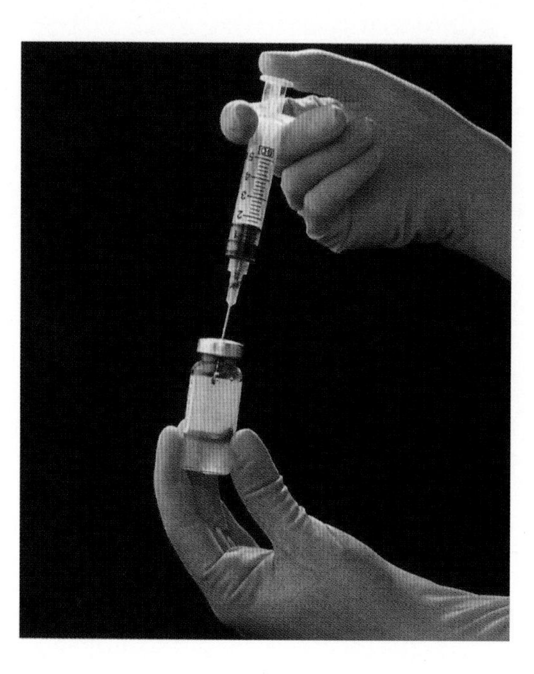

Figure 2-3 Anaerobic transport vial for liquid specimen. Inject specimen without introducing air into the vial (Caution: inject slowly to ensure that specimen remains on top of agar) (From Finegold, Baron, and Wexler, *A Clinical Guide to Anaerobic Infections,* 1992. Star Publishing Co.)

Figure 2-4 Anaerobic pouch is an ideal system for transporting tissue specimens. (From Finegold, Baron, and Wexler, *A Clinical Guide to Anaerobic Infections,* 1992. Star Publishing Co.)

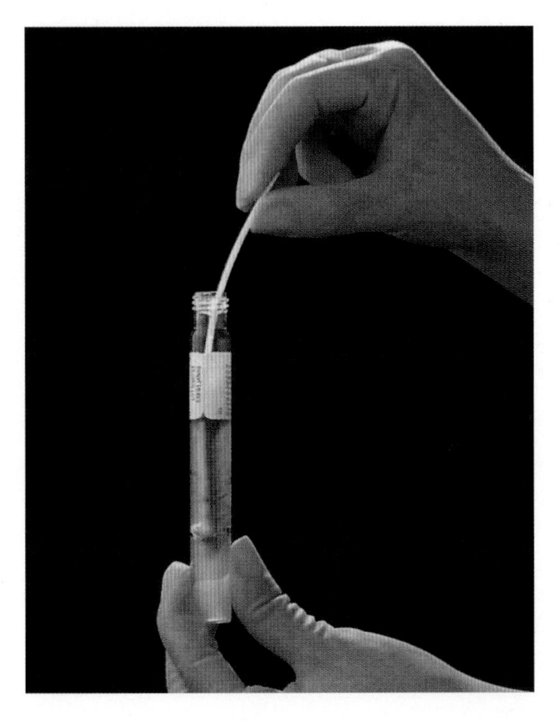

STEP 1. Insert swab deep into agar (contains reducing substances to remove residual oxygen).

STEP 2. Break off contaminated end of swab shaft.

STEP 3. Immediately replace cap.

Figure 2-5 Anaerobic transport tube with agar deep for transporting swabs when no other specimen can be obtained. (From Finegold, Baron, and Wexler, *A Clinical Guide to Anaerobic Infections,* 1992. Star Publishing Co.)

shaft only) should be placed deep into the agar butt, the swab stick broken off high where held, and the cap replaced quickly. Two newer swab systems containing Amies medium and gel have been found to satisfactorily preserve viability of most anaerobes for at least 24 h (72;258). No organisms in swab specimens seem to survive well for longer than 48 h (260).

ANAEROBIC CULTURE TECHNIQUES

Anaerobic methods include the use of anaerobic jars or boxes, plastic anaerobic bags, the roll-tube method of Hungate (157) (and modifications), and the anaerobic chamber or glove box. The roll-tube procedures are rather time consuming, require complex equipment, and utilize PRAS media for culture and identification. Anaerobic chambers, especially the glove-free models, utilize standard methods and are convenient and cost-effective when larger numbers of specimens are processed. They are essential for work on anaerobes of the normal flora as these organisms are more oxygen-sensitive than clinical isolates. Comparative studies have shown that when clinical specimens are collected, transported, and processed properly, recovery of clinically significant anaerobes is as good with anaerobic jars as with the more complex methods (321).

We do not recommend the large diameter Gas Pak-150 jar for incubation of primary plates or the large freestanding anaerobic incubators for routine work since they do not minimize oxygen exposure. Small jars opened one at a time are more satisfactory for bench work. Small incubators inside anaerobic chambers are particularly useful, as the plates can be inspected at 24 h or earlier for rapidly growing anaerobes such as the *B. fragilis* group and *Clostridium* species, if necessary. In anaerobic chambers that require the use of desiccants, store the plates in a gas-permeable plastic bag while incubating, to avoid drying out the media.

Use of a liquid medium as the only anaerobic culture technique is not satisfactory except in the case of blood cultures. Anaerobic infections usually contain multiple organisms and it might be difficult to isolate the individual components of the mixture from a liquid medium. Facultative organisms may overgrow the anaerobes in broth, making quantitation impossible. Accordingly, liquid media should be used only as a back-up medium to other anaerobic culture methods.

Blood Culture Techniques

Liquid Media

A number of commercially available media appear to be satisfactory for recovery of anaerobic bacteria, but considerable inoculum size–dependent differences in performance of individual products may exist (74;281;355). Some contain sodium polyanethol sulfonate (SPS), which has been reported to enhance the recovery of some anaerobes but may be inhibitory to *Peptostreptococcus anaerobius*. This inhibitory effect can be overcome by addition of 1.2% gelatin. Some of the media available in unvented blood culture bottles, with or without SPS, and prepared under vacuum with CO_2 added, are tryptic soy broth, thiol broth, and Columbia broth (BBL), trypticase soy broth (B-D), thioglycolate medium 135C (B-D), Septi-Chek

(Roche), Wilkins Chalgren broth (Hemoline; bioMerieux, Marcy l'Etoile, France), and modified Schaedler broth (BIO ANAER; Sanofi Pasteur, Narnes-La-Coquette, France).

Automated Systems

The automated systems that will be discussed are the BACTEC (Becton-Dickinson) nonradiometric systems and the BacT/Alert (Organon Teknika Corp.). The BACTEC systems are based on growing organisms' production of CO_2, which causes a pH change in the medium. A change in a fluorescence indicator, which is quenched in the presence of acidity, is measured by a sensor in the instrument. When the CO_2 concentration reaches a specific threshold value the culture is considered positive. Previous BACTEC systems used radioactive indicators of CO_2 which led to problems such as potential hazards of storage, monitoring, and disposal of radioactive material. The next generation system used infrared detection of bacterial metabolism. Studies performed indicated no significant difference in the recovery of anaerobic bacteria or in the mean time for detection of anaerobes between the radiometric and infrared system (173). The newest fluorescence-based system seems to have equally good recovery (381).

Another automated system designed to detect anaerobic growth in blood culture bottles is the BacT/Alert (BTA; Organon Teknika). Like the BACTEC fluorescence monitoring system, the BTA uses minimal sample invasion and maintains virtually continuous monitoring of all bottles. A semipermeable membrane and a colored CO_2 sensor are located at the bottom of each culture bottle. If bacteria are present, they generate CO_2 which diffuses through the membrane and reacts with the sensor to produce hydrogen ions. The sensor registers the subsequent drop in pH by a gradual color change from blue-green to yellow. This, in turn, alters the amount and type of light reflected from the sensor. The sensor is illuminated by an LED light source. Reflected light is converted by a photodiode to a signal that is proportional to the concentration of CO_2 in the bottle. The instrument is highly automated; once a bottle is entered into the system, it need not be manipulated until it is removed. Positivity depends on changes in CO_2 level, total level of CO_2, or initial CO_2 level. Unless the bottle is read as positive, it is not handled by a technologist. This system performs essentially as well as BACTEC for recovery of anaerobes (340;375;381).

Standard Blood Culture Processing

After properly cleaning and disinfecting the skin, draw blood from the patient and inoculate it immediately into the medium at a ratio of 1 ml of blood to 10–20 ml of medium. It is advisable to inoculate two bottles (containing different media compositions), venting one for maximal recovery of strict aerobes (these should also be checked for anaerobes) and leaving the other unvented for recovery of anaerobes. When the bottle is inoculated with the patient's sample, air may be inadvertently introduced into the system (although, for most systems, it shouldn't be if the proper precautions are taken); therefore, to avoid further aeration, do not shake the anaerobic bottles. Examine cultures daily for turbidity, colonies, hemolysis, or gas. Gram-stained smears and blind aerobic subcultures should be made routinely after the first 6–12 h of incubation. All bottles showing macroscopic growth should be subcultured anaerobically and aerobically on blood agar plates. Negative culture bottles should be held for 5–7 d, based on the laboratory-validated protocol.

Anaerobic Incubation Methods (Table 2-2)

Jar Techniques

Anaerobic jars (Almar, BBL, EM Science, Key Scientific, Mitsubishi, and Oxoid) are used primarily with plated media, both primary plates and subcultures (Figure 2-6). Reliable results from any anaerobic jar require adequate replacement of the oxygenated environment with an anaerobic atmosphere. This is achieved by introducing a gas mixture containing H_2 into the jar. In the presence of a catalyst, the oxygen combines with the H_2 to produce water, thus establishing anaerobiosis. A catalyst composed of palladium-coated alumina pellets (Coy Manufacturing Co., Anaerobe Systems, and Engelhard Industries) held in a wire mesh basket is preferred because it is convenient to use and this "cold" catalyst poses no explosion hazard. The catalysts must be regenerated after each use, as excess moisture and H_2S will inactivate the pellets. This process requires heating the catalysts to 160 °C in a drying oven for 2 h; a toaster oven set to 320 °F is a convenient alternative method. After cooling, the reactivated catalysts can be stored in a dry area until further use. After extended use, the catalyst pellets may be boiled for 2 h in mild acid to remove surface deposits (see Appendix C). Reduced conditions in the jar are monitored by the addition of methylene blue or resazurin indicators. These indicators, initially blue and pink (respectively), become colorless with low concentrations of oxygen. A biologic indicator may also be used (see "Anaerobic Chamber Techniques," this chapter).

Anaerobic jars can be set up by different methods. One method employs commercially available gas-generating envelopes (Figure 2-7). In some systems, H_2 and CO_2 are released when 10 ml of water is added (GasPak, BBL; Anaerobic Gas generator, Oxoid). The envelope is placed into the jar (equipped with a catalyst and indicator), water is added, and the jar is sealed. Production of heat within a few minutes (detected by touching the lid or the outside of the jar near the catalyst) and subsequent development of moisture on the walls of the jar are indications that the catalyst and generator envelope are functioning properly. Anaerobiosis is achieved in 1–2 h, although the indicators may take longer to decolorize (321). New catalyst-free systems (Anaero-Pack, Key Scientific, Mitsubishi; Anaerogen, Oxoid) activate in air, and no catalysts or water are needed. Another catalyst-free system uses iron filings packaged in an envelope, which is saturated with water and placed inside the jar (Anaerocult, Remel).

Alternatively, the evacuation-replacement system may be used. Air is removed from the sealed jar by drawing a vacuum of 25 in of mercury. This process is repeated two times, filling the jar with an oxygen-free gas such as N_2 between evacuations. The final fill of the jar is made with a gas mixture containing 80–90% N_2, 5–10% H_2, and 5–10% CO_2. Once again, an indicator is included to indicate anaerobiosis. Anaerobiosis is achieved more quickly by the evacuation-replacement method. Both methods give comparable yields of anaerobes from clinical specimens, if the specimen is properly transported and set up in jars immediately after plates are streaked.

The Anoxomat (Mart BV^R Laboratorium, Holland) is an automated method that employs the evacuation-replacement technique to create an anaerobic environment in the jar (35;321) (Figure 2-8). This fully automated system is similar to the technique already described but it also provides an internal quality assurance program that executes additional tests when selected. These tests include leak checks, a test for sufficient

(a)

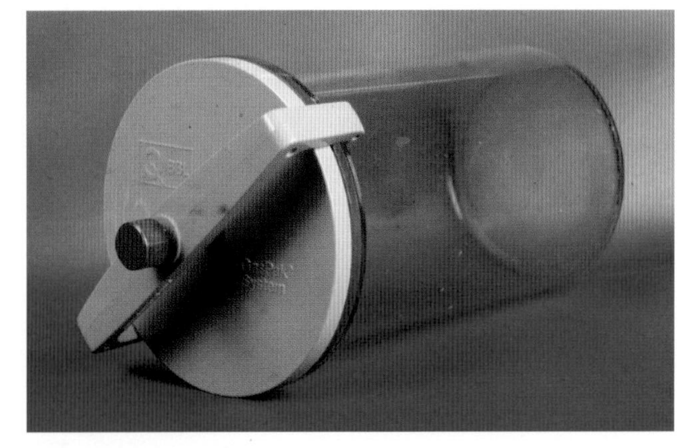

(b)

(c)

Figure 2-6. Anaerobic jar systems: BBL (a), Oxoid (b), and Mitsubishi box (c).

table 2-2. Comparison of culture methods

Factors	PRAS roll tubes	Anaerobic chamber
Initial cost	Moderate	High
Continuing cost	Moderate	Moderate
Time factor	Moderate	Moderate
Space required	Moderate	Moderate-Large
Recovery of significant anaerobes	Satisfactory	Satisfactory
Suitability for specimen processing[a]	Not suitable	Allows processing without exposure to oxygen
Principal advantages	Tubes can be inspected at any time without disturbing anaerobic conditions	Plates can be inspected at any time under anaerobic conditions; conventional isolation procedures are employed
Principal disadvantages	Technique requires more training time; cumbersome to isolate and purify strains; colonial morphology frequently not distinctive	Space requirement; high initial cost

[a] Inoculation, tissue grinding, specimen homogenization with a blender.

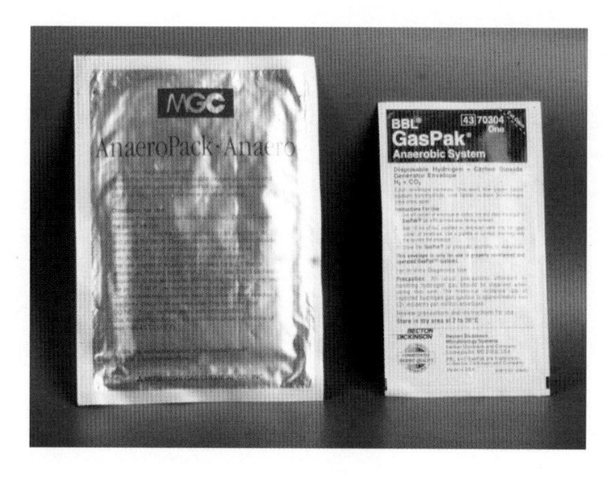

Figure 2-7. Gas-generating envelopes.

catalyst activity, and a test for adequate addition of replacement gas. Studies in our laboratories comparing the incubation of primary plates in the Anoxomat jars, the conventional anaerobe jars, and the anaerobic chamber have shown a slight increase in the recovery of anaerobes from clinical specimens including the oral cavity with the Anoxomat system compared to the other two systems (117;321).

for isolation of anaerobes from clinical specimens

Anaerobic jar	Anoxomat	Anaerobic Bag Technique
Moderate	High	Low
Moderate	Low	Moderate
Low	Low	Low
Moderate	Moderate	Small
Satisfactory	Satisfactory	Satisfactory
Not suitable	Not suitable	Not suitable
Uncomplicated and convenient; conventional techniques employed	Uncomplicated and convenient; conventional techniques employed. Automated, offers quality assurance for jar performance	Plates can be inspected at any time without disturbing anaerobic conditions; uncomplicated and convenient; conventional techniques employed
Possibility of prolonged exposure of plates to oxygen during inspection and subculture	Possibility of prolonged exposure of plates to oxygen during inspection and subculture	Holds only a few plates; possibility of prolonged exposure of plates to oxygen during subculture

Figure 2-8. The Anoxomat system. (Courtesy of Mart BV, Holland.)

Some laboratories may modify the jar technique to create a type of holding jar for multiple primary plates. Essentially, the holding jar is equipped with a vented lid. After the plates have been placed into the jar, the jar is filled and maintained with a continuous slow stream of oxygen-free gas. Nitrogen gas is preferred over CO_2 as the latter lowers the pH of the agar. A holding jar is left at room temperature until it is completely filled. We are skeptical about survival of delicate anaerobes in such a jar. If you must utilize this type of system, we suggest that you use an oxygen-sensitive organism such as *Fusobacterium nucleatum* to evaluate anaerobiosis and that you close and incubate the jar within 2 h to facilitate anaerobic recovery. We prefer batching properly transported specimens and culturing them

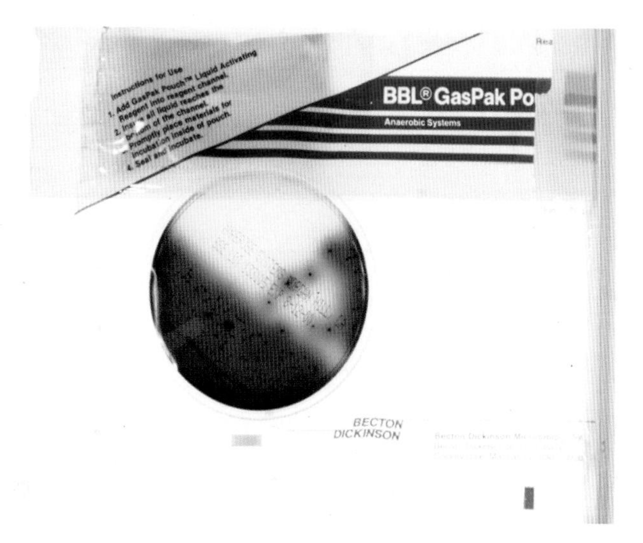

Figure 2-9. Anaerobic bag (BBL GasPack Pouch) containing a BBE plate growing *B. fragilis*.

on one occasion at a convenient time. Single specimens could be processed and individually incubated in anaerobic pouches or jars, depending on the number of plates set up.

Anaerobic Bag Techniques

The anaerobic bag systems commercially available include the GasPak Pouch (BBL Microbiology Systems), AnaeroPouch (Mitsubishi), AnaerogenCompact (Oxoid), Anabag (Hardy Diagnostics), and Anaerocult P (Merck). The GasPak Pouch consists of a transparent bag (bar or heat sealed), a catalyst-free liquid reagent packet (anaerobic atmosphere), and a methylene blue indicator strip. Anaerocult P utilizes wetted (rusting) iron filings to remove oxygen from the bag atmosphere. AnaeroPouch and Anaerogen utilize a system that activates with air; no water or reagents need to be added. The Anabag accommodates a 150 mm plate used for testing susceptibility with the Etest or multiple 100 mm plates.

It has been shown that if an anaerobic chamber is not available for use, an anaerobic bag can be used to incubate single plates (primary and subculture) with good anaerobic recovery (90) (Figure 2-9).

Roll-Tube Technique

The roll-tube method, developed originally by Hungate (157) and popularized for clinical work by Holdeman and Moore at the Virginia Polytechnic Institute (154), is useful for recovery of strict anaerobes when an anaerobic chamber is not available. All preparations and manipulations of the prereduced, anaerobically sterilized (PRAS) media–containing tubes are carried out under a constant flow of oxygen-free gas, usually introduced through the mouth of the tube via a bent metal cannula or blunt-ended needle. After autoclaving, the tubes of molten agar are spun rapidly in a special automated holder. Centrifugal force causes the agar to solidify evenly in a thin layer on the sides of the tube. Specimen inoculation and colony picking are performed with a bent wire, inserted through the mouth of

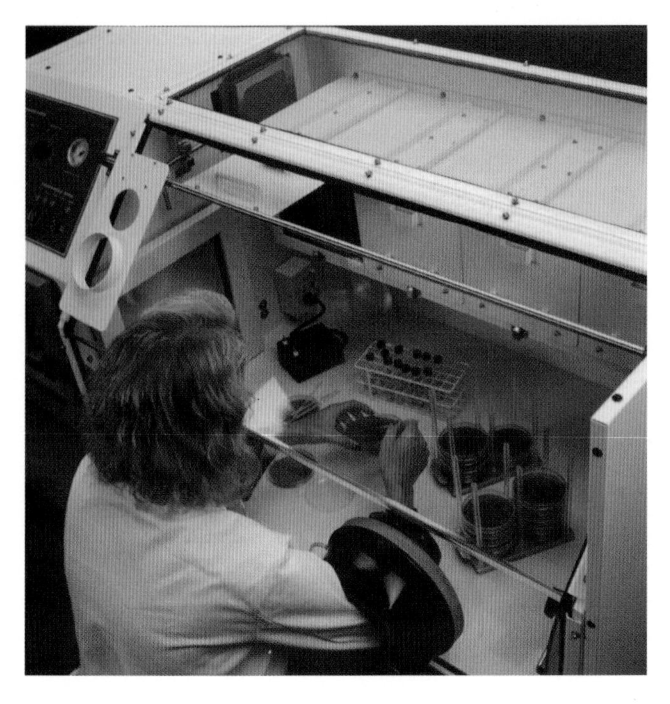

Figure 2-10. Microbiologist working in a gloveless anaerobic chamber.

the tube while oxygen-free gas flows into the tube. The need for specialized equipment, the high level of patience and skill required by users, and the inability to recognize subtle differences in colonial morphology through the agar have discouraged most microbiologists from employing this technique. In addition, most clinically significant anaerobes are not extremely oxygen-sensitive and can be recovered with less stringent methods.

Anaerobic Chamber Techniques

Either flexible plastic bags or rigid gas-tight cabinets can be used as anaerobic chambers. In the first system, access to the inside of the chamber is achieved by means of gloves sealed to the chamber (Coy Laboratory Products, Inc.; Forma Scientific, Inc.). The gloved chamber allows the microbiologist the convenience of entering and exiting the chamber without atmospheric replacement; however, loss of dexterity may be experienced since the gloves tend to be cumbersome. The glove-free chamber (Sheldon Manufacturing, Inc., U.S.A; Ruskinn Technology Limited and Don Whitley, England) utilizes sleeves with cuffs that seal around the forearm and requires evacuation of air in the sleeve by means of a foot pedal switch prior to entering the chamber. This leaves the hands unencumbered by gloves (Figure 2-10).

In both of these systems material is passed in and out of the chamber through an interchange, but the glove-free system has the advantage of permitting passage of small materials through the cuff (sleeve). Concept 300 (Ruskinn Technologies) also has a single-plate entry system allowing immediate transfer of small items. The interchange is a rigid compartment with an inner and outer door and is attached to the chamber by a gas-tight seal. To minimize expense the interchange initially may be evacuated and refilled by a series of flushes with an inexpensive oxygen-free gas such as N_2. The final fill is made with a gas mixture containing 5% H_2, 5–10% CO_2, and 85–90% N_2, which should be safe and nonexplosive (80). The number of evacuations required depends upon the porosity of the material being transported into the chamber. Three evacuations and replacements are usually sufficient. The anaerobic environment of the chamber is maintained by the palladium catalysts and H_2. The H_2 is needed to rid the chamber of any oxygen that may slowly leak through the chamber walls and it also can serve as an electron donor to stimulate the growth of certain anaerobes such as the *B. ureolyticus*–like group (336). The CO_2 is included since many anaerobes require it for growth. The balance of N_2 is chosen for its low cost and low flammability and explosion properties.

A new methylene blue or resazurin indicator strip should be opened within the chamber every 1–2 d or one may open a tube of PRAS peptone yeast extract broth (PY) with resazurin indicator and cap it loosely with aluminum foil. A biologic indicator such as Simmon's citrate slant inoculated with *Pseudomonas aeruginosa* can also be used. Conversion of green agar to deep blue indicates oxygen exposure. In the gloved chambers moisture produced by the catalytic conversion of H_2 and O_2 to water is controlled by its absorption into silica gel crystals with indicator. The desiccant turns from bright blue to pink when water-logged and is recharged by drying in a hot oven. The newer chambers have a cold spot that condenses excess humidity and allows the water formed to be removed through an external drain. In routine use, the catalysts should be rejuvenated on a regular basis, such as every other day. After extended use, the catalyst pellets may be boiled for 2 h in mild acid to remove surface deposits (see Appendix C). Anatox is a charcoal-like substance that absorbs volatile substances produced by growing anaerobes in the anaerobic chamber. It protects the catalysts from being poisoned by H_2S-producing organisms. It can be rejuvenated like the catalysts, and reused for many years.

Common Pitfalls in Specimen Collection and Culture Techniques

- Failure to bypass normal flora in collecting specimen.
- Failure to set up anaerobic cultures promptly from clinical specimens or to keep these under anaerobic conditions pending culture.
- Failure to minimize air exposure during processing.
- Failure to use a good anaerobic jar. Check for cracks and leaks in plastic lids, O-rings, and vent openings.
- Failure to use active catalysts when using hydrogen in anaerobic systems. Catalysts must be reactivated after each use by heating to 160 °C for 2 h.
- Failure to use redox indicator or known fastidious anaerobe in the anaerobic systems used.
- Using toxic gas (such as methane from certain municipal sources) in displacement procedure.
- Failure to include CO_2 in anaerobic gas. Carbon dioxide is essential for the growth of some anaerobes.

chapter 3

Processing Clinical Specimens and Isolation Procedures

Initial Processing Procedure

Initial processing includes direct examination and inoculation of the specimen onto appropriate media as represented in Figure 3-1.

Direct Examination

Direct examination involves macroscopic and microscopic examination of the specimen, and may in selected cases include gas-liquid chromatography, darkfield microscopy, DNA probe screening techniques, or direct nucleotide sequencing (266). The direct microscopic examination provides immediate semiquantitative information about the types of organisms present and presumptive evidence about the presence of anaerobic bacteria (sometimes specific anaerobes). Since culture results may not be available for several days, this information is important in helping to decide the initial therapy.

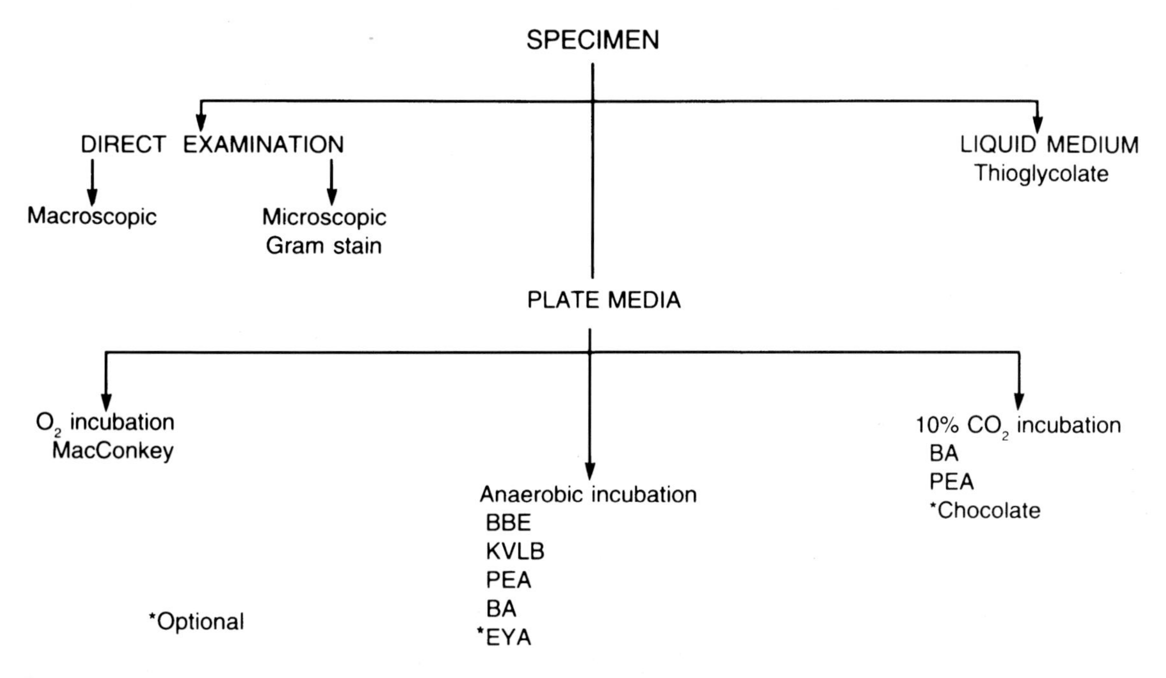

Figure 3-1. Initial specimen processing scheme.

Macroscopic Examination

The macroscopic characteristics for which the specimen is inspected are (1) foul odor (due to the volatile fatty acid and amine endproducts of metabolism and H_2S produced by anaerobic organisms); (2) brick-red fluorescence due to pigmented *Porphyromonas* species or *Prevotella* species; (3) black necrotic tissue or discharge which may be due to the pigmented gram-negative rods; (4) "sulfur" granules (Figure 3-2); (5) purulence; and (6) blood.

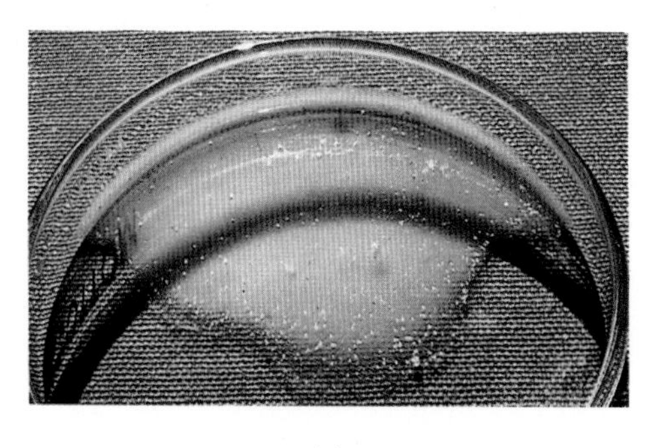

(a)

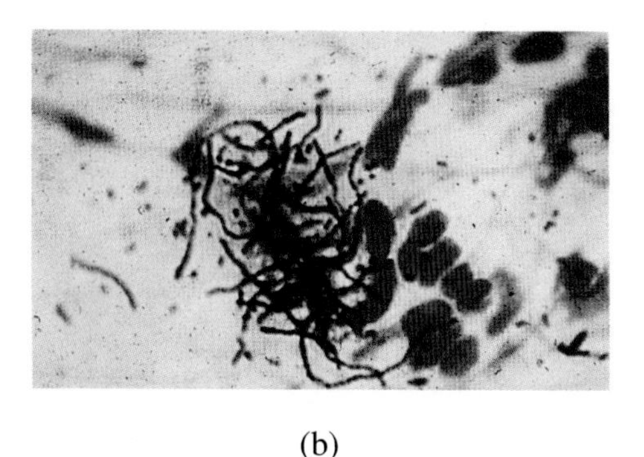

(b)

Figure 3-2. Pus with sulfur granules aspirated from chest mass of patient (a), and Gram stain showing characteristic morphology of *Actinomyces* (b). (From Finegold, Baron, and Wexler, *A Clinical Guide to Anaerobic Infections,* 1992. Star Publishing Co.)

Microscopic Examination

The microscopic examination should always include a Gram stain (Figure 3-3). Fix the direct smear in methanol for 30 s to preserve red and white blood cell morphology (214). Standard Gram stain procedures and reagents are used except that 0.1% basic fuchsin is substituted for safranin as the counterstain. Basic fuchsin enhances the staining of gram-negative anaerobes. The Gram stain reveals the types and relative numbers of microorganisms and host cells present, and serves as a quality control measure for the microbiologist. It can also suggest using special media. Failure to recover all the morphotypes seen on the Gram stain smear may indicate a problem in anaerobic techniques in specimen collection, transportation, or processing; occasionally it may be due to the inhibition of the organisms by residual antibiotics in the specimen. In thick films from exudates and body fluids the recognition of organisms may be facilitated by staining with

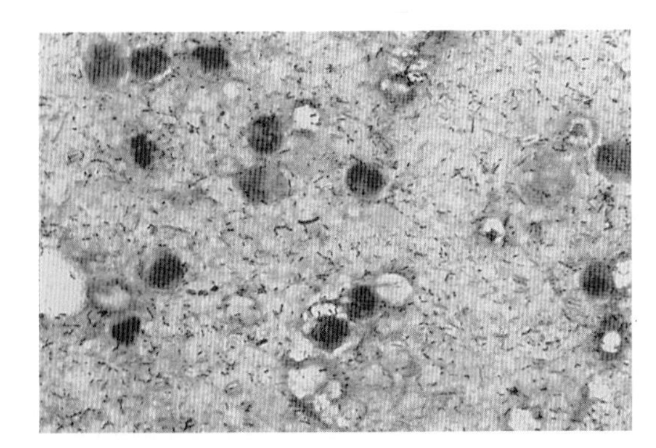

Figure 3-3. Gram stain from peritoneal fluid in a case of perforated appendicitis. Numerous bacterial morphologies are present.

acridine orange. Darkfield and phase contrast microscopy may be helpful in the detection of small, poorly stained organisms (*Dialister pneumosintes*), for direct observation of motility (*Campylobacter* spp.), for noting spores (*Clostridium* spp.), and for the recognition of morphotypes not cultivable on ordinary media (spirochetes) (172). It is always important to correlate the specimen source with microscopic findings.

Pale, irregularly stained organisms are frequently anaerobic. Gram-negative, coccobacillary forms are suggestive for pigmented *Prevotella* or *Porphyromonas* species, but also for *Haemophilus* and *Actinobacillus* species. A thin gram-negative bacillus with pointed ends suggests *Fusobacterium nucleatum* or the microaerobic *Capnocytophaga* species. A very large gram-negative bacillus with one pointed and one square end and intracellular gram-positive stained granules is suggestive of *Leptotrichia buccalis*. *F. mortiferum* typically is an extremely pleomorphic gram-negative rod with filaments containing swollen areas, round bodies, and irregular staining. The same type of morphology can be seen occasionally with *F. necrophorum*.

A small (<5 μm) gram-negative coccus is suggestive of *Veillonella* species or *Dialister pneumosintes*. Observation of gram-positive rods with spores indicates *Clostridium* species; however, many clostridia, such as *C. clostridioforme* (see Figure 6-23), appear gram-negative and spores are not always seen. A large gram-positive rod with boxcar-shaped square cells and no spores can presumptively be identified as *C. perfringens*. A branching gram-positive bacillus may be *Actinomyces*, *Propionibacterium*, or *Bifidobacterium*. (See figures in Chapters 4 and 6 for different cellular morphologies.)

Molecular methods, such as nucleic acid probes and PCR amplification, used in direct demonstration of some oral (85;307;352) and non-oral (264;266;315) gram-negative bacilli, as well as amplification and direct demonstration of anaerobes by 16S rRNA sequencing from clinical specimens, are not yet standardized or produced for commercial distribution. While they are not yet commonly used in clinical laboratories, in the future they will undoubtably offer an option when accurate identification and rapid diagnosis are indicated (172). Preliminary trials have already been carried out (266).

Direct gas-liquid chromatographic analysis of clinical specimens or blood culture bottles for metabolic endproducts may provide rapid presumptive evidence for anaerobic infection together with Gram stain and macroscopic findings (133). Our experience is that it seldom adds to information gained from the Gram stain and macroscopic findings.

Specimen Preparation and Inoculation

After direct examination, the prepared specimen is inoculated onto appropriate anaerobic and aerobic plating media, and into liquid medium (see Figure 3-1).

Initial specimen processing should be done in an anaerobic chamber to minimize aeration. If a chamber is not available, process the specimen within 15 min.

- Mix grossly purulent material by vortexing it under anaerobic conditions to ensure even distribution of the organisms.
- Homogenize pieces of tissue or bone fragments with approximately 1 ml of liquid anaerobic medium using a tissue grinder or large pieces in a tissue tearer (985-370 Biospec Products, Inc.) to make a thick mixture.
- Extract swab specimens in about 0.5 ml of anaerobic broth and then treat them as a liquid specimen.
- Using a Pasteur pipette, add the prepared specimen as follows:

  (1) One drop per plate for purulent material,

  (2) Two to three drops per plate for nonpurulent material,

  (3) One drop onto a slide for Gram stain; spread evenly.

  (4) Put the rest into liquid medium (such as thioglycolate) for enrichment culture.

Media

Use a combination of enriched, nonselective, selective, and differential media for isolation and presumptive identification of bacteria from clinical material. To determine the appropriate media required for the recovery of aerobic and facultative anaerobic bacteria from various sites, refer to the Manual of Clinical Microbiology (242) or any other standard reference source. Sheep blood, chocolate, MacConkey, and phenylethyl alcohol blood agar plates can be used as a minimum set for primary inoculation for recovery of aerobic bacteria.

The following media are used in the Wadsworth and KTL Anaerobic Bacteriology Laboratories for the isolation of the obligate anaerobic bacteria:

(1) Brucella agar (BA) supplemented with 5% sheep blood and vitamin K_1 (1 µg/ml) and hemin (5 µg/ml) as a nonselective medium to support the growth of all bacteria.

(2) Bacteroides bile esculin (BBE) agar for the selective isolation of the *Bacteroides fragilis* (200) group and *Bilophila* (19).

(3) Kanamycin vancomycin laked blood (KVLB) agar for the selection of pigmented and other *Prevotella* species and *Bacteroides* species (309).

(4) Phenylethyl alcohol sheep blood agar (PEA) to inhibit facultative gram-negative rods and to inhibit swarming of certain clostridia.

(5) Thioglycolate medium without indicator, supplemented with vitamin K_1, hemin, and a marble chip, for enrichment and back-up culture.

The primary plating media ideally should be prereduced or freshly prepared. Prereduced, anaerobically sterilized (PRAS) plate media are available commercially (Anaerobe

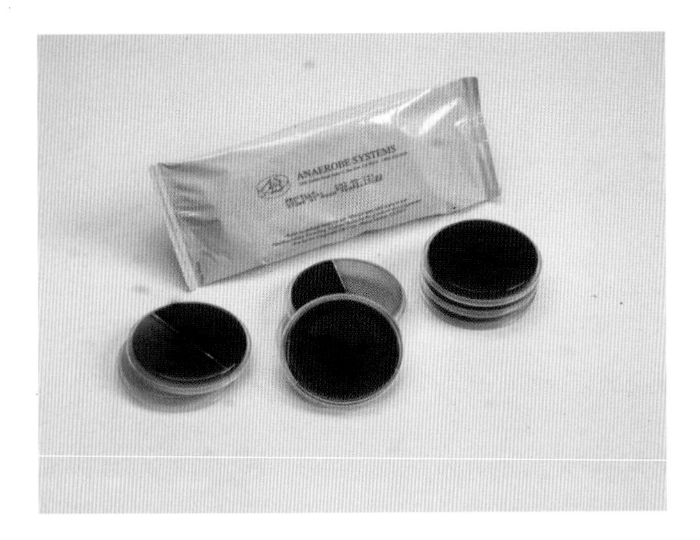

Figure 3-4. Prereduced Anaerobe Systems plates.

Systems) (Figure 3-4). Plate media used for subsequent subculturing usually do not need to be reduced (309). Unless it is purchased in oxygen-free packages (Anaerobe Systems), boil or steam liquid medium for 5 min to drive off the dissolved oxygen; use it the same day. Use of prereduced or fresh media stored under anaerobic conditions will enhance and expedite recovery of anaerobes (215). Prereduced selective media CCFA (cycloserine cefoxitin fructose agar) has increased isolation of toxigenic *C. difficile* three-fold (260). Any one-day reduction in the turnaround time of laboratory results will contribute to considerable savings in patient care (estimated at $2,000 per day) (260).

In addition to Brucella agar base, brain heart infusion supplemented with 0.5% yeast extract, CDC, Columbia, FAA (fastidious anaerobe agar, Lab M, England), or Schaedler agars can be used as a base for anaerobic blood agar plates. The following supplements must be added to the base medium to enhance the growth of anaerobic bacteria (309): (1) 5% sheep, horse, or rabbit blood; (2) vitamin K_1 (1 µg/ml); and (3) hemin (5 µg/ml). The growth of anaerobes may vary according to the basal medium used (215;241;260). It has been reported that Brucella agar supports the growth of anaerobic gram-negative bacilli better than CDC (trypticase soy) or Schaedler, and that gram-positive cocci grow in higher numbers on CDC agar than on Brucella agar (231;296). The FAA medium allows fusobacteria and some formate-fumarate–requiring species to grow luxuriantly (32;33;172). Brain heart infusion agar is superior to trypticase soy in isolation efficiency for *Eubacterium* species, but far inferior for pigmented gram-negative rods from subgingival and other samples (231).

BBE agar is useful for rapid isolation and presumptive identification of the *B. fragilis* group. The medium contains 100 µg/ml of gentamicin to inhibit most aerobic organisms and 20% bile to inhibit most other anaerobes. As differential agents, esculin and iron have been incorporated into the agar to aid in detecting the esculin-positive *B. fragilis*–group organisms. BBE agar is also a useful medium for *Bilophila* species, which usually produce black-centered colonies on this medium due to H_2S production. Other organisms that may grow on this medium are occasional strains of *Fusobacterium mortiferum* and *F. varium,* gentamicin-resistant *Enterobacteriaceae,* enterococci, pseudomonads, staphylococci, and yeast. However, the colony size of the facultative anaerobic organisms is usually less than 1 mm in diameter.

KVLB agar is useful for the rapid isolation of *Prevotella* species. Of course, *B. fragilis*–group strains also grow well on this medium. The medium contains 75 μg/ml kanamycin, which inhibits most aerobic, facultative, and anaerobic gram-negative rods except for *Bacteroides* and *Prevotella,* and 7.5 μg/ml vancomycin, which inhibits most gram-positive organisms and *Porphyromonas* species. KVLB with reduced vancomycin concentration (2 μg/ml) may be used when *Porphyromonas* species are looked for (172). The laked blood allows earlier pigmentation of *Prevotella* and *Porphyromonas* species. Yeast and other kanamycin-resistant organisms, such as *Capnocytophaga* species, sometimes grow on KVLB; therefore, one should perform a Gram stain and determine the aerotolerance of all isolates.

PEA should be inoculated for purulent specimens and when mixed infections are suspected. PEA inhibits facultative gram-negative rods, preventing *Enterobacteriaceae* from overgrowing the anaerobes, and inhibits swarming of *Proteus.* It also prevents certain clostridia, such as *C. septicum,* from swarming, which will facilitate the isolation of other colonies. Most gram-positive and gram-negative anaerobes will grow on primary PEA medium, especially in mixed culture, and the morphology of the colonies is similar to that on blood agar plates; however, a longer incubation time may be necessary to detect the more slowly growing and pigmented anaerobes. However, pigment develops more rapidly than on nonselective Brucella agar.

Thioglycolate broth serves as a back-up source of culture material, which may be needed in the case of failure of anaerobiosis or growth inhibition on plates due to antibiotic or other antibacterial factors (e.g., polymorphonuclear leukocytes) or small numbers of bacteria in the specimen.

If clostridia are suspected clinically or from the Gram stain of the clinical material, a primary egg yolk agar (EYA) plate can be inoculated to check for the production of lipase and lecithinase. If boxcar-shaped cells are seen in the Gram stain, a direct reverse-CAMP test can be performed on extra BA (see Appendix C); however, care must be taken to interpret the results if culture yields mixed growth. If culturing for *C. difficile,* a CCFA or CCEY plate should be inoculated.

For other selective and nonselective media that may be useful, see Tables A-1, A-3, and A-4 in Appendix A.

Incubation of Cultures

Streak inoculated plates in a four-quadrant fashion to yield isolated colonies and immediately place them in an anaerobic environment (jars, chambers, anaerobic pouches). If plates are incubated in jars, the cultures should not be exposed to air until after 48 h of incubation, since anaerobes are most sensitive to oxygen during their logarithmic phase of growth. Plates incubated in chambers or anaerobic pouches can be inspected for growth after 24 h or earlier, as they do not need to be removed from the anaerobic environment. Selective media, such as BBE for the *Bacteroides fragilis* group or EYA for clostridia, may be examined after 24 h of incubation since the selected organisms grow rapidly; however, one should not open an anaerobic pouch or a jar containing the nonselective blood agar plate before 48 h.

Hold negative culture plates for a minimum of 7 d before final examination. Research laboratories should hold plates for 2 weeks. In our experience, with shorter incubation times some anaerobic species, such as *Porphyromonas* and *Actinomyces* species, may not be detected. Inspect the thioglycolate broth accompanying a set of negative plates daily; if it becomes turbid, subculture a few drops onto a blood agar plate (anaerobic incubation) and onto a chocolate or blood agar plate for CO_2 incubation.

EXAMINATION OF CULTURES AND ISOLATION OF ANAEROBES

Anaerobes are usually present in mixed culture with other anaerobes and/or aerobic or facultative bacteria (Figure 3-5). It is important to study all colony morphotypes, since facultative and anaerobic bacteria sometimes have very similar colony appearances. Clues to the presence of anaerobes in a culture include: (1) a foul odor upon opening an anaerobic jar or pouch, (2) more colony types present on the anaerobically incubated BA plate than on the CO_2 incubated plate, (3) growth on BBE or other highly selective agar, and (4) red fluorescing or black pigmented colonies on KVLB or BA.

Examination of Primary Plates

Examine primary anaerobic plates with a stereoscopic microscope or a hand lens. The use of a dissection microscope in the inspection of primary plates is highly recommended because reliance on mere visual impression leads to overlooking small colony types that may represent the predominant growth. Describe colonies from the various media, as listed below, and semiquantitate the growth. The quantity may be described as light, moderate, or heavy, or by the number of quadrants in which the growth occurs (1+, 2+, 3+, 4+ growth).

All different colony types observed should be described, semiquantitated, and subcultured for further testing. A detailed colony description includes (Figure 3-6) (1) size; (2) shape (circular, etc.); (3) edge (entire, irregular, lobate, etc.); (4) profile (convex, flat, umbonate, etc.); (5) color; (6) opacity (transparent, translucent, opaque); (7) any other noteworthy characteristics such as pigment, fluorescence, pitting, hemolysis, or distinct odor.

Pitting colonies on anaerobic BA plates but not on the CO_2 plate are suggestive of the *Bacteroides ureolyticus*–like group. Large colonies with a double zone of β-hemolysis are presumptively *C. perfringens.* These colonies should be Gram stained and subcultured on an EYA plate for the demonstration of lecithinase and on Brucella agar for the reverse CAMP test. Bread crumb–like or speckled colonies that are subsequently shown to contain slender fusiform gram-negative bacilli suggest *F. nucleatum.* A molar tooth–shaped colony of gram-positive branching rods may be *Actinomyces* species or *Propionibacterium propionicus.* The BA plate should also be examined for pigmented organisms and fluorescence under UV light (366 nm). (See figures in Chapters 4 and 6 for different colonial morphologies on various agar plates.)

Figure 3-5. Odontogenic abscess; polymicrobial finding with several colonial types.

From BBE agar, subculture the *B. fragilis*–group colonies, which are typically >1 mm in diameter, circular, entire, convex, and usually surrounded by a dark grey zone (*B. vulgatus* is esculin-negative) (Figure 3-7). A granular precipitate around the colonies is characteristic of *B. fragilis* species. Occasional colonies of fusobacteria that grow on BBE agar are also >1 mm in diameter, but are flat and irregular. Convex colonies on BBE with black centers and translucent margins are suggestive of *Bilophila* and should be subcultured (Figure 3-8). Smaller, translucent colonies may represent *Sutterella wadsworthensis.*

Subculture all colony types from the KVLB plate (Figure 3-9) and check them carefully for pigment in bright light and for red fluorescence under UV light. Touching colonies with a swab or smearing on white blotting paper may help to recognize weak pigment. Pigment production often takes >5 d to develop (167).

Subculture any colonies from PEA that have colonial morphologies not found on the BA plate; subculture all colonies from PEA if the BA plate is overgrown by swarming *Proteus,* clostridia, or other organisms.

Pleomorphic yellow ground-glass colonies with a horse stable odor on CCFA are characteristic of *C. difficile* (Figure 3-10).

For descriptions of typical colony morphologies on selective media, see Appendix A.

Subculture of Isolates

Gram stain each different colony type and subculture onto the following media (Figure 3-11):

(1) BA—anaerobic incubation (extra BA provides material for rapid enzyme tests).
(2) Chocolate agar—CO_2 incubation (aerotolerance).
(3) BA—O_2 incubation (aerotolerance).
(4) EYA for suspected *Clostridium, Fusobacterium,* and *Prevotella* species—anaerobic incubation (optional) for detection of lecithinase, lipase, and proteolytic activities of the isolates.
(5) Rabbit blood agar (RBA) for rapid demonstration of pigmented organisms—anaerobic incubation (optional).

Use only one colony for all subculturing, aerotolerance testing, and Gram stains. Use a sterile wooden stick for transferring, because it picks up the colony and retains enough

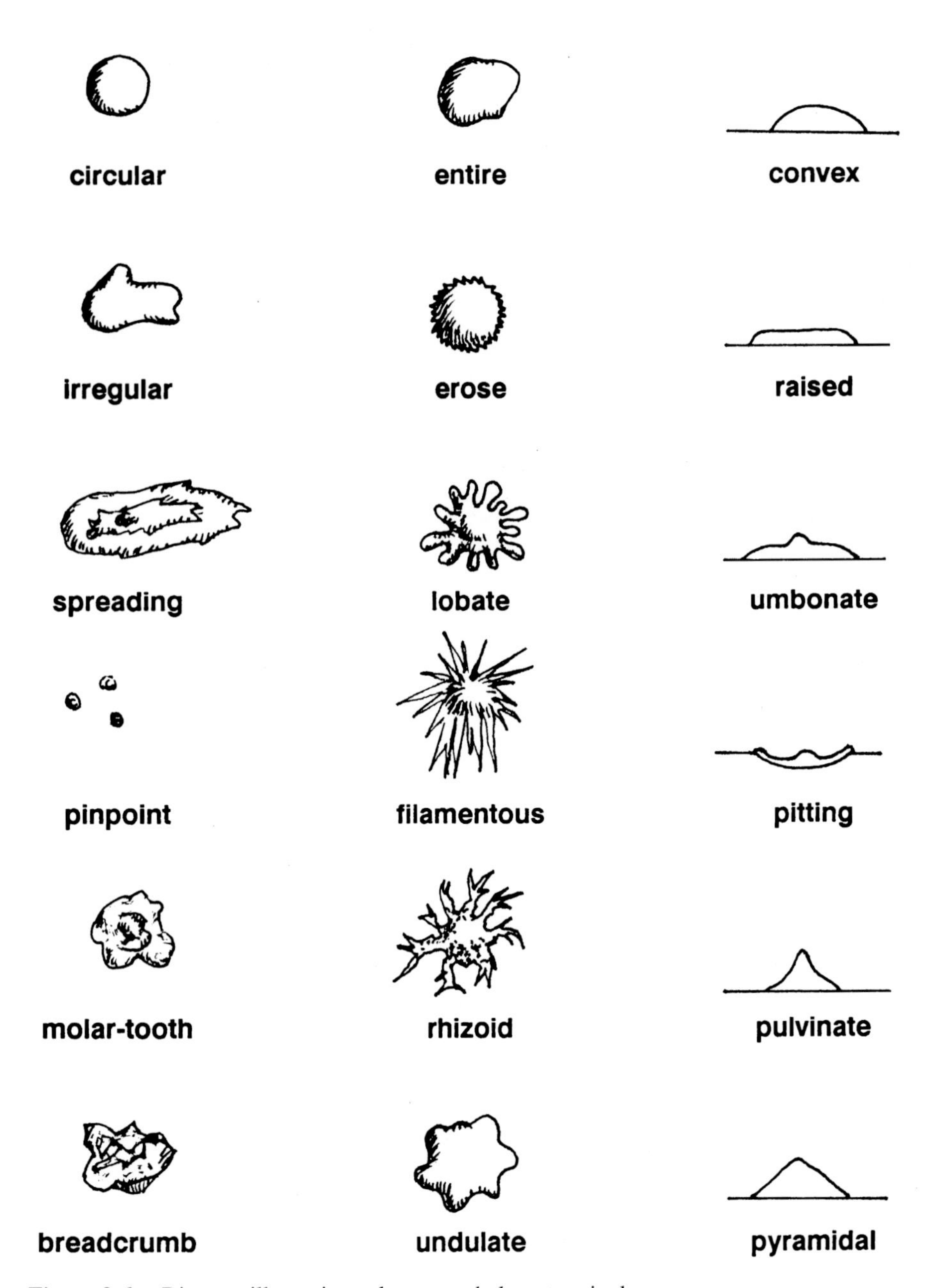

Figure 3-6. Diagram illustrating colony morphology terminology.

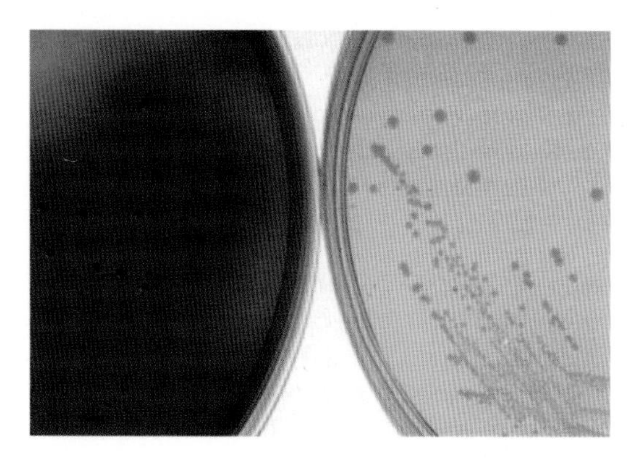

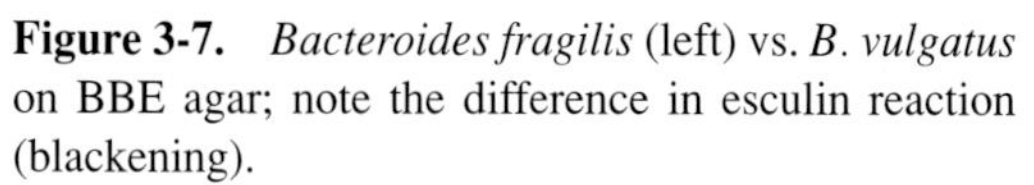

Figure 3-7. *Bacteroides fragilis* (left) vs. *B. vulgatus* on BBE agar; note the difference in esculin reaction (blackening).

Figure 3-8. *Bilophila wadsworthia* on BBE agar; note the black-centered colonies.

material for inoculation of multiple plates and for a Gram stain. If the original colony is not large enough to inoculate all the plates, inoculate only the anaerobic BA plate and use that subculture for subsequent testing. EYA, RBA, and aerotolerance plates can be divided so that several isolates can be inoculated onto each plate.

Streak the first and second quadrants of the anaerobic BA plate back and forth several times to ensure an even lawn of heavy growth; streak the other two quadrants for isolated colonies. Place the special-potency disks (vancomycin, 5 μg; kanamycin, 1,000 μg; and colistin, 10 μg) on the first quadrant (Figure 3-12) well apart from each other. These disks serve as an aid in grouping the anaerobic bacteria and for confirming the Gram reaction, but do not imply susceptibility of an organism for antibiotic therapy (325). A 10 μg sodium polyanethol sulfonate (SPS) disk may be placed near the colistin disk for rapid identification of *Peptostreptococcus anaerobius* (363), and a nitrate disk may be placed on the second quadrant for the detection of nitrate reduction (362). Alternatively, the latter two disks may be used on selected isolates at a later stage.

If processing is done on the open bench, incubate all anaerobic plates promptly (within 20 min), since some isolates may die after relatively short exposure to oxygen (e.g., *F. necrophorum*, *Porphyromonas* spp.). Always reincubate the primary plates along with the subcultures for an additional 48 h, and inspect them again for new morphotypes and pigment production.

A chocolate agar plate incubated in 5%–10% CO_2 atmosphere is required for aerotolerance testing to detect organisms that require CO_2, especially slow-growing, fastidious facultative or microaerophilic species that do not grow alone on blood agar plates (such as *Haemophilus* and *Actinobacillus*). Use of a BA plate alone for CO_2 incubation may yield false negative results. An additional BA plate incubated in air will further define the atmospheric requirements and hemolytic properties of facultative organisms.

Many gram-positive nonsporeforming rods, such as most *Actinomyces* species, many *Propionibacterium* species, *Lactobacillus* species, and some *Bifidobacterium* species can grow in CO_2-enriched ambient atmosphere and may be aerotolerant enough to grow

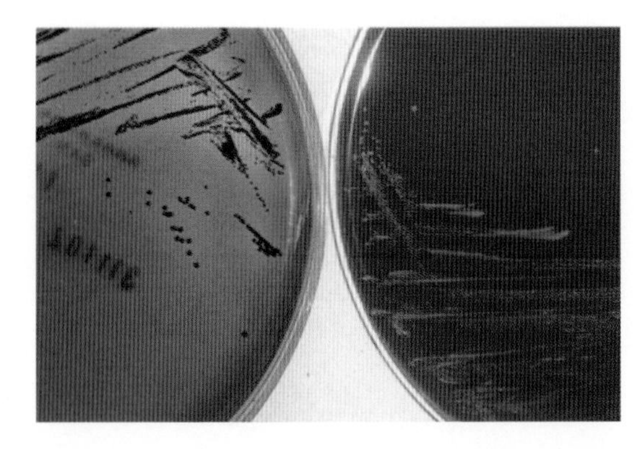

Figure 3-9. *Prevotella intermedia* on KVLB (left) and Brucella blood agar (right); note the stronger pigment production on KVLB.

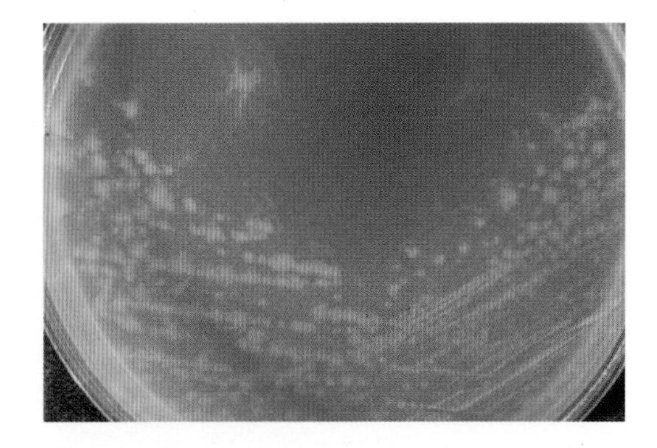

Figure 3-10. *Clostridium difficile* on CCFA agar; note the yellow, ground-glass appearance of colonies and pale pink color of uninoculated medium.

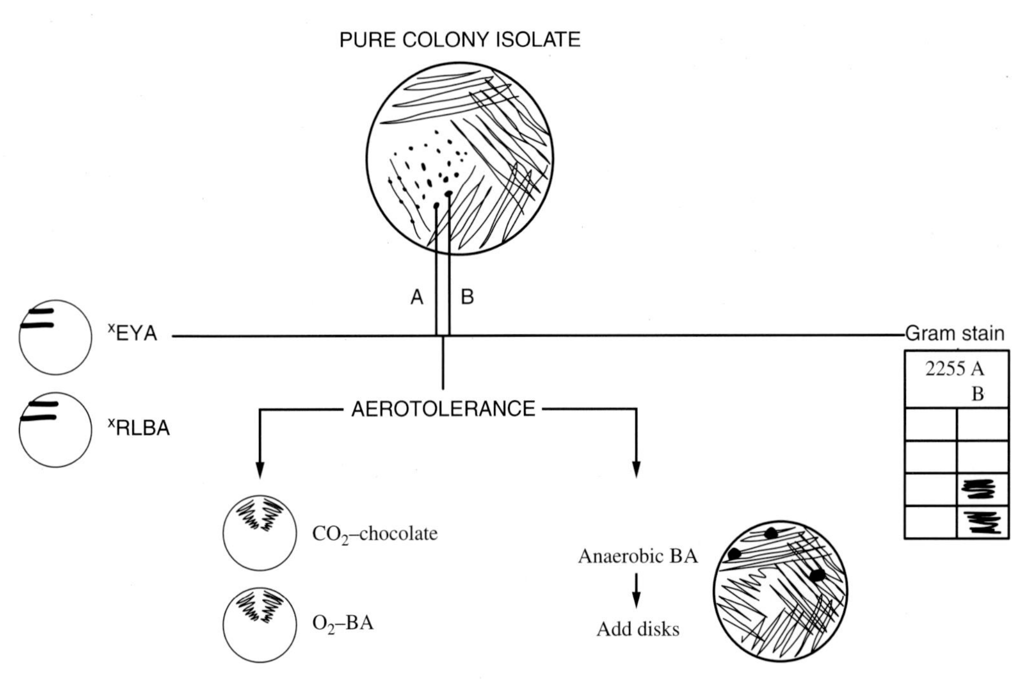

Figure 3-11. Flow chart illustrating scheme for subculturing isolated colonies

poorly in air, but are considered anaerobic bacteria. Several species of *Clostridium*, including *C. carnis, C. histolyticum,* and *C. tertium,* can grow, but not sporulate, in air. Furthermore, some microaerophilic cocci require several subcultures before they can grow in a CO_2 atmosphere.

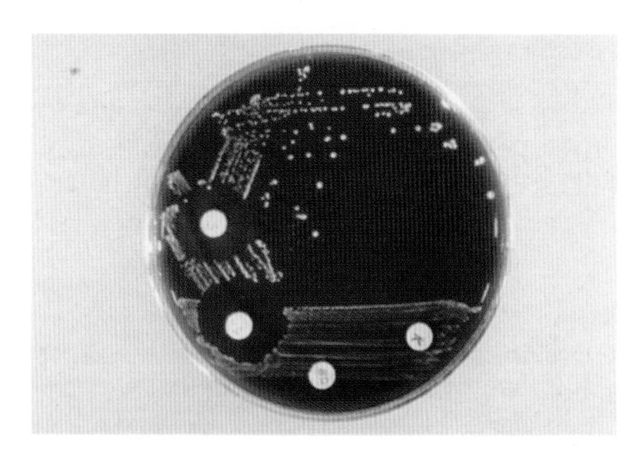

Figure 3-12. Special potency disks arranged on subculture of single colony from primary plate.

Common Pitfalls in Specimen Processing and Isolation Procedures

- Failure to prepare the Gram stain directly from the clinical specimen. The Gram stain alerts one to the possible presence of organisms requiring special media or conditions of incubation and indicates that the techniques are inadequate if the organisms seen on the smear fail to grow.
- Use of thioglycolate or other liquid medium as the only system for growing anaerobes. A number of anaerobes will not grow in thioglycolate medium, even when it is enriched. Solid media are required to separate various organisms in a mixed culture since rapidly growing organisms may overgrow the anaerobes in liquid medium, making recovery difficult.
- Failure to check liquid medium if growth is recovered only from aerobic plates. A Gram stain of the broth often alerts one to the presence of additional anaerobes.
- Use of inadequate media. Failure to use fresh media.
- Failure to do quality control on media and reagents for performance.
- Failure to use supplements in media. Vitamin K_1 and hemin are required by many anaerobic organisms.
- Failure to use selective media. Some anaerobes may be overgrown by more rapidly growing organisms and overlooked if selective media are not used.
- Failure to hold cultures for extended periods. Some fastidious organisms, especially if present in small numbers, may require 5–7 d to grow on a plate.
- Failure to minimize exposure of colonies on plates to air. For some fastidious anaerobes, >15 min exposure can cause loss of viability.

chapter 4

Preliminary Identification Methods

4

General Considerations

Isolates in pure culture are further processed for identification as shown in Figure 4-1. The Wadsworth and KTL Anaerobic Bacteriology Laboratories, as research facilities, are committed to performing as complete an identification as possible. However, this philosophy is not practical for all laboratories for a variety of reasons, such as time, expense, and material and personnel considerations.

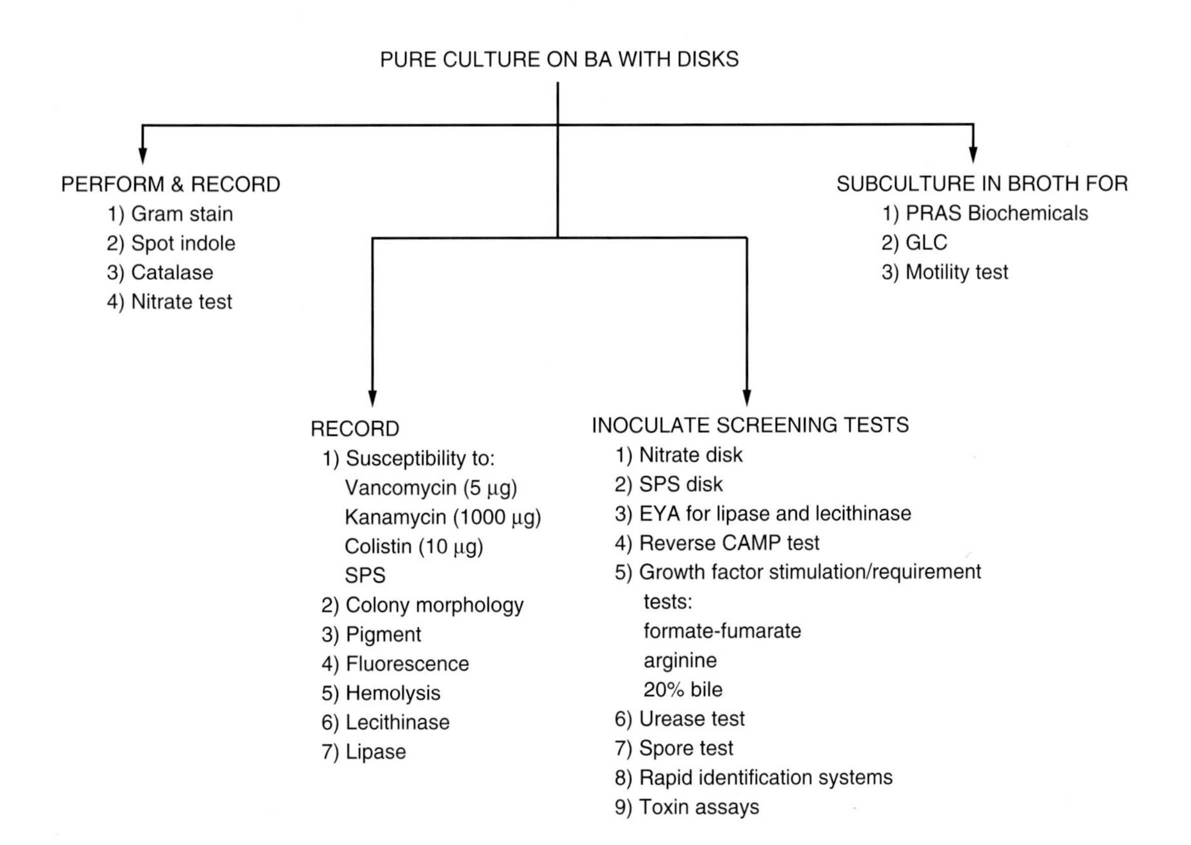

Figure 4-1. Flow chart illustrating scheme for initial identification of anaerobes from pure cultures.

In this manual we describe different approaches laboratories can take with regard to anaerobic workup. All laboratories should be capable of presumptive identification from primary plates, isolation, and maintenance of an anaerobe in pure culture so that the isolate can be sent to a reference laboratory for identification and susceptibility testing if necessary or desirable. Simple tests to further group and identify anaerobes are presented in Chapters 4 and 5. Chapter 6 describes tests for laboratories to identify isolated anaerobes as completely as possible using a variety of techniques that may include PRAS biochemicals, miniaturized biochemical systems (e.g., API), rapid enzyme detection panels (e.g., RapID ANA, Vitek, Rapid ID 32A, Crystal) and disks (Rosco, WEE-TAB), gas-liquid chromatography, MIS fatty acid methyl ester analysis, and toxin assays.

Key to the tables:

Cell morphology:
B = bacillus
C = coccus
CB = coccobacillus

Spore characteristics:
O = oval
R = round
S = subterminal
T = terminal

Susceptibility:
R = resistant
S = sensitive

PRAS Carbohydrates:
pH ≤ 5.5 positive (+)
pH 5.6-5.8 weak (W)
pH ≥ 5.9 negative (–)
PY = peptone yeast extract
PYG = glucose
ARA = arabinose
FRU = fructose
GLU = glucose
CEL = cellobiose
LAC = lactose
MAL = maltose
MANO = mannose
MANIT = mannitol
RAF = raffinose
RHA = rhamnose
RI = ribose
SAL = salicin
SORB = sorbitol
SUC = sucrose
TREH = trehalose
XYL = xylose
XLAN = xylan
LT = lactate
TH = threonine
ESC = esculin hydrolysis
GEL = gelatin hydrolysis
IND-NIT = indole-nitrate
ST = starch

GLC = gas-liquid choromatography:
A = acetic
P = propionic
IB = isobutyric
B = butyric
IV = isovaleric
V = valeric
IC = isocaproic
C = caproic
L = lactic
S = succinic
PA = phenylacetic acid

Note: 1) capital letters indicate major metabolic products, 2) small letters, minor products, and 3) parentheses, variable reaction.

Reactions:
– = negative
+ = positive
W = weak positive
V = variable reaction
$+^{-}$ = most strains positive, some negative
$+^{W}$ = most strains positive, some weakly positive
W^{-} = most strains weakly positive, some negative
$-^{+}$ = most strains negative, some positive
$-^{W}$ = most strains negative, some weakly positive
F/F REQUIRED = formate/fumarate required
α-FUC = α-fucosidase
α-GAL= α-galactosidase
α-GLU = α-glucosidase
β-GAL= β-galactosidase
β-NAG = β-N-acetylglucosaminidase
β-XYL = β-xylosidase
ONPG = O-nitrophenyl-β-D-galactopyranoside
PGUA= β-glucuronidase

Milk reactions:
C = clot formed
D = digested

table 4-1. Preliminary identification of clinically most commonly isolated groups

	Cell shape	Gram reaction	Aero-tolerance	Distinguishing characteristics
B. fragilis group	B	–	–	Growth on BBE with colony size >1 mm in diameter
Bilophila sp.	B	–	–	Growth on BBE with black-centered colonies <1 mm in diameter
Pigmented *Prevotella* spp. or *Porphyromonas* spp.	CB, B	–	–	Black/brown pigmented or red fluorescing colony
Campylobacter spp. / *B. ureolyticus*	B	–	–	Pitting colonies
F. nucleatum (presumptive)	B	–	–	Slender bacillus with pointed ends; "breadcrumb" or speckled colony
Anaerobic gram-negative bacillus	B	–	–	
Anaerobic gram-negative coccus	C	–	–	
Anaerobic gram-positive coccus	C	+	–	
C. perfringens (presumptive)	B	+	–	Double-zone β-hemolysis; boxcar-shaped cells
Other *Clostridium* spp.	B	+	–	Spores seen on Gram stain
Anaerobic gram-positive bacillus	B	+	–	No boxcar-shaped cells, no spores

PRESUMPTIVE IDENTIFICATION

Information from the primary plates, in conjunction with the atmospheric requirements, Gram stain, and colonial morphology of a pure isolate, provides presumptive identification of anaerobic organisms. Table 4-1 summarizes the extent to which isolates can be identified using this information.

Preliminary grouping of anaerobes and even complete identification of some organisms are based on colonial and cellular morphology; Gram reaction; susceptibility to the special-potency antibiotic disks; catalase, indole, and nitrate reactions; and possibly other simple tests (Tables 4-2, 4-3). Appendix C describes the test procedures and principles.

Preliminary Examination of All Isolates

The following characteristics are recorded on a worksheet and certain direct tests are performed from a pure culture on a BA plate for all isolates:

1. Describe colonial morphology in detail; this includes colony size, shape, edge, internal appearance, color, profile, opacity, general appearance (e.g., pitting, mucoid, dull, breadcrumb–like), and any other distinctive characteristic (see Figures 3-6 to 3-10 and those in this chapter and Chapter 6). The pitting characteristic may be easier to detect after 4 d of incubation. Also, recording a peculiar smell of colonies may give a valuable clue to the presence of some organisms such as the *B. ureolyticus*–like group.

table 4-2. Group identification of

	Vancomycin (5 µg)	Kanamycin (1000 µg)	Colistin (10 µg)	Growth in 20% bile	Catalase	Indole	Lipase
B. fragilis group	R	R	R	+	V	V	–
Other *Bacteroides* spp.	R	R	V	$-^{+}$	$-^{+}$	V	
Pigmented species	V	R	V	–	$-^{+}$	V	V
Porphyromonas spp.	S	R	R	–	V	$+^{-}$	$-^{+}$
Prevotella spp.	R	R^{s}	V	–	–	V	V
P. intermedia-P. nigrescens	R	R^{s}	S	–	–	+	$+^{-}$
P. loescheii	R	R	V	–	–	–	V
Other *Prevotella* spp.	R	R	V	–	$-^{+}$	$-^{+}$	$-^{+}$
Campylobacter spp./*B. ureolyticus*	R	S	S	–	$-^{+}$	–	–
B. ureolyticus	R	S	S	–	–	–	–
Campylobacter spp.	R	S	S	–	–	–	–
C. gracilis	R	S	S	–	–	–	–
Sutterella sp.	R	S	S	+	–	–	–
Bilophila sp.[a]	R	S	S	+	+	–	–
Desulfomonas pigra[b]	R	S	R	V	V	–	–
Desulfovibrio spp.[b]	R	S	R	V	V	–	
Fusobacterium spp.	R	S	S	V	–	V	V
F. nucleatum	R	S	S	–	–	+	–
F. necrophorum	R	S	S	$-^{+}$	–	+	$+^{-}$
F. varium/*F. mortiferum*	R	S	S	+	$-^{+}$	V	$-^{+}$
Gram-negative cocci	R	S	S	–	V	–	–
Veillonella spp.	R	S	S	–	V	–	–

[a] Growth stimulated by 1% pyruvate (final concentration).

[b] Desulfoviridin positive. *Depigra sulfomonas* was reclassified as *Desulfovibrio piger.*

anaerobic gram negative organisms

Pigment	Brick-red Fluorescence	Slender cells with pointed ends	Growth stimulated by formate-fumarate	Nitrate reduction	Urease	Motility	Pitting of agar
	−						
+	$+^-$						
$+^-$	$+^-$						
+	+						
+	+						
+	+						
−	−						
			+	+	V	V	V
			+	+	+	−	V
			+	+	−	$+^-$	V
			+	+	−	−	V
			+	+	−	−	V
			−	+	$+^-$	−	−
			−	$-^+$	$-^+$	−	−
			−	V	$-^+$	+	−
		V					
		+		−			
		−		−			
		−					
				V			
				+			

table 4-3. Group identification of anaerobic

	Cellular morphology	Spore test	Vancomycin (5µg)	Kanamycin (1000 µg)	Colistin (10µg)	Sodium polyanethol sulfonate (SPS)	Indole	Nitrate
Anaerobic gram-positive cocci	C	–	S	V	R	V	V	$-^{+}$
P. anaerobius	C/CB	–	S	R^{S}	R	S	–	–
P. asaccharolyticus	C	–	S	S	R	R	+	–
M. micros[a]	C	–	S	S	R	V	–	–
Clostridium species	B	+	S^{R}	V	R		V	V
C. perfringens	B	+	S	S	R		–	V
C. bifermentans	B	+	S	S	R		+	–
C. sordellii	B	+	S	S	R		+	–
C. difficile[b]	B	+	S	S	R		–	–
C. septicum[c]	B	+	S	S	R		–	V
C. sporogenes[c]	B	+	S	S	R		–	–
C. tetani[c]	B	+	S	S	R		V	–
Nonsporeforming bacilli	CB/B	–	S^{R}	S	R		V	V
P. acnes	B	–	S	S	R		$+^{-}$	+
E. lenta[d]	B	–	S	S	R		–	+
Actinomyces sp.	B	–	S	S	R		–	$+^{-}$

[a] "Halo" around colonies.

[b] Yellow ground glass colonies on CCFA; chartreuse fluorescence.

[c] Swarming.

[d] ADH positive; red fluorescence.

2. Describe cellular morphology including size, shape, spores, granules, Gram reaction, and any other distinctive characteristic (see figures in this chapter and Chapter 6).
3. Examine colonies for hemolysis on blood agar. Use transmitted light to look for β-hemolysis or a double zone of β-hemolysis (clear zone of hemolysis extending just beyond the colonies and zone of partial hemolysis extending well beyond the colony edge) (Figure 4-2). Also look for greening of the agar due to H_2O_2 production around colonies (Figure 4-3), which is apparent after exposure to air. This is most frequently seen with fusobacteria, lactobacilli, and some clostridia. Clostridial colonies are often pyramidal or rhizoid (Figure 4-4).
4. Examine colonies for pigment; colors may vary from light tan to black with *Prevotella* species and *Porphyromonas* species (Figure 4-5), from pink to red with *Actinomyces odontolyticus* (Figure 4-6), from mustard yellow to speckled blue with *Capnocytophaga* species (see Figure 6-11, blue sapphire), and from speckled

gram-positive organisms

Catalase	Fluorescence	Lecithinase	Reverse CAMP test	Boxcar shape	Double zone β-hemolysis	Urease
V						
–						
$-^{+}$						
–						
–	$-^{+}$	V				
–		+	+	+	+	–
–		+	–	–	–	–
–		+	–	–	–	$+^{-}$
–	+	–				
–	–	–				
–	–	$-^{+}$				
–	–	–				
V						
+	–					
+	+					
$-^{+}$	$-^{+}$					

green to pinkish with *Eubacterium yurii*. To detect pigment, it is often helpful to pick up several colonies with a swab and look at the cell mass on the swab.

5. Examine colonies for fluorescence (Figure 4-7). A variety of colors may be seen such as red, orange, pink, and chartreuse (yellow-green) (30;34;299).
6. Perform a catalase test (Figure 4-8).
7. Perform a spot indole test (Figure 4-9).
8. Determine susceptibility to special-potency antibiotic disks (vancomycin 5 μg, kanamycin 1,000 μg, and colistin 10 μg) (see Figure 3-12). The disks are used as an aid in determining the Gram reaction and in separating different anaerobic species and genera. Generally, gram-positive organisms are sensitive to vancomycin and resistant to colistin, whereas the gram-negative organisms are resistant to the vancomycin disk. The vancomycin susceptibility is especially helpful with those clostridia that consistently stain gram-negative. Occasionally, however, some clostridia (e.g., *C. innocuum*)

4

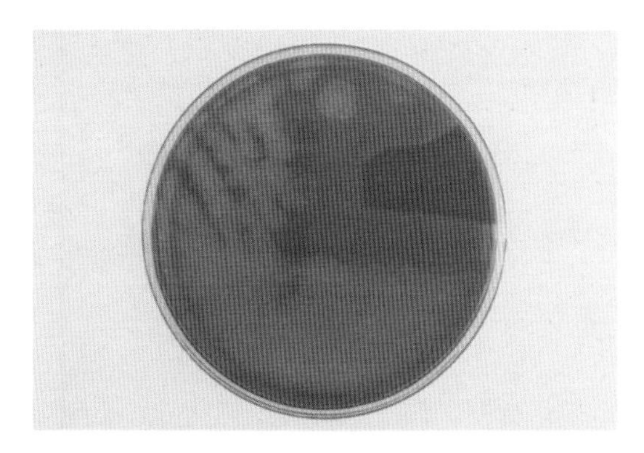

Figure 4-2. *Clostridium perfringens* on blood agar, showing double zone of β-hemolysis.

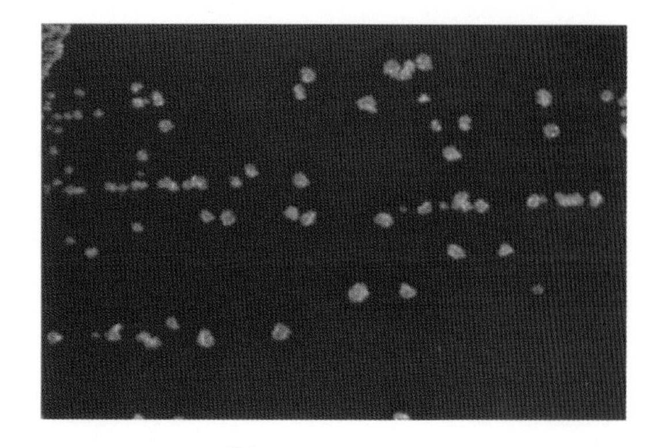

Figure 4-3a. *Fusobacterium nucleatum* subsp. *fusiforme*: dry, irregular, white breadcrumb-like colony morphology; note greening of the agar.

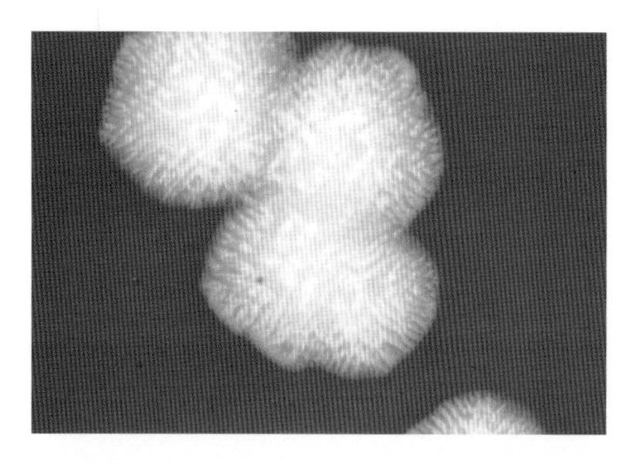

Figure 4-3b. *Fusobacterium nucleatum* subsp. *polymorphum*: circular, entire, speckled colony morphology.

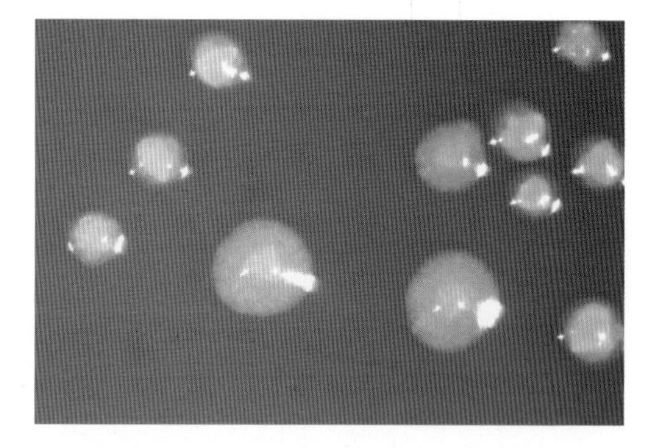

Figure 4-3c. *Fusobacterium nucleatum* subsp. *nucleatum*: circular, entire, smooth "nucleus" centered colony morphology.

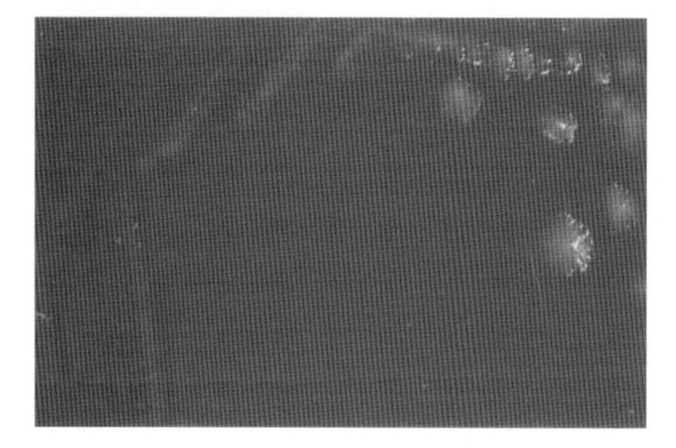

Figure 4-4a. Colony typical of clostridia on blood agar; note pyramidal shape and erose edge.

Figure 4-4b. Colony typical of clostridia on egg yolk agar; note rhizoid edge.

Figure 4-5. *Prevotella intermedia* on blood agar with special potency disks.

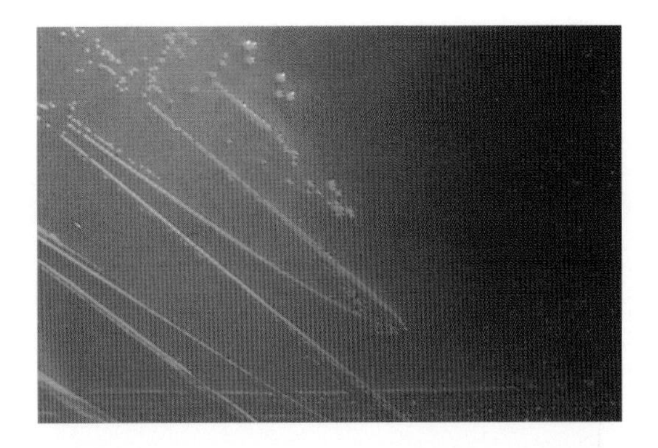

Figure 4-6. *Actinomyces odontolyticus* on blood agar; note pale orange/pink pigment.

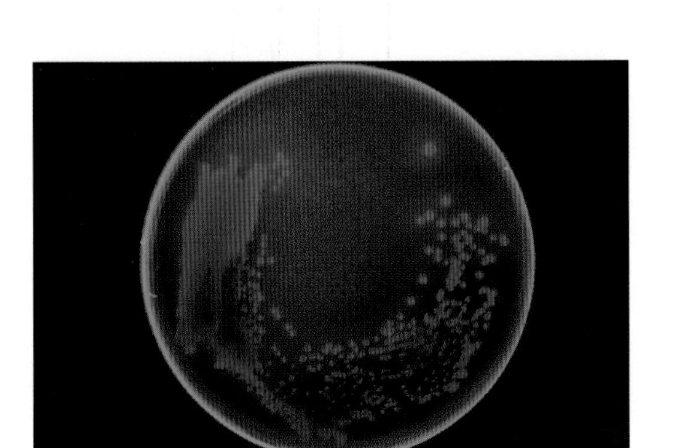

Figure 4-7. Brick-red fluorescence under long-wave UV light of pigmented anaerobic gram-negative bacillus. (Photograph courtesy of Dr. John Brazier.)

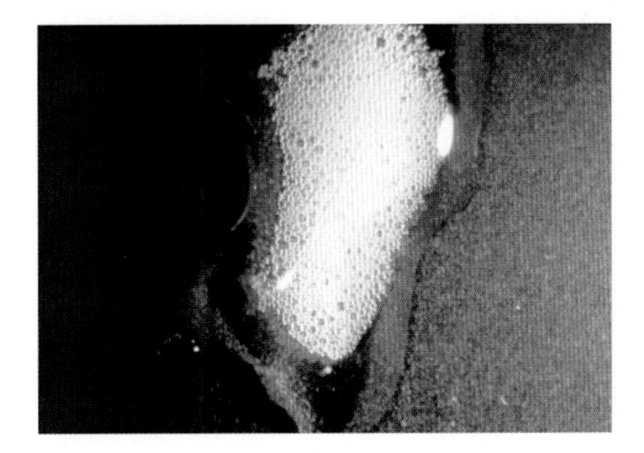

Figure 4-8. Catalase test; strong reaction.

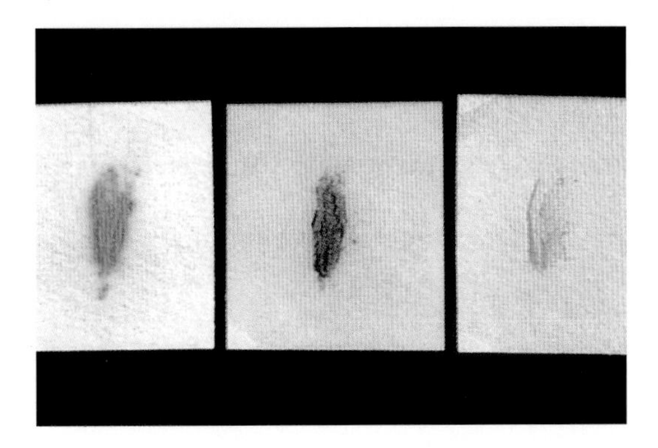

Figure 4-9. Spot indole test showing positive result on left (blue), negative result on right (pink), and slight masking of positive result by pigment of *Prevotella intermedia* (center).

box 4-1. Fluorescence of selected anaerobes

Organism	Color
P. gingivalis, P. catoniae, P. levii	No fluorescence
Other *Porphyromonas* spp.	Red, orange
Pigmented *Prevotella* spp.	Red
Non-pigmented gram-negative bacilli	No fluorescence or pink, orange, yellow
Fusobacterium spp.	Chartreuse
Veillonella spp.	Red or no fluorescence
Eggerthella lenta	red or no fluorescence
Clostridium difficile	Chartreuse
C. innocuum	Chartreuse
C. ramosum	Red

box 4-2. Special potency disk patterns

	Vancomycin 5 μg	Kanamycin 1000 μg	Colistin 10 μg	SPS 5%
Gram-negative	R[1]	V	V	
Gram-positive	S[2]	V	R	
Bacteriodes fragilis group	R	R	R	
Campylobacter spp./B. *ureolyticus*	R	S	S	
Fusobacterium spp.	R	S	S	
Bilophila/Sutterella spp.	R	S	S	
Desulfomonas/Desulfovibrio spp.	R	S	R	
Porphyromonas spp.	S	R	R	
Veillonella spp.	R	S	S	
Capnocytophaga spp.	R	S	R	
Leptotrichia spp.	R	S	S	
Selenomonas spp.	R	S	R	
Peptostreptococcus anaerobius				S[3]
Other gram-positive cocci				R

[1] *Porphyromonas* spp. are vancomycin-sensitive

[2] Some strains of *Lactobacillus* and most *Clostridium innocuum* strains are vancomycin-resistant.

[3] Some strains of *Micromonas micros* are SPS-sensitive.

and lactobacilli can appear resistant to the vancomycin 5 μg disk. On the other hand, *Porphyromonas* species usually are susceptible to vancomycin.

Screening Tests for Selected Isolates

1. Perform a nitrate disk test (Figure 4-10). Test (a) indole-negative gram-positive rods (not clostridia), (b) gram-negative rods that form small transparent or pitting colonies and are susceptible to kanamycin and colistin (*B. ureolyticus*–like organisms), and (c) gram-negative cocci.
2. Determine susceptibility to sodium polyanethol sulfonate (SPS) using the disk test. Test gram-positive cocci.
3. Determine ability to produce lecithinase (Figure 4-11). Test all clostridia.
4. Determine ability to produce lipase (Figure 4-12). Test pigmented gram-negative rods; suspected fusobacteria (sensitive to kanamycin and colistin, umbonate or

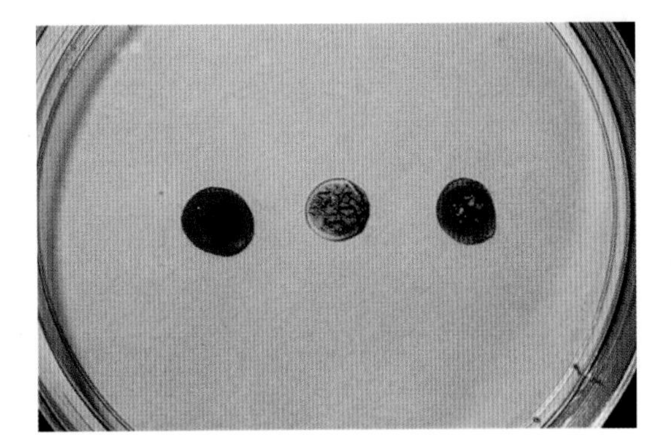

Figure 4-10. Nitrate disk test showing positive test result with reagents only (left), positive test result with reagents plus zinc (center), and negative test result with reagents plus zinc (right).

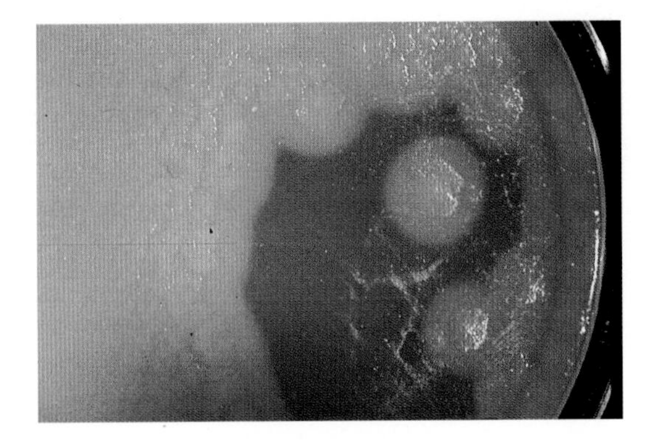

Figure 4-11. Lecithinase reaction on egg yolk agar; note opacity of agar surrounding colonies due to precipitation of complex fats.

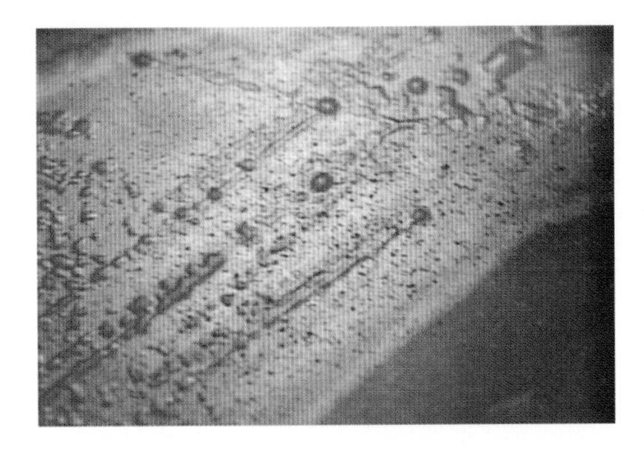

Figure 4-12. Lipase reaction on egg yolk agar; note mother-of-pearl sheen on agar surface.

convex, speckled colony, foul odor); catalase-positive, indole-negative, aerotolerant gram-positive rods (*A. neuii*); and clostridia.

5. Perform a reverse-CAMP test (55;141) (Figure 4-13). Test all suspected *C. perfringens* and *A. neuii.*
6. Determine ability to produce urease. Test lecithinase-positive clostridia, *B. ureolyticus*–like organisms, suspected *Bilophila, Actinomyces* species, and *Peptostreptococcus* species.

Figure 4-13. *Clostridium perfringens* (cross-streak) and *Streptococcus agalactiae* showing positive reverse CAMP test (arrow-shaped synergistic enhancement of hemolysis).

7. Determine whether the organism produces spores. Test gram-positive rods and occasional gram-negative rods when needed to detect *Clostridium* species.
8. Determine resistance to bile. This test is useful for separating gram-negative bacilli: the *B. fragilis* group, *Bilophila,* some fusobacteria, and *Leptotrichia* are not inhibited by 20% bile, while *Prevotella* species are. Some *Actinomyces* species, lactobacilli, and bifidobacteria may also be bile-resistant. If colonies were isolated from BBE agar, this test is not necessary as BBE contains 20% bile. However, all bile-resistant organisms do not grow on BBE agar due to other inhibitors.
9. Perform a formate and fumarate (F/F) growth requirement test. Test *B. ureolyticus*–like organisms.
10. Perform an arginine growth stimulation test. Test small indole-negative, gram-positive bacilli not resembling diphtheroids or clostridia.
11. Examine for motility. Test any curved gram-negative bacillus. Test *B. ureolyticus*– like bacilli and other gram-negative bacilli that do not key out if assumed to be nonmotile.

Tables 4-2 and 4-3 group anaerobes by Gram reaction, and list characteristics that identify groups and certain anaerobes. **It is important for all characteristics to be present to assign identification.**

Gram-Negative Organisms

Gram-Negative Bacilli

The gram-negative bacilli are divided into the following major categories based on microscopic morphology, special-potency antibiotic disk patterns, and other simple tests: (1) *B. fragilis* group, (2) *Prevotella* species, (3) *Porphyromonas* species,

(4) *B. ureolyticus*–like group, (5) *Bilophila* species, (6) *Sutterella* species, (7) *Fusobacterium* species, and (8) other gram-negative bacilli. Organisms assigned to other than the above mentioned genus designations are discussed in Chapter 6 under advanced identification.

The *B. fragilis* group can be identified by a special-potency antibiotic disk pattern showing resistance to all three disks and by resistance to 20% bile in a tube test, the bile disk test, or on BBE agar (see Figure 3-7; Figures 4-14, 4-15) (94;354).

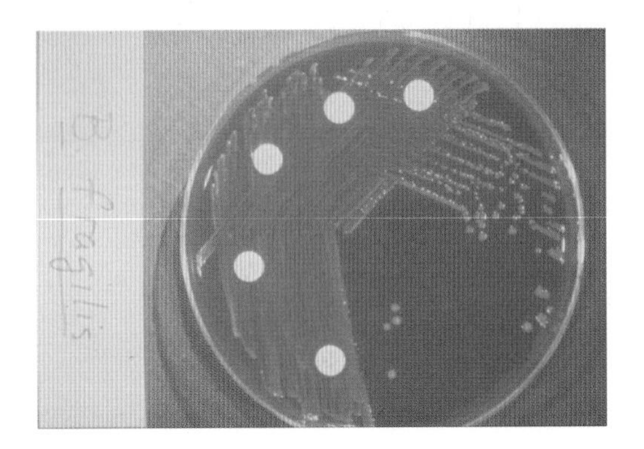

Figure 4-14a. *Bacteroides fragilis* on blood agar with special potency disks.

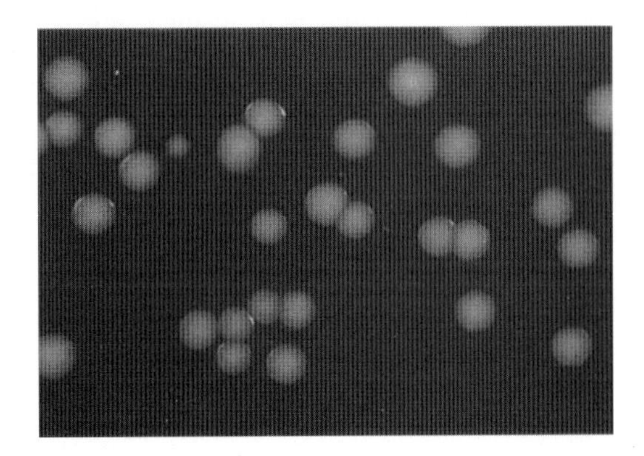

Figure 4-14b. *Bacteroides fragilis* colonies on blood agar.

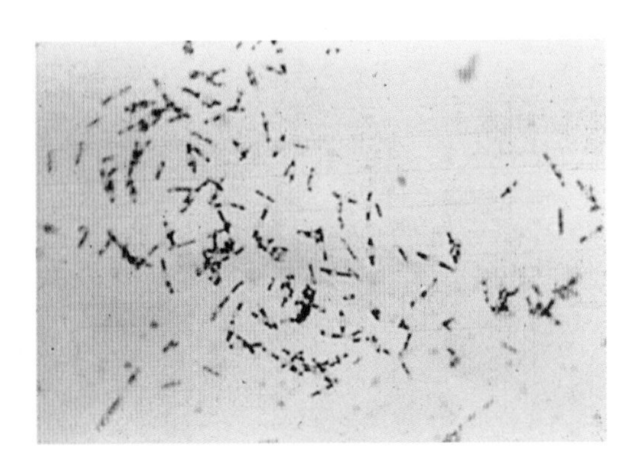

Figure 4-15. Gram stain of *Bacteroides fragilis*; note bipolar staining.

Figure 4-16. *Bacteroides ureolyticus* on blood agar; note pitting colony.

Anaerobic gram-negative bacilli that require formate and fumarate for growth in broth culture may be identified presumptively as belonging to the *B. ureolyticus*–like group. *B. ureolyticus, Campylobacter gracilis* and other "anaerobic" *Campylobacter* species, and *Sutterella* species are thin gram-negative rods with rounded ends that are sensitive to kanamycin and colistin special-potency disks. The colonies are small and translucent or transparent and may produce greening of the agar. Three colony morphotypes exist: smooth and convex, pitting (Figure 4-16), and spreading. All colony types can occur in the same culture. These organisms are asaccharolytic, nitrate or nitrite reducing, and they require supplementation of broth media with formate and fumarate for growth. *C. rectus*,

C. curvus, C. concisus, and *C. showae* are motile and oxidase-positive; unlike the former three, *C. showae* is catalase-positive. *C. gracilis* is nonmotile and urease-negative, and thus differentiated from urease-positive *B. ureolyticus. Bilophila* (Figures 4-17, 4-18), which phenotypically resembles *B. ureolyticus,* is distinguished by its resistance to bile and strong catalase reaction. *Sutterella* is also resistant to bile, but is urease- and catalase-negative, characteristics that differentiate this genus from *Bilophila* (19;359).

Organisms that fluoresce brick-red or produce brown to black colonies on blood-containing medium are placed into pigmented *Prevotella* species or *Porphyromonas* (Figure 4-19). Indole-positive and lipase-positive pigmented coccobacilli can be identified as *P. intermedia/P. nigrescens* (see Figure 4-5), and indole-positive but lipase-negative weakly pigmented rods as *P. pallens* (185;187). Indole-negative but lipase-positive strains are either *P. loescheii* or, rarely, *P. melaninogenica.* Lipase-negative strains must be identified using other biochemical tests. An indole-positive pigmenter sensitive to the vancomycin special-potency disk can be identified as *Porphyromonas.*

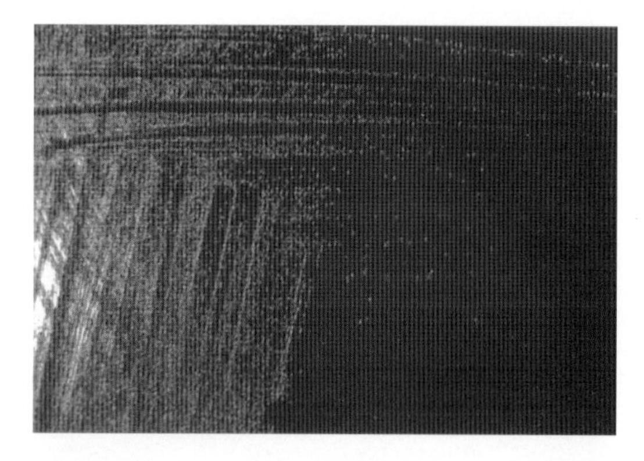

Figure 4-17. *Bilophila wadsworthia* on blood agar; colonies are small and translucent.

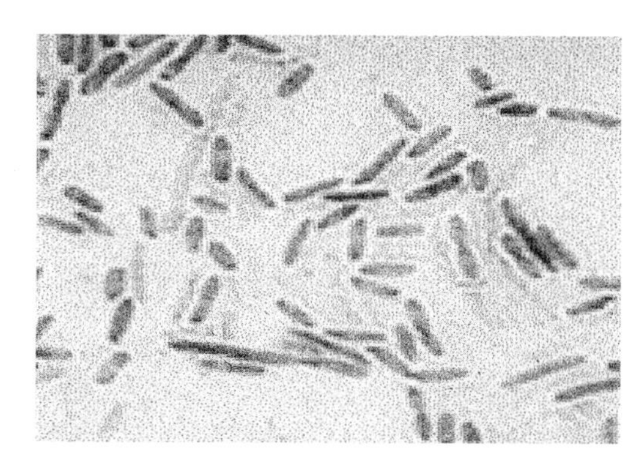

Figure 4-18. Gram stain of *B. wadsworthia* from blood agar; note straight bacillus of uniform width and internal vacuoles.

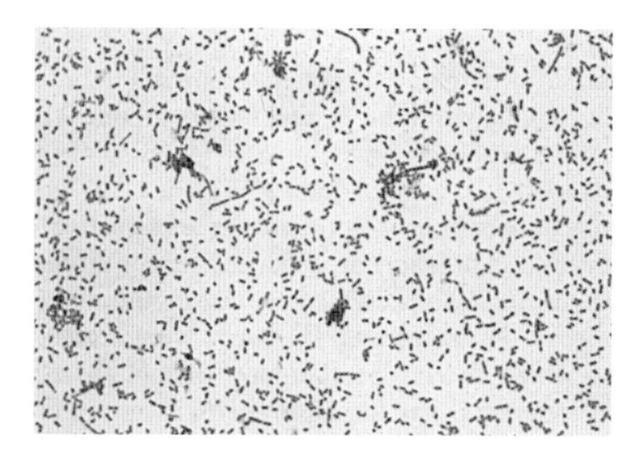

Figure 4-19. Gram stain of pigmented *Prevotella* sp.; many coccobacillary forms with some rod-shaped forms interspersed.

Fusobacterium species are sensitive to the special-potency disks kanamycin and colistin, are catalase- and nitrate-negative, and produce a characteristic rancid butter odor due to butyric acid production. *F. nucleatum* is a thin rod with tapering ends (Figure 4-20) and is indole-positive. The needle-shaped morphology is shared with the microaerophilic, indole-negative *Capnocytophaga* and *Leptotrichia* species. *F. nucleatum* fluoresces chartreuse under UV light and often produces greening of blood agar after exposure to air. At least three different colony morphotypes of *F. nucleatum* exist. Due to considerable phenotypic and genotypic heterogeneity and the uncertainty of valid criteria to separate the *Fusobacterium nucleatum* subspecies, it is hard to judge whether the reported colonial morphologies consistently coincide with the subspecies designations. *F. nucleatum* subspecies *fusiforme* colonies may be small (<0.5–1 mm), granular, and irregular (breadcrumb) (see Figure 4-3a); *F. nucleatum* subspecies *polymorphum* colonies often are large, speckled, smooth, translucent, and butyrous (see Figure 4-3b); and *F. nucleatum* subspecies *nucleatum* colonies may be small, greyish white, and smooth (see Figure 4-3c). *F. necrophorum* subspecies *necrophorum* is lipase-positive and usually bile sensitive. It produces indole, fluoresces chartreuse, produces greening of blood agar, and often demonstrates β-hemolysis around the greyish to yellow dull, umbonate colonies (Figure 4-21). It is a pleomorphic long rod with round ends and often has bizarre forms (Figure 4-22). Lipase-negative strains require further biochemical testing. *Fusobacterium mortiferum* is indole-negative and extremely pleomorphic, with filaments containing swollen areas with large, round bodies and exhibiting irregular staining (Figure 4-23). *F. necrophorum* may have similar morphology but usually fewer round bodies. A bile-resistant fusobacterium isolated from BBE agar may be identified presumptively as *F. mortiferum* or *F. varium;* however, other fusobacteria (e.g., *F. necrophorum* and *F. ulcerans*) may grow in 20% bile. Therefore, further testing is required to confirm the presumptive identification.

Gram-Negative Cocci

A gram-negative coccus that reduces nitrate is *Veillonella;* cells are <0.5 μm (Figure 4-24), and the colonies are small and transparent or opaque and yellowish grey, like a silver coin. About 35% are catalase-positive and 99% sensitive to kanamycin and colistin special-potency antibiotic disks (168). All other anaerobic gram-negative cocci should be called "anaerobic gram-negative cocci."

Figure 4-20. Gram stain of *Fusobacterium nucleatum*; note thin gram-negative bacillus with pointed ends.

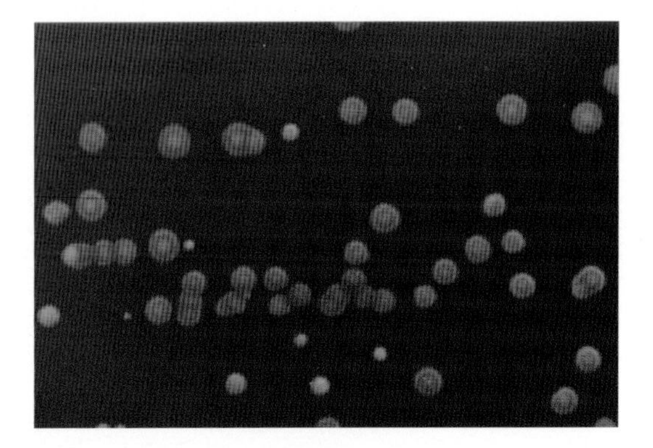

Figure 4-21. *F. necrophorum* colonies; note umbonate profile and irregular shapes.

Figure 4-22. Gram stain of *F. necrophorum*; note pleomorphism and swollen forms that may be similar to *F. mortiferum*.

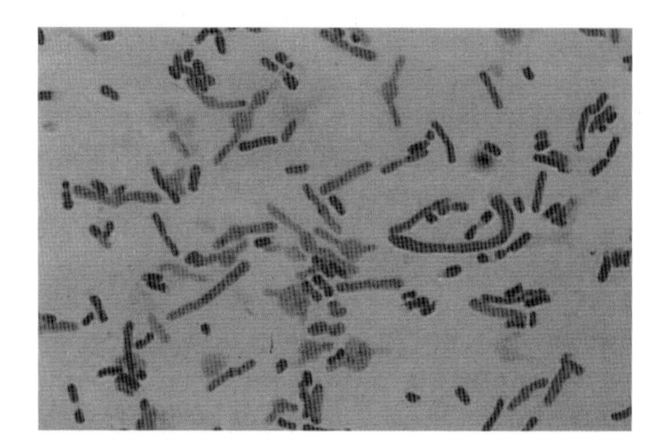

Figure 4-23. Gram stain of methanol-fixed *F. mortiferum*. Note pleomorphic, swollen forms. The cell wall is so fragile that morphology may not be visualized on Gram stains prepared on heat-fixed smears.

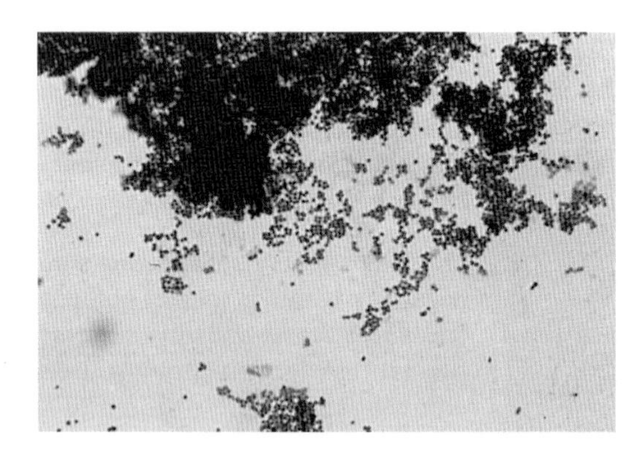

Figure 4-24. Gram stain of *Veillonella* sp.; note small gram-negative coccus.

Gram-Positive Organisms

Gram-positive organisms are divided into the following three major categories based on microscopic morphology and the presence of spores: (1) anaerobic gram-positive cocci, which include *Peptostreptococcus* and several other genera (see Table 1-4); (2) *Clostridium,* the anaerobic sporeforming bacilli; and (3) anaerobic nonsporeforming bacilli which include *Actinomyces, Bifidobacterium, Eubacterium, Lactobacillus, Propionibacterium,* and several other newly described or reclassified genera (see Table 1-4).

Gram-positive organisms that stain gram-negative can be separated from the true gram-negatives with the special-potency antibiotic disks. Gram-positive organisms are generally resistant to colistin and susceptible to vancomycin, whereas the gram-negatives are resistant to vancomycin with the exceptions noted previously. Biochemical tests and endproduct analysis are required to separate the genera of anaerobic gram-positive bacilli. If spores are not observed in a Gram stain of the organism, the ethanol or heat test for spores will help separate clostridia from the anaerobic nonsporeforming bacilli. Catalase and nitrate are helpful for grouping the nonsporeformers; however, endproduct analysis is often required for full identification.

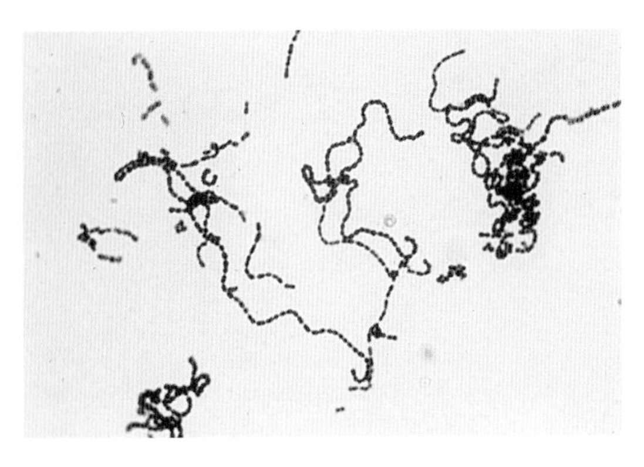

Figure 4-25. Gram stain of *Peptostreptococcus anaerobius*; note large coccobacillary cells that usually form chains.

Figure 4-26. *M. micros* on blood agar; note milky halo around white colonies.

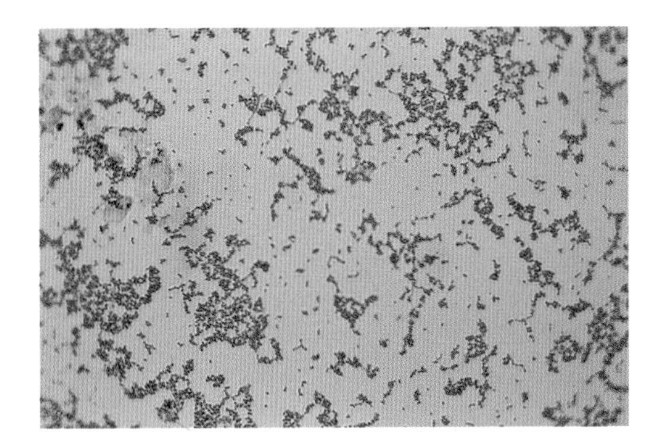

Figure 4-27. Gram stain of *M. micros*; note very small cell size and chain formation.

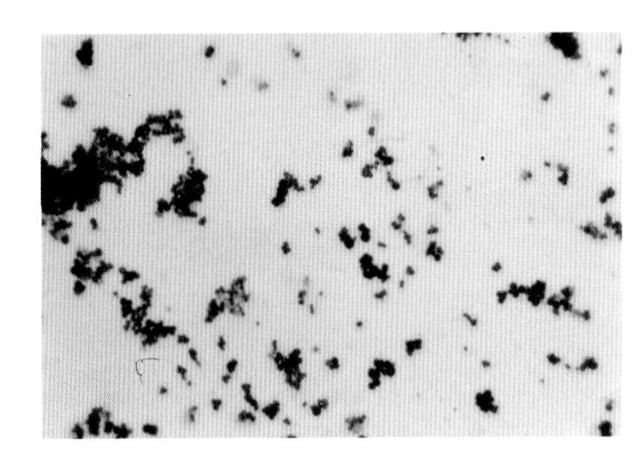

Figure 4-28. Gram stain of *F. magna*; note large size and tetrad formation.

Gram-Positive Cocci

A gram-positive coccus sensitive (exhibiting a zone of inhibition of ≥ 12 mm about the disk) to sodium polyanethol sulfonate (SPS) is *P. anaerobius*. It is a large elongated coccus that occurs in pairs and chains (Figure 4-25). The colonies on blood agar are usually larger than most anaerobic cocci, ranging in size from 0.5 to 2 mm in diameter. They are grey-white and opaque in appearance. A sweet odor is associated with this organism due to the production of isocaproic acid. An anaerobic coccus with a milky halo around the colonies (Figure 4-26) and small cells (<0.6 μm) (Figure 4-27) can be identified as *Micromonas* (*Peptostreptococcus*) *micros*. *M. micros* can exhibit a zone of inhibition with SPS; however, the zone is usually <12 mm in size. *Finegoldia* (*Peptostreptococcus*) *magna* cells are larger than those of most peptostreptococci (Figure 4-28). The most common indole-positive gram-positive coccus isolated from clinical specimens is *P. asaccharolyticus*. *P. indolicus*, another indole-positive coccus, is usually seen in animal specimens. The latter can be differentiated from *P. asaccharolyticus* by the ability of *P. indolicus* to reduce nitrate and produce coagulase.

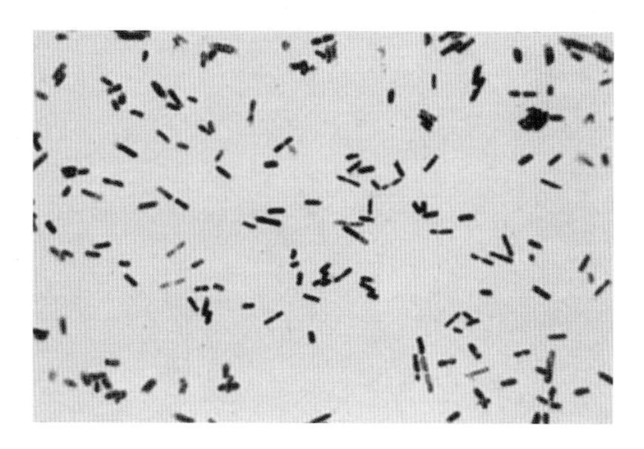

Figure 4-29. Gram stain of *C. perfringens*; note boxcar-shaped cells and absence of spores.

Clostridia

C. perfringens is a boxcar-shaped gram-positive rod (Figure 4-29) that is lecithinase-positive. Most strains show a typical double zone of β-hemolysis which can be enhanced by exposure to cold. *C. perfringens* is usually reverse-CAMP–positive (see Figure 4-13), whereas other clostridia have been reported to be negative (55;141). It is not necessary to demonstrate the presence of spores in *C. perfringens*. *C. sordellii* and *C. bifermentans* are similar in that they are both indole- and lecithinase-positive. *C. sordellii* is usually urease-positive, whereas *C. bifermentans* is always urease-negative and produces chalk-white colonies on EYA. These clostridia and *C. baratii* are Nagler-test–positive, but the antiserum used in that test is no longer commercially available (for details, see the fifth edition of this book). *Clostridium novyi* type A is lecithinase- and lipase-positive.

Nonsporeforming Gram-Positive Bacilli

P. acnes is an indole-positive, catalase-positive, gram-positive pleomorphic coryneform rod (Fig 4-30). It is usually nitrate-positive. The colonies are initially small and white and become larger and more yellowish-tan with age. *P. acnes* may grow in a 5–10% CO_2 environment. The indole- or catalase-negative strains must be identified with more extensive biochemical tests. The two aerotolerant subspecies of *A. neuii* may resemble the other propionibacteria since they are catalase-positive and indole-negative; the positive reverse-CAMP test of *A. neuii* may be helpful in differentiation.

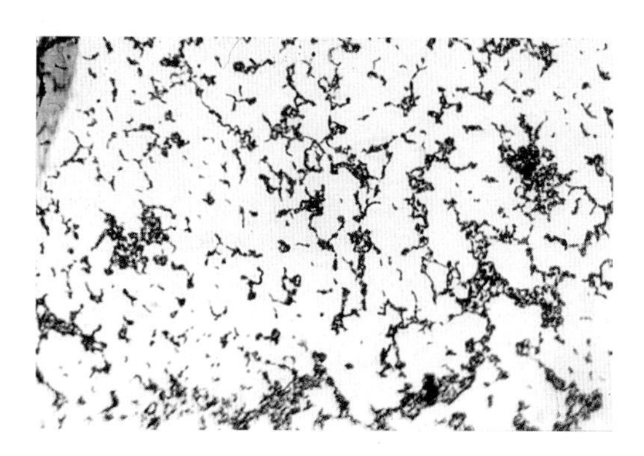

Figure 4-30. Gram stain of *Propionibacterium acnes*; note pleomorphic, branching gram-positive bacilli.

Eggerthella lenta is a nitrate-positive, small gram-positive bacillus whose growth is stimulated by arginine. The strains exhibit a red fluorescence under long-wave UV light and are catalase-positive and bile-resistant.

Commonly Encountered Errors in Identification

- Failure to obtain a pure culture before performing biochemical tests.
- Failure to determine whether an organism is a true anaerobe (aerotolerance testing). A frequent error is inadequate aerotolerance testing of an organism. Many of the fastidious unusual aerobic gram-negative bacilli grow slowly, requiring $\geq$ 48 h to appear on a plate. Testing of metronidazole susceptibility may help; true anaerobes are generally susceptible, microaerophiles are resistant. Thus the following misidentifications have been made:

  1. *Actinobacillus actinomycetemcomitans* as *Bacteroides* species.
  2. *Haemophilus aphrophilus* as *B. ureolyticus* group.
  3. *Eikenella corrodens* as *B. ureolyticus* group.
  4. *Capnocytophaga* species as *F. nucleatum.*
  5. *Mycoplasma hominis* as *Dialister pneumosintes.*
  6. *Abiotrophia* and *Granulicatella* (nutritionally variant *Streptococcus*) species as anaerobic coccus.
  7. *H. influenzae* as anaerobic gram-negative bacillus (failure to use chocolate agar for aerotolerance testing).

- Failure to determine the Gram reaction correctly (many anaerobes overdecolorize and appear gram-negative). An example of this is misidentification of gram-negatively staining *Clostridium* species, especially *C. clostridioforme,* as *Bacteroides.* These *Clostridium* species grow well in 20% bile. Also, coccobacillary gram-negative rods easily can be misinterpreted as anaerobic cocci (*Dialister pneumosintes* as *Veillonella*). Use of special-potency antibiotic disks for presumptive identification will, with some exceptions, differentiate a gram-negative (vancomycin-resistant) organism from a gram-positive (vancomycin-sensitive). Gas chromatography will also provide differentiation.
- Failure to confirm the viability and adequate growth of an organism when doing biochemical tests. Errors can result from using a nonviable organism to inoculate biochemical or spore tests and then reading the results as valid negatives; always include viability control.
- Failure to perform confirmation tests when using commercial micromethod identification systems.
- Inaccurate or incomplete identification. Common errors due to inadequate testing include:

  1. *Clostridium tertium* (catalase-negative, spores present in anaerobic culture) as *Lactobacillus* species (no spores) or *Bacillus* species (usually catalase-positive).
  2. *Bilophila wadsworthia* (catalase-positive) as *B. ureolyticus* (usually catalase-negative).

3. *B. wadsworthia* (nonmotile, colistin sensitive) as *Desulfomonas* species (nonmotile, colistin resistant) or *Desulfovibrio* species (motile, colistin-resistant).
4. *Sutterella wadsworthensis* (bile-resistant) as *Campylobacter gracilis* (bile sensitive).

- Failure to inoculate bile has resulted in misidentification of members of *Prevotella* species as *B. fragilis*–group organisms and vice versa. This may be a particular problem when using rapid identification kits, which typically do not include the bile reaction. Several *Prevotella* species, especially *P. bivia, P. disiens,* members of the *P. oralis* group, and *Bacteroides tectus* and *B. pyogenes,* biochemically appear quite similar to *B. fragilis.* Growth in 20% bile or a bile disk susceptibility test on the purity plate and catalase are key reactions for separating the *Prevotella* species from the latter two (119).

chapter 5

Rapid, Cost-Efficient Identification Methods Using Preformed Enzyme Tests

5

General Considerations

This chapter is specifically designed to assist in the preliminary and presumptive identification of the main groups of clinically occurring anaerobic bacteria with the aid of cost-efficient tests including Gram staining; sensitivity to special-potency antibiotic disks; bile, catalase, and indole tests; and individually available rapid enzyme tests that detect preformed (constitutive) bacterial enzymes. The enzyme substrates used are either chromogenic, yielding color reactions with or without reagent addition, or fluorogenic, detected by long-wave UV light (366 nm) as blue-white fluorescence. These tests will produce results ranging in times from a few minutes (4-methylumbelliferyl substrates, Sigma) (210;216;227;264;289) to 2–4 h (WEE-TAB system, Key Scientific Products, Round Rock, TX; Rosco Diagnostic Tablets, Taastrup, Denmark) (5;99;156;172;289). Commercial identification kits based on this principle are available from several manufacturers (see Chapter 6), but they are much more expensive and lack the flexibility to tailor optimal sets of reactions from single, dual, or triple enzyme substrates for certain bacterial groups.

The identification approach presented in this chapter may be used in hospital laboratories that usually do not aim at advanced-level identification of anaerobes. However, it is important for these laboratories, too, to be familiar with the situations when isolates must be identified further and sent to a reference laboratory (see Chapter 1). Academic centers and teaching hospitals may apply the schemes presented here as the preliminary approach to further workup (if necessary) and as a basis for interim reports.

Identification Using Preformed Enzyme Tests

The detailed procedures to perform the tests using either 4-methylumbelliferyl substrates (4-MU) or individually available tablets or disks and guidelines for interpretation of reactions are presented in Appendix C.

Tables 5-1–5-8 contain the reactions used to identify the common clinically occurring anaerobic bacteria using preformed enzymes. Note that identifications in these tables are always based on testing aerotolerance, Gram-staining properties, and all the reactions presented, not only the enzyme tests.

table 5-1. Rapid identification of *Bacteroides fragilis* group

BBE+ or Oxgall^{R}, Vanco^{R} Kana^{R}, Colistin^{R}

Bacterium	Catalase	Indole	Esculin hydrolysis	α-fucosidase	l-arabinose
B. fragilis[a]	+	–	+	+	–
B. caccae	$-^{+}$	–	+	+	+
B. distasonis[b]	+	–	+	–	$+^{-}$
B. merdae[b]	–	–	+	–	–
B. vulgatus	$-^{+}$	–	–	+	+
B. thetaiotaomicron[c]	+	+	+	+	+

[a] Suggested reporting: *B. fragilis* group, MCR *B. fragilis*. If indicated, perform: API 20A or short set of fermentation tests.

[b] *B. distasonis* is β-glucuronidase-negative and catalase-positive, *B.merdae* is β-glucuronidase-positive and catalase-negative.

[c] Report other indole-positive organisms as indole-positive *B. fragilis* group.

table 5-2. Rapid identification of nonpigmented *Prevotella* spp.

BBE– or Oxgall^{S},Vanco^{R}, Kana^{R}, Colistin$^{R/S}$

Bacterium	Indole	Esculin hydrolysis	β-NAG	α-fucosidase	β-xylosidase	Colistin
P. buccae	–	+	–	–	+	S
P. oris	–	+	+	+	+	R
P. heparinolytica[a]	+	+	+	+	+	R/S
P. zoogleoformans[a]	–	+	+	+	+	R/S
P. dentalis [b] (*Mitsuokella dentalis*) (*Hallella seregens*)	–	+	+	–	V	R/S
P. oralis group	–	+	+	+	–	R/S
P. enoeca [c]	–	V	+	+	–	R
P. bivia [c]	–	–	+	+	–	R
P. disiens	–	–	–	–	–	R/S

[a] Colonies adhere to agar, viscous mass in liquid broth. Phylogenetically these species cluster within *B. fragilis* group, although they are sensitive to bile.

[b] *Prevotella dentalis* forms water drop-like colonies.

[c] *P. enoeca* is esculin variable *P .bivia* is -negative; *P. enoeca* recovered from oral sites, *P. bivia* elsewhere.

table 5-3. Rapid identification of pigmented (fluorescence red-coral) gram-negative rods

Prevotella: BBE-, Oxgall^{S},Vanco^{R}, Kana$^{R(S)}$, Colistin$^{S(R)}$
Porphyromonas: BBE-, Oxgall^{S},Vanco^{S}, Kana^{R}, Colistin^{R}

Bacterium	Indole	α-fucosidase	α-glucosidase	ONPG	β-NAG	Esculin hydrolysis	Lipase	Note
Prevotella denticola/melaninogenica/loescheii[a]*/tannerae*[b] group	–	+	+	+	+	–/+	+[a]	Brown colony
P. intermedia-nigrescens group	+	+	+	–	–	–	+	Dark colony
P. pallens	+	+	+	–	–	–	–	Pale beige colony
P. corporis[c]	–	–	+	–	–	–	–	Brown colony
P. bivia	–	+	+	+	+	–	–	Slow pigment
P. disiens[c]	–	–	+	–	–	–	–	Slow pigment
Porphyromonas asaccharolytica	+	+	–	–	–	–	–	
P. gingivalis	+	–	–	+[d]	+	–	–	Trypsin+; no fluorescence
P. endodontalis	+	–	–	–	–	–	–	

[a] *P. loescheii* is often lipase-positive.
[b] *P. tannerae* is esculin- and sucrose-negative.
[c] *P. corporis* is similar to *P. disiens* but pigments faster.
[d] Positive by Rosco test; negative by API ZYM.

table 5-4. Rapid identification of fusobacteria

Vanco^{R}, Kana^{S}, Colistin^{S}, Nitrate–
(*F. varium/mortiferum,* BBE+, Oxgall^{R})

Bacterium	Indole	Lipase	Esculin hydrolysis	BBE growth	ONPG	Gram
F. nucleatum	+	–	–	–	–	Pointed ends
F. necrophorum	+	+	–	–	–	Pleomorphic
F. varium	+	$-^{+}$	–	+	–	Regular shaped rods
F. mortiferum	–	–	$-^{+}$	+	+	Pleomorphic; swollen forms

5

table 5-5. Rapid identification of *B. ureolyticus*–like group

Vanco[R], Kana[S], Colistin[S], Nitrate +
(*Bilophila/Sutterella* BBE+, Oxgall[R])

Bacterium	Urease	Motility[a]	Catalase	Oxgall	CO_2 growth
Campylobacter gracilis (*Bacteroides*)	–	–	–	S	–[b]
Bacteroides ureolyticus	+	–	–[+]	S	–[b]
Campylobacter rectus (*Wolinella*)	–	+	–	S	–[b]
Bilophila wadsworthia	+	–	+	R	–
Sutterella wadsworthensis	–	–	–	R	–[b]
Eikenella corrodens [c]	–	–	–[+]	S	+

[a] Suspend colonies from warm plate in a drop of water, place cover slip and examine under high power field (400×).
[b] These organisms may grow microaerophilically in the presence of 2–6% O_2 (Campy atmosphere).
[c] Oxidase-, LDC-, and ODC-positive.

table 5-6. Rapid identification of gram-positive cocci

Vanco[S], Kana[V], Colistin[R], Oxgall[S]

Bacterium	Size (μm)	SPS	Catalase	Indole	Urease	α-glucosidase	Note
P. anaerobius	1.0	S	–	–	–	+	IC—sweet odor
P. asaccharolyticus	0.8	R	–[+]	+	–	–	Butyrous odor
M. micros	≤ 0.6	R/S	–	–	–	–	Milky “halo”
F. magna	1.2	R	–[+]	–	–	–	Clumping cells
P. tetradius	1.2	R	–	–	+	+	Clumping cells, tetrads

IC=isocapronic acid; sweet odor, “halo”= milky haze around the colonies

table 5-7. Rapid identification of common lecithinase-positive *Clostridium* spp.

Vanco^{S}, Kana^{S}, Colistin^{R}

	Lecithinase	Indole	Urease	Note
C. novyi type A	+	$-^{+}$	−	Swarming, lipase-positive, strong β-hemolysis
C. perfringens	+	−	−	Double-zone hemolysis, reverse CAMP-positive
C. bifermentans	+	+	−	Chalkwhite colonies
C. sordellii	+	+	+	

table 5-8. Rapid identification of some swarming clostridia

Vanco^{S}, Kana^{S-R}, Colistin^{R}

Bacterium	Lipase	Indole	α-glucosidase	Note
C. novyi type A	+	$-^{+}$	+	Lecithinase-positive, strong β-hemolysis
C. septicum	−	−	+	
C. sporogenes	+	−	−	
C. tetani	−	+	−	Tennis racket-shaped cells

Common Pitfalls Associated with Preformed Enzyme Tests

- Inoculum was too dilute.
- Inoculum was not prepared from a fresh culture.
- Reagents were not added prior to interpretation (aminopeptidase tests).
- Mineral oil overlay was not used before incubation when indicated (ADH, ODC, etc.).
- Failure to detect weak reactions under UV light (positive and negative controls will help).

chapter 6

Advanced Identification Methods

General Considerations

Differentiation of genera of anaerobes is shown in Table 6-1. In addition to the tests performed in preliminary and rapid approaches, identification of many anaerobes to species or even genus level requires additional biochemical tests and metabolic end-product analysis by gas-liquid chromatography. In this chapter, anaerobes are divided and discussed as follows: (1) gram-negative bacilli (Tables 6-2 to 6-6 and Figures 6-1 to 6-16), (2) gram-negative cocci (Table 6-7), (3) gram-positive cocci (Tables 6-8 to 6-9 and Figures 6-17 to 6-18), (4) *Clostridium* species (Table 6-10 and Figures 6-19 to 6-23), and (5) nonsporeforming gram-positive bacilli (Tables 6-11 to 6-16 and Figures 6-24 to 6-30).

Advanced identification systems are discussed at the end of this chapter and gas-liquid chromatography is covered in Appendix B. Tables and flow charts show key characteristics for the listed organisms based on reactions using PRAS biochemicals, gas-liquid chromatography, and preformed enzyme tests such as Rosco Diagnostic Tablets, WEE-TAB disks, and API ZYM (Identification Systems, below). Table 6-17 suggests a battery of tests to be inoculated for each anaerobe type.

Advanced Identification

Gram-Negative Bacilli

Figure 6-1 presents a flow diagram for grouping gram-negative bacilli. *Bacteroides, Prevotella, Porphyromonas, Fusobacterium, Bilophila, Campylobacter,* and *Sutterella* are encountered most commonly in clinical specimens. Other gram-negative bacilli listed in Table 6-1, such as *Leptotrichia, Capnocytophaga, Selenomonas,* and *Anaerobiospirillum,* are rare; they may occur as contaminating normal flora or occasionally as causes of infection. Bacteremia in an immunocompromised host is the most common form of infection with this latter group of organisms (excluding *Selenomonas*) (25;116). Also, *Anaerobiospirillum* has been isolated from fecal specimens of patients with diarrhea (213), and a zoonotic role for this organism has been proposed (212). Initial differentiation of these latter genera is based on motility, flagellar arrangement, and analysis of metabolic endproducts (155). All gram-negative bacilli that are curved, or metabolically fastidious, or that form spreading or pitting colonies, should be tested for motility and identified according to Table 6-1. (A flagella stain modified from the method of Ryu (182) is described in Appendix C.)

table 6-1. Differentiation

Characteristic	Genus
GRAM-NEGATIVE BACILLI	
Rod-Shaped Cells or Coccobacilli	
I. Nonmotile or peritrichous flagella	
A. Produce butyric acid (without isobutyric and isovaleric acids)	*Fusobacterium*
B. Produce major lactic acid	*Leptotrichia*
C. Produce acetic acid and hydrogen sulfide; reduce sulfate	*Desulfomonas*
D. Not as above (A, B, or C)	*Anaerorhabdus*
	Bacteroides
	Bilophila
	Campylobacter (gracilis)
	Catonella
	Dialister
	Dichelobacter
	Fibrobacter
	Johnsonella
	Megamonas
	Mitsuokella
	Porphyromonas
	Prevotella
	Rikenella
	Ruminobacter
	Sebaldella
	Sutterella
	Tissierella
II. Polar flagella, motile	
A. Fermentative	
1. Produce butyric acid	*Butyrivibrio*
2. Produce succinic acid	
a. Spiral-shaped cells	*Succinivibrio*
b. Ovoid cells	*Succinimonas*
3. Produce propionic and acetic acids	*Anaerovibrio*
B. Nonfermentative	
1. Produce succinic acid from fumarate	*Campylobacter*
2. Produce hydrogen sulfide; reduce sulfate	*Desulfovibrio*
III. Tufts of flagella on concave side of curved cells; fermentative	*Selenomonas*
	Centipeda
IV. Bipolar tufts of flagella	*Anaerobiospirillum*
GRAM-NEGATIVE COCCI	
Spherical- or Kidney Bean-Shaped Cells	
I. Produce propionic and acetic acids	*Veillonella*
II. Produce butyric and acetic acids	*Acidaminococcus*
III. Produce isobutyric, butyric, isovaleric, valeric and caproic acids	*Megasphaera*

of genera of anaerobes

Characteristic	Genus
GRAM-POSITIVE COCCI	
I. Fermentable carbohydrate required or stimulatory	
A. Produce butyric (plus other acids)	*Coprococcus*
B. No butyric produced	*Ruminococcus*
II. Do not require a fermentable carbohydrate	
A. Lactic acid sole major product	*Abiotrophia*[a]
	Atopobium[b]
	Gemella
	Granulicatella[a]
	Streptococcus
B. Not as above	*Finegoldia*
	Micromonas
	Peptococcus
	Peptostreptococcus
	Staphylococcus
GRAM-POSITIVE SPOREFORMING BACILLI	*Clostridium*
	Filifactor
GRAM-POSITIVE NON-SPOREFORMING BACILLI	
I. Produce propionic and acetic acids	*Propionibacterium*
	Corynebacterium matruchotii
II. No propionic acid produced	
A. Produce acetic and lactic acid (A≥L)	*Bifidobacterium*
B. Produce lactic acid as sole major end product	*Atopobium*[b]
	Lactobacillus
C. Produce moderate acetic acid plus one of the following:	
1. Major succinic and lactic acids	*Actinomyces, Mobiluncus*
2. Major succinic acid	*Actinomyces*
	Arcanobacterium
3. Moderate or minor succinic acid	*Actinobaculum*
	Bulleidia
4. Major lactic acid	*Collinsella*
	Cryptobacterium
	Holdemania
	Rothia
5. Major caproic acid	*Pseudoramibacter*
6. Phenylacetic acid	*Mogibacterium*
D. Other: butyric ± others, acetic or no major acids	*Eggerthella*
	Eubacterium
	Slackia

[a]Nutritionally variant streptococci (NVS) often isolated in anaerobic culture only.
[b]Displays coccal and rod shaped morphologies by different species.

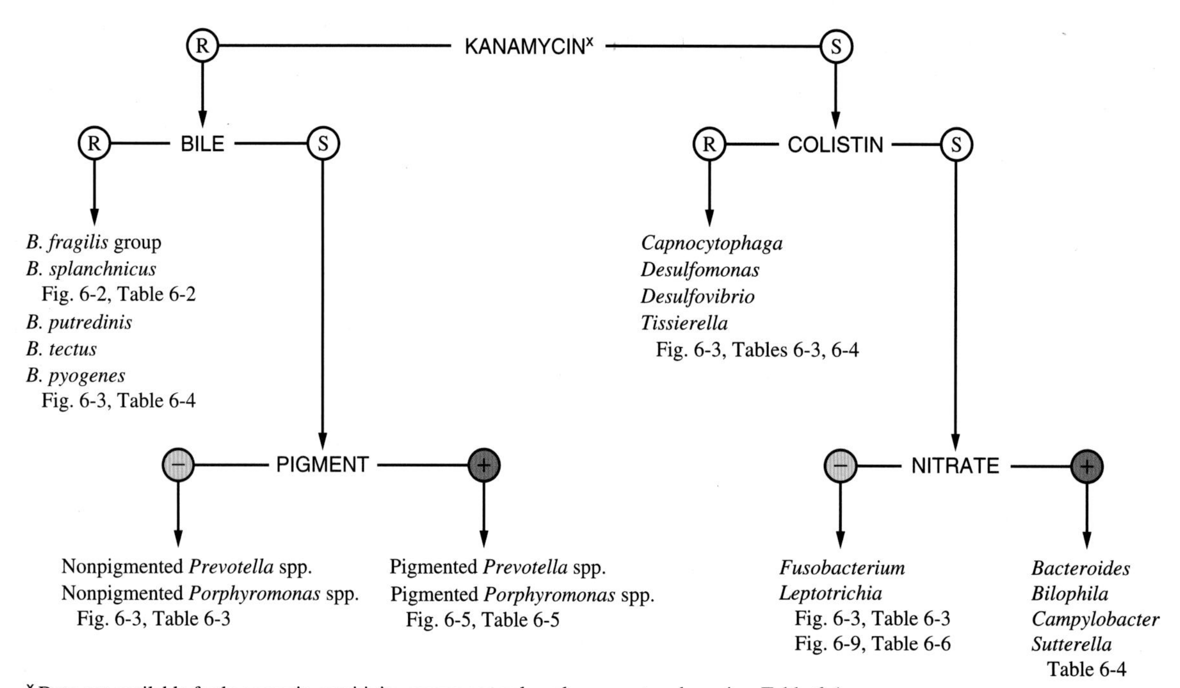

Figure 6-1. Flow chart illustrating scheme for identification of anaerobic gram-negative bacilli.

Bacteroides fragilis Group

Table 6-2 (Figure 6-2; see also Table 5-1 for rapid preliminary identification) presents the key characteristics for identifying the *B. fragilis* group. Good growth on BBE medium and other 20% bile (2% oxgall)–containing media is characteristic of the *B. fragilis* group, with the exception of some *B. uniformis* strains that may grow poorly in the presence of bile. Some non–*B. fragilis* group organisms are bile resistant; included are *B. splanchnicus, B. tectus, B. putredinis, B. pyogenes, Mitsuokella multiacida, Bilophila, Sutterella, Leptotrichia,* and some fusobacteria. Not all of these, however, will grow on BBE agar. Morphologic characteristics, special-potency disks, and some biochemicals will differentiate these species.

Colonies of the *B. fragilis* group on a blood agar (BA) plate are 2–3 mm in diameter, circular, entire, convex, and grey to white in color (see Figure 4-14). On BBE agar, they hydrolyze esculin, blackening the agar except for most strains of *B. vulgatus,* which are esculin-negative (see Figure 3-7). The cells are uniform, pleomorphic, or vacuolized, traits that are medium and age dependent (see Figure 4-15). Some of the species within the group are very similar biochemically and therefore difficult to differentiate. Xylan fermentation is used to separate *B. ovatus* (positive) from *B. thetaiotaomicron* (negative); additionally, ovoid cells suggest *B. ovatus*. Arabinose and cellobiose are useful in identifying *B. stercoris* and *B. uniformis*; both carbohydrates are usually fermented by *B. uniformis,*

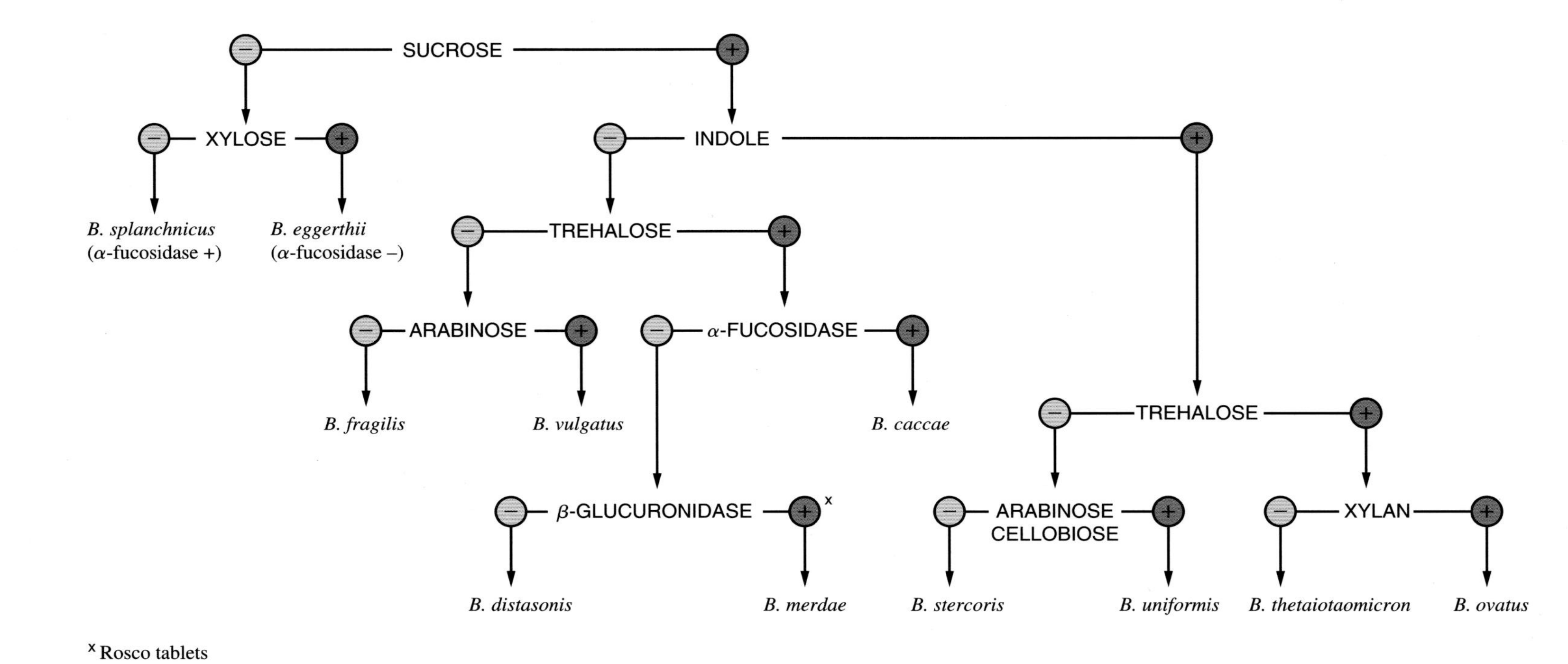

Figure 6-2. Flow chart illustrating scheme for identification of the bile-resistant *Bacteroides fragilis* group and B. *splanchnicus*.

6

table 6-2. Characteristics of *B. fragilis* group and *B. splanchnicus*

Species	Growth in 20% bile	Indole	Catalase	Esculin hydrolysis	Arabinose	Cellobiose	Rhamnose	Salicin	Sucrose	Trehalose	Xylose	Xylan	α-fucosidase	Fatty acids from PYG
B. caccae	+	–	$-^{+}$	+	+	$+^{-}$	$+^{-}$	$-^{+}$	+	+	+	–	+	A,p,S(iv)
B. distasonis[a]	+	–	$+^{-}$	+	$-^{+}$	+	V	+	+	+	+	–	–	A,p,S(pa,ib,iv,l)
B. eggerthii	+	+	–	+	+	$-^{+}$	$+^{-}$	–	–	–	+	+	–	A,p,S(ib,iv,l)
B. fragilis	+	–	+	+	–	$+^{-}$	–	–	+	–	+	–	+	A,p,S,pa(ib,iv,l)
B. merdae[a]	+	–	$-^{+}$	+	$-^{+}$	V	+	+	+	+	+	–	–	A,p,S(ib,iv)
B. ovatus	+	+	$+^{-}$	+	+	+	+	+	+	+	+	+	$+^{-}$	A,p,S,pa(ib,iv,l)
B. stercoris	+	+	–	$+^{-}$	$-^{+}$	$-^{+}$	+	$-^{+}$	+	–	+	V	V	A,p,S(ib,iv)
B. thetaiotaomicron	+	+	+	+	+	$+^{-}$	+	$-^{+}$	+	+	+	–	+	A,p,Spa(ib,iv,l)
B. uniformis	W^{+}	+	$-^{+}$	+	+	+	$-^{+}$	$+^{-}$	+	$-^{w}$	+	V	$+^{-}$	a,p,l,S(ib,iv)
B. vulgatus	+	–	$-^{+}$	$-^{+}$	+	–	+	–	+	–	+	$-^{+}$	+	A,p,S
B. splanchnicus	+	+	–	+	+	–	–	–	–	–	–	–	+	A,P,S,ib,b,iv(l)

[a] *B. merdae is* β-glucuronidase-positive with Rosco disks, *B. distasonis* is negative.

but not by most strains of *B. stercoris*. *B. caccae* and *B. merdae* closely resemble *B. distasonis*. *B. merdae* and most *B. caccae* are catalase-negative, whereas *B. distasonis* is usually positive. α-Fucosidase differentiates *B. distasonis* and *B. merdae* from *B. caccae: B. distasonis* and *B. merdae* are α-fucosidase-negative and *B. caccae* is positive. Preliminary tests in our laboratories suggest that β-glucuronidase may distinguish *B. distasonis* (negative) from *B. merdae* (positive). *B. eggerthii* and *B. splanchnicus,* which resemble the *B. fragilis* group and are generally considered with this group, do not ferment sucrose; *B. eggerthii* is xylose-positive and α-fucosidase-negative; *B. splanchnicus* is xylose-negative and α-fucosidase-positive.

As mentioned earlier, major taxonomic changes have occurred with the anaerobic gram-negative rods other than the *B. fragilis* group. This chapter will divide these organisms into pigmented and nonpigmented gram-negative rods.

Nonpigmented *Prevotella* and Other Gram-Negative Rods

Saccharolytic

The saccharolytic, nonpigmented gram-negative rods are listed in Table 6-3 (Figure 6-3). These organisms fall into three categories: pentose fermenters and nonfermenters

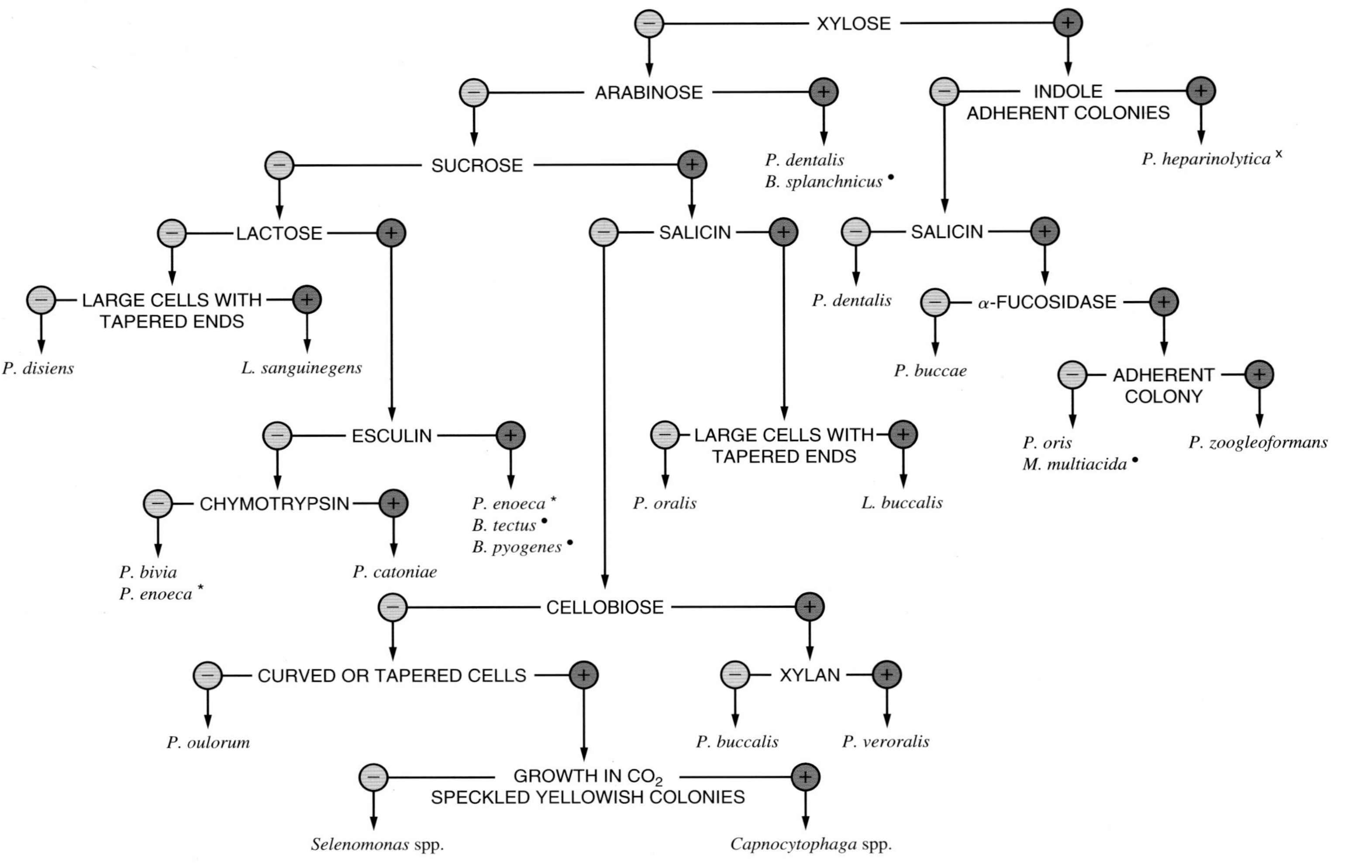

ˣ Indole production difficult to demonstrate; rare indole-negative colonies may be *P. zoogleoformans*
• Bile-resistant
* *P. enoeca* is esculin variable; isolated from oral sites

Figure 6-3. Flow chart illustrating scheme for identification on nonpigmented, saccharolytic gram-negative bacilli.

table 6-3. Characteristics of nonpigmented

Subgroup and species	Growth in 20% bile	Indole	Esculin hydrolysis	Arabinose	Cellobiose	Glucose	Lactose	Salicin
Pentose fermenters								
M. multiacida	+	–	+	+	+	+	+	+
P. buccae	–	–	+	+	+	+	+	+
P. dentalis	–	–	V	+	+	+	+	–
P. heparinolytica[a]	–	+	+	+	+	+	+	+
P. oris	–	–	+	$+^{-}$	+	+	+	+
P. zoogleoformans[a]	–	–	+	V	+	+	+	V
Not pentose fermenters								
P. buccalis	–	–	+	–	+	+	+	–
P. enoeca	–	–	V	–	–	+	+	–
P. oralis	–	–	+	–	+	+	+	+
P. oulorum[c]	–	–	+	–	–	+	+	–
P. veroralis	–	–	+	–	+	+	+	–
Proteolytic								
P. bivia	–	–	–	–	–	+	+	–
P. disiens	–	–	–	–	–	+	–	–
Other								
Capnocytophaga spp.	–	–	$+^{-}$	–	$-^{+}$	+	V	–
Leptotrichia buccalis	$+^{-}$	–	+	$-^{+}$	$+^{-}$	+	$+^{-}$	$+^{-}$
Leptotrichia sanguinegens		–	+	–	+	+	–	+
Selenomonas spp.	–	–	V	–	–	$+^{-}$	V	–

[a] Produces viscous sediment in broth and colonies usually adhere to agar.

[b] Positive by 4–methyl–umbelliferyl substrate.

[c] Catalase and lipase positive.

(xylose and arabinose are usually tested; suboptimal growth may give false negative results, so testing of β-xylosidase with a heavy inoculum is recommended) and proteolytic (gelatin and milk are usually tested) (172). The other key carbohydrates useful for differentiating these organisms are cellobiose, lactose, salicin, and sucrose. Certain strains of pigmented *Prevotella* can take up to 21 d to develop visible pigment and can therefore be difficult to differentiate from the nonpigmented species. Fluorescence is

saccharolytic *Prevotella* spp. and other genera

Sucrose	Xylose	Xylan	α-Fucosidase	ONPG (β-galactosidase)	N-Acetyl-β-glucosaminidase	β-Xylosidase	Fatty acids from PYG
+	+		+	+	+	+	A,L,S
+	+		−	+	−	+	A,S(p,ib,b,iv,l)
W	V		−	+	+	V	A,S
+	+		+	+	+	+	A,p,S(iv)
+	+		+	+	+	+	A,S(p,ib,iv)
+	V		+	+	+	+	A,P,S(ib,iv)
+	−	−	+	+	+	−	a,iv,S
−	−		+	+	+	−	a,S
+	−	−	+	+	+	−[b]	A,S(l)
+	−	−	−	+	+	−	A,S
+	−	+	+	+	+	−	a,S
−	−		+	+	+	−	A,iv,S(ib)
−	−		−	−	−	−	A,S(p,ib,iv)
+	−						A,S
+	−						L(a,s)
−	−						L(a)
$+^-$	−						A,P

useful for recognizing these strains. Also cell morphology and ability to grow in 5% CO_2 are crucial differential reactions. β-N-acetyl-glucosaminidase and α-fucosidase tests differentiate the phenotypically similar pentose fermenters *P. oris* (positive) from *P. buccae* (negative). Additionally, *P. buccae* often appears sensitive to the special-potency colistin disk, while *P. oris* is resistant. *Prevotella dentalis* (a water drop–like colony; Figure 6-4) can be differentiated from *P. oris* (white or grey opaque) by colony morphology and by salicin fermentation and α-fucosidase reaction (*P. dentalis* is negative for both; *P. oris* is positive). *Mitsuokella multiacida* can be differentiated from *P. oris* since it is resistant to bile and positive for nitrate, and further by its peculiar special-potency antibiotic disk pattern: sensitive to both kanamycin and colistin. *P. zoogleoformans* and *P. heparinolytica* form a highly viscous, tenacious (zoogleal) mass in broth cultures and adherent colonies on solid media. Indole is the key reaction to distinguish *P. zoogleoformans* from *P. heparinolytica* (indole-positive) and also from other pentose fermentors. Indole production is often difficult to demonstrate; a heavy inoculum from a pure culture on EYA or an old culture (>5 d) may facilitate detection. *P. heparinolytica* is often isolated from oral specimens of humans and animals whereas *P. zoogleoformans* is infrequently isolated from clinical specimens of humans. The other pentose fermenters of animal origin, *P. ruminicola, P. brevis, P. bryantii,* and *P. albensis,* are rarely, if ever, isolated from clinical samples of humans. Their differential identification is based on demonstration of extracellular enzyme activities such as DNase, carboxymethylcellulase, and xylanase (12). Key reactions in the differentiation of the non–pentose fermenting *P. oralis* group (*P. oralis, P. buccalis, P. veroralis,* and *P. oulorum*) are salicin, cellobiose, xylan, and sucrose. Xylan is used to differentiate *P. veroralis. P. oulorum* is lipase- and catalase-positive. *P. bivia, P. disiens,* and *P. enoeca*

table 6-4. Characteristics of nonpigmented weakly

Species	Growth in 20% bile	Glucose	Catalase	Indole	Nitrate	Motility
Anaerorhabdus furcosus	$+^{-}$	W	–	–	–	–
Bilophila wadsworthia[a]	+	–	+	–	+	–
Bacteroides capillosus	$-^{+}$	W^{-}	–	–	–	–
B. coagulans	+	–	–	+	–	–
B. forsythus[b]	–	–	–	–	–	–
B. putredinis	$+^{-}$	–	$+^{-}$	+	–	–
B. pyogenes	+	W	–	–	–	–
B. tectus	+	W	–	–	–	–
B. ureolyticus	–	–	$-^{+}$	–	+	–
Campylobacter spp.	–	–	–	–	+	$+^{-}$
C. gracilis	–	–	–	–	+	–
Desulfomonas pigra[a, c]	V	–	V	–	$-^{+}$	–
Desulfovibrio spp.[a]	V	–	V	–	V	+
Dialister pneumosintes	–	–	–	–	–	–
Sutterella wadsworthensis	+	–	–	–	+	–
Tissierella praeacuta	+	–	–	–	V	+

[a]*Bilophila* sp. is colistin sensitive; *Desulfomonas* sp. and *Desulfovibrio* spp. are colistin resistant.

[b]Reclassified as *Tannerella forsythensis* (Reclassification of *Bacteroides forsythus* (Tanner et al. 1986) as Tannerella forsythensis corrig. gen. nov., comb nov. M. Sakamoto, M. Suzuki, M. Umeda, I. Ishikawa, and Y. Benno. IJSEM, In press, published online 29 November 2001.)

[c]Reclassified as *Desulfovibrio piger.* (Recalssification of *Desulfomonas pigra* as *Desulfovibrio piger* comb. nov. and suppression of the genus *Desulfomonas,* by J. Loubinoux, F. M. A. Valente, I. A. C. Pereira, A. Costa, P. A. D. Grimont, and A. E. Le Faou. IJSEM, In press.)

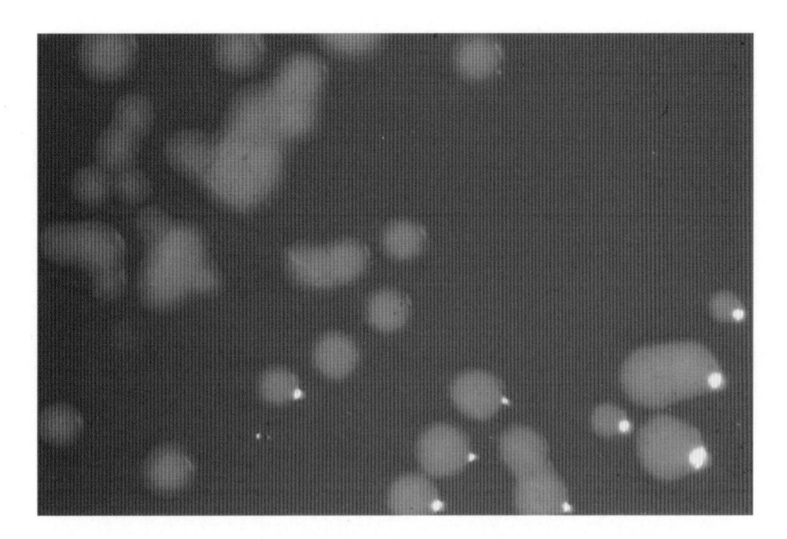

Figure 6-4. *P. dentalis* colony; note water drop-like slimy appearance.

saccharolytic or asaccharolytic gram-negative bacilli

F/F required	Desulfoviridin	Urease	Esculin hydrolysis	Gelatin hydrolysis	Fatty acids from PYG
–	–	–	+	$-^{W}$	a,l(s)
–	W^{-}	$+^{-}$	–	–	A(s)
–	–	–	+	$-^{W}$	a,s(p,l)
–	–	–	–	+	a(p,l,s)
–	–	–	+	+	A,S,pa
–	–	–	–	+	a,P,ib,b,IV,S(l),pa
–	–	–	+	+	a,P,ib,b,IV,S(l)
–	–	–	+	+	A,p,iv,S,pa
+	–	+	–	–	A,S
+	–	–	–	–	a,S
+	–	–	–	–	a,S
–	+	$-^{+}$	–	–	A
–	+	$-^{+}$	–	–	A
–	–	–	–	$-^{W}$	a(l,s)
+	–	–	–	–	a,S
–	–	–	–	+	A,p,ib,B,IV,s(l)

are strongly proteolytic and do not ferment sucrose. Their colonies may fluoresce pink (coral) under UV light and should not be confused with the pigmented *Prevotella* species. Lactose fermentation is used to differentiate *P. bivia* and *P. enoeca* (lactose-positive) from *P. disiens* (lactose-negative). *P. enoeca* may be differentiated from *P. bivia* by a positive esculin test and by the fact that it is an oral organism.

A simplified approach to the identification of nonpigmented *Prevotella* species using enzyme reactions is presented in Chapter 5.

Asaccharolytic

Asaccharolytic, nonpigmented gram-negative rods (Table 6-4) other than the *B. ureolyticus*–like organisms and *Bilophila* are occasionally isolated from clinical specimens. The recognition of *B. ureolyticus*–like organisms and *Bilophila* was discussed in Chapters 4 and 5. *Anaerorhabdus furcosus, Bacteroides capillosus, B. tectus,* and *B. pyogenes* may ferment glucose and other carbohydrates weakly, and are esculin-positive. Gram stain and gas-liquid chromatography are useful for differentiating them. In addition, *B. capillosus* coagulates milk and often grows better with polysorbate (Tween-80)-supplemented media. *B. tectus* and *B. pyogenes* are resistant to 20% bile and are isolated only from animal sources or dog and cat bite wound infections (119). *B. coagulans* and *B. putredinis* are both indole-positive, and can be separated by gas-liquid chromatography and the catalase test. *B. forsythus,* which is a recognized periodontal pathogen, is a fusiform-shaped bacillus and requires N-acetylmuramic acid (NAM) for growth in monoculture, but grows well as satellite colonies in mixed cultures (378); it also shows a wide variety of enzyme activities including trypsinlike activity, β-N-acetyl-glucosaminidase, β-glucuronidase, and α-fucosidase and is esculin-positive. Visualization of *D. pneumosintes* colonies may require magnification, and negative staining may be necessary to see the cells (which are coccoid and very minute, <0.3 μm).They have often been misidentified as *Veillonella.* However, *Veillonella* is nitrate-positive, whereas *D. pneumosintes* is nitrate-negative and resistant to colistin, while the vast majority of *Veillonella* strains are susceptible (168). *Desulfomonas* and *Desulfovibrio* are sensitive to the special-potency disk kanamycin and resistant to colistin, and are sulfate-reducing; the desulfoviridin test described in Appendix C can be used to confirm the identification. Although the latter is motile and the former is not, the two genera are genetically closely related.

Pigmented Gram-Negative Rods

Table 6-5 and Figure 6-5 show characteristics for identifying the pigmented gram-negative rods. Pigmented *Prevotella* and *Porphyromonas* species vary greatly in the degree and rapidity of pigmentation, depending primarily on the type of blood and basal medium used. Laked rabbit blood agar (LBA) is considered most rapid and reliable (155;167). The pigmentation ranges from buff to tan to black, and may take several days to develop (Figures 6-6 to 6-8). As mentioned earlier, the identification of those strains that usually do not show pigment until after 21 d of incubation must be established by other biochemical tests. The UV fluorescence of the pigmented

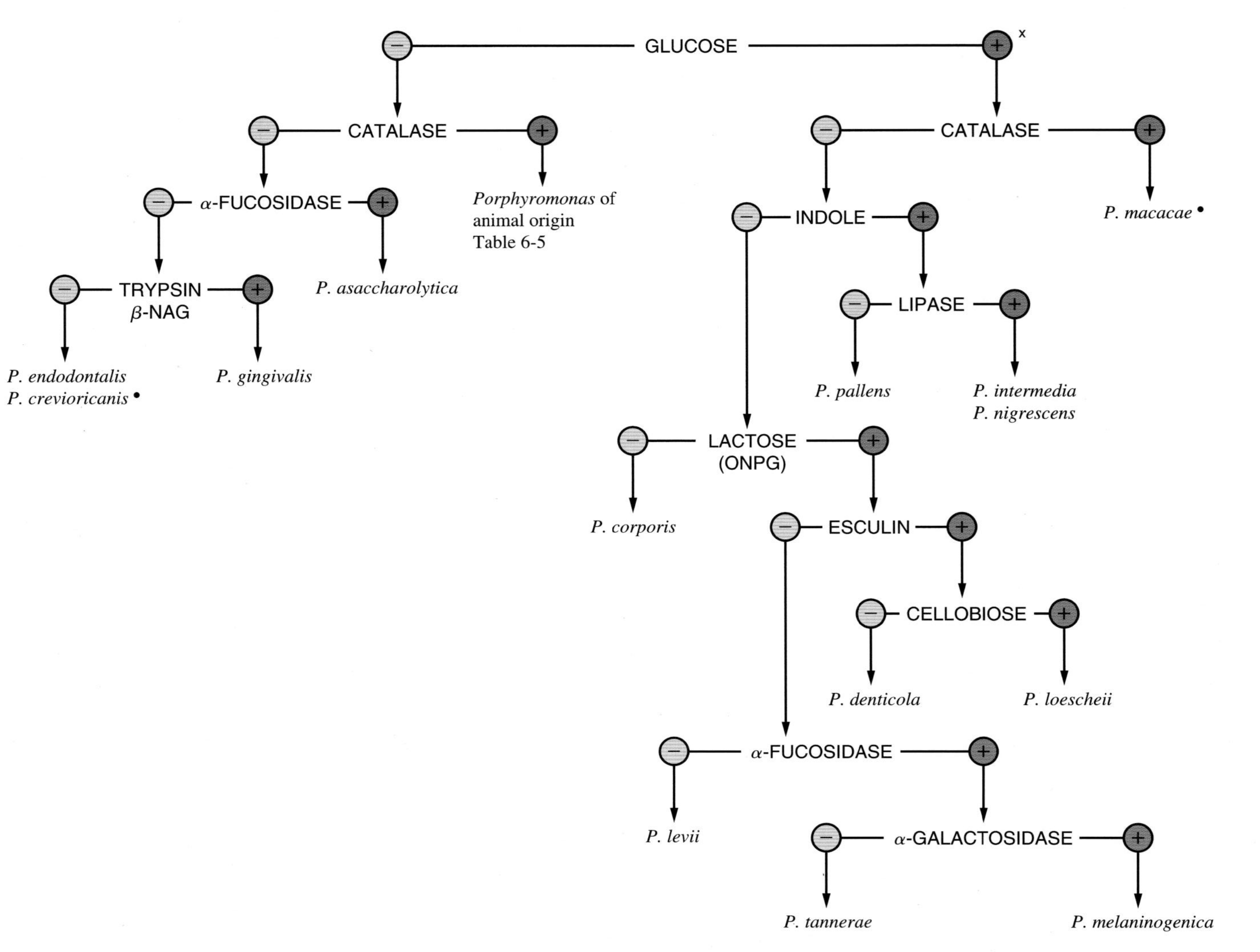

x *Porphyromonas levii* and *P. macacae* weakly positive
• *Porphyromonas* of animal origin

Figure 6-5. Flow chart illustrating scheme for identification of pigmented gram-negative bacilli.

table 6-5. Characteristics of pigmented

	Indole	Lipase	Catalase	Fermentation of Glucose	Cellobiose	Lactose	Sucrose
Porphyromonas: Asaccharolytic or weakly saccharolytic							
P. asaccharolytica	+	–	–	–	–	–	–
P. canoris[b]	+	–	+	–	–	–	–
P. cangingivalis[b]	+	–	+	–	–	–	–
P. cansulci[b]	+	–	$+^{W}$	–	–	–	–
P. catoniae[c]	–	–	–	W	–	W	–
P. circumdentaria[b]	+	–	+	–	–	–	–
P. crevioricanis[b]	+	NA	–	–	–	–	–
P. endodontalis	+	–	–	–	–	–	–
P. gingivalis[d]	+	–	–	–	–	–	–
P. gingivicanis[b]	+	NA	+	–	–	–	–
P. levii[b,d]	–	–	–	W	–	W	–
P. macacae[b]	+	+	+	W	–	W	–
Prevotella: Saccharolytic							
P. corporis	–	–	–	+	–	–	–
P. denticola	–	–	–	+	$-^{+}$	+	+
P. intermedia	+	$+^{-}$	–	+	–	–	$+^{-}$
P. loescheii	–	V	–	+	+	+	+
P. melaninogenica	–	–	–	+	$-^{+}$	+	+
P. nigrescens[f]	+	$+^{-}$	–	+	–	–	+
P. pallens	+	–	–	+	–	–	+
P. tannerae	–	–	–	+	–	+	V

[a]Reaction by API ZYM System or Rosco Diagnostic tablets. Reactivity in these systems is not always identical (see footnote e).

[b]Animal origin; may be isolated from bite infections; *P. macacae* presently includes former *P. salivosa*; NA information not available.

[c]Nonpigmented.

[d]*P. gingivalis* does not show fluorescence; isolates phenotypically resembling *P.levii* (PLLO) may show weak or no fluorescence. Animal strains of *P. gingivalis* (*Porphyromonas gulae;* 119a) are catalase-positive, human strains are catalase-negative.

[e]Negative by the API ZYM System; positive with the Rosco O-nitrophenyl-ß-D-galactopyranoside test.

[f]*P. intermedia* shares the same characteristics. Differentiation based on enzyme electrophoresis, oligonucleotide probe or AP-PCR.

6

Porphyromonas spp. and *Prevotella* spp.

Esculin hydrolysis	α-Fucosidase[a]	α-Galactosidase[a]	β-Galactosidase[a]	N-Acetyl-β-glucosaminidase[a]	Trypsin[a]	Chymotrypsin[a]	Fatty acids from PYG
–	+	–	–	–	–	–	A,P,ib,B,IV,s
–	–	–	+	+	–	+	A,P,ib,b,IV,s
–	–	–	–	–	–	+	A,p,ib,B,IV
–	–	–	–	–	–	–	A,P,ib,B,IV,S,pa
–	+	$-^{+}$	+	+	$-^{+}$	$+^{-}$	a,P,iv,l,S
–	–	–	–	–	–	–	A,P,ib,b,IV,s,pa
–	–	NA	NA	–	–	NA	A,p,ib,B,IV,s
–	–	–	–	–	–	–	A,P,ib,B,IV,s
–	–	–	–[e]	+	+	–	A,P,ib,B,IV,s,pa
–	–	NA	NA	–	–	NA	A,p,ib,B,IV,s
–	–	–	+	+	–	+	A,P,ib,B,IV,s
–	–	+	–[e]	+	+	+	A,P,ib,B,IV,s,pa
–	–	–	–	–	–	$+^{-}$	A,ib,iv,S(b)
$+^{-}$	+	+	+	+	–	–	A,S(ib,iv,l)
–	+	–	–	–	–	–	A,iv,S(p,ib)
$+^{-}$	+	+	+	+	–	–	a,S(l)
$-^{+}$	+	+	+	+	–	–	A,S(ib,iv,l)
–	+	–	–	–	–	–	A,iv,S(p,ib)
–	+	–	–	–	–	–	A,S(p,ib)
–	+	–	+	+	–	–	A,iv,S(ib)

6

Prevotella and *Porphyromonas* species varies from pink to orange to red, and is masked by visual pigment production. Brick-red fluorescence is the only reliable color for presumptive identification of these organisms since some nonpigmented *Prevotella* species can fluoresce coral or pink. Gram stain will easily differentiate the other organisms that either produce pigment or fluoresce: the gram-positive rod, *Actinomyces odontolyticus,* usually forms red pigment; the gram-positive rod *E. lenta* and the gram-negative coccus *Veillonella* may both fluoresce red.

The unusual special-potency disk pattern (sensitive to vancomycin) and asaccharolytic or weakly saccharolytic nature separate *Porphyromonas* from *Prevotella* species. A positive catalase reaction will separate *Porphyromonas* species (except *P. crevioricanis*) of animal origin from those of human origin (167). Trypsinlike activity and β-N-acetylglucosaminidase activity (see Appendix C) will separate *P. gingivalis* from other *Porphyromonas* species; α-fucosidase activity separates *P. asaccharolytica* from *P. endodontalis.* In addition, *P. gingivalis* does not fluoresce under UV light. *P. levii*–like organisms have been isolated from clinical specimens, particularly from chronic foot infections of diabetics or other patients with vascular insufficiency (171). Satellite-like growth in primary culture, weakly saccharolytic nature, and negative indole and catalase are key characteristics in species differentiation. Demonstration of pigment (often weak in young cultures) is crucial to differentiate *P. levii* from *P. catoniae,* which is nonpigmented but closely related by other phenotypic characteristics (Figure 6-6). The latter is an oral organism, but *P. levii* is usually isolated from extraoral sources. *P. macacae* of animal origin shares similar characteristics but is positive for indole, catalase, and lipase (172). The indole-positive *Prevotella intermedia* and *P. nigrescens* produce dark brown to black pigment, but *P. pallens'* pigment is pale (Figure 6-7). Furthermore, *P. pallens* is lipase-negative (185;187). Routine biochemical tests do not separate *P. intermedia* from *P. nigrescens;* arbitrarily primed PCR, oligonucleotide probes, or electrophoretic enzyme mobilities must be used (187;221;294). Lactose and cellobiose fermentation and esculin hydrolysis are the key reactions in species differ-

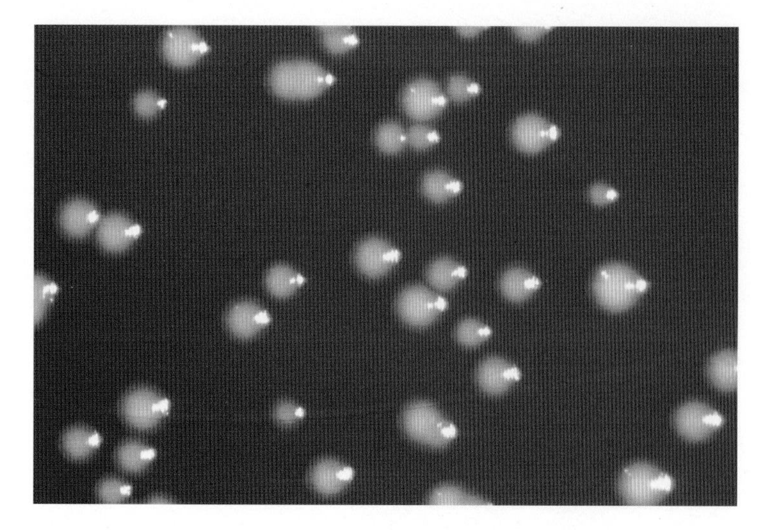

Figure 6-6. *P. catoniae* colony; note absence of pigment.

entiation of the indole-negative, pigmented *Prevotella* species. *P. corporis* is lactose-negative. The esculin-positive *P. denticola* is separated from *P. loescheii* by not fermenting cellobiose and by being lipase-negative. *P. melaninogenica* and *P. tannerae* are usually esculin-negative; *P. melaninogenica* is α-galactosidase positive and *P. tannerae* is negative. Although all the pigmented gram-negative rods presented above are susceptible to bile, a bile-resistant pigment-producing organism, probably belonging to a new genus and species, has been described recently (270;271).

Chapter 5 presents a simplified approach to the identification of pigmented gram-negative bacilli using enzyme reactions.

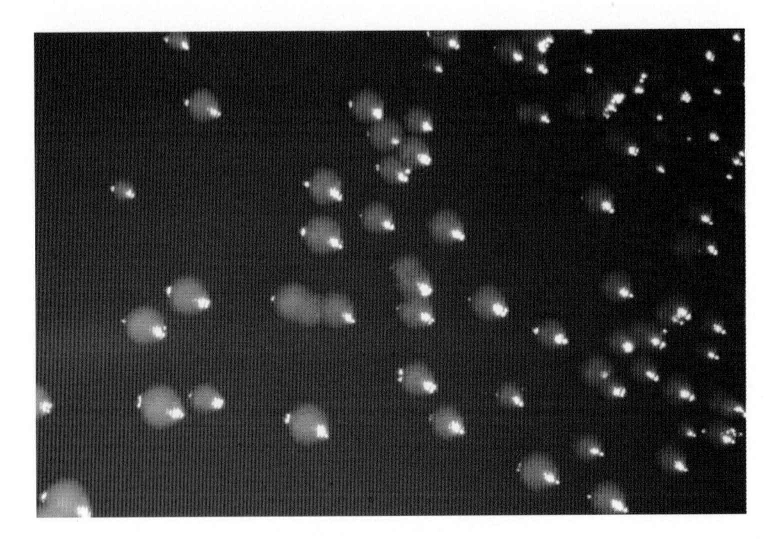

Figure 6-7. *P. pallens* colony; note pale pigment.

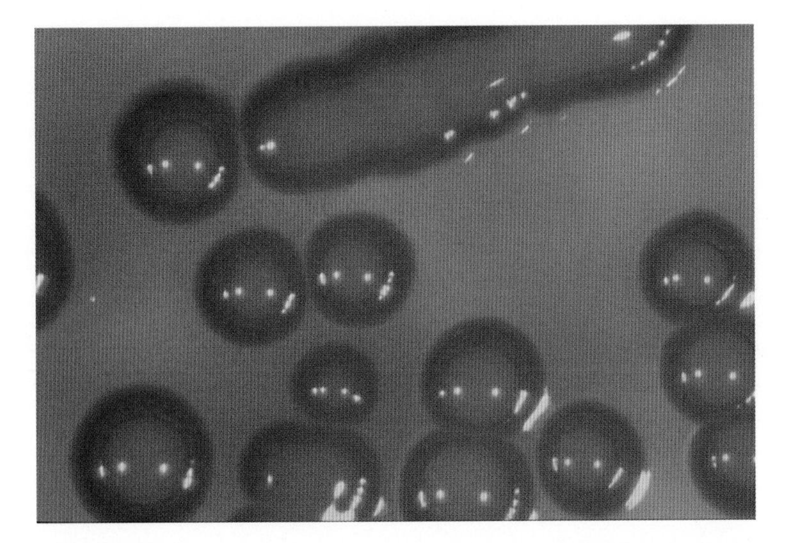

Figure 6-8. *Porphyromonas* sp. colony; note brown pigment and mucoid appearance.

Fusobacteria

Fusobacterium species are resistant to the vancomycin special-potency disk and sensitive to colistin and kanamycin. Most species are indole-positive. They are weakly saccharolytic or nonfermentative and produce butyric acid without isobutyric or isovaleric acids. The colonial morphology of fusobacteria varies greatly, but Chapter 4 describes the characteristic. Furthermore, many strains fluoresce chartreuse under UV light and most strains produce greening of blood agar when exposed to air due to H_2O_2 production.

Table 6-6 lists the fusobacteria and the key reactions for differentiating them (Figure 6-9). The most common clinical isolates, *F. necrophorum* and *F. nucleatum,* are discussed in Chapters 4 and 5. Lipase-negative and nonhemolytic strains of *F. necrophorum* were transferred into a new species, *F. pseudonecrophorum,* and were subsequently reclassified as *F. varium* (13). Bizarre, pleomorphic rods with large round bodies are characteristic of *F. mortiferum* (see Figure 4-23). This organism may grow on BBE agar. Another fusobacterium that can be isolated from BBE agar is *F. varium,* which is esculin-negative and can be indole- and lipase-positive. Lipase production can take up to 10 d to develop. *F. ulcerans,* isolated from tropical ulcers, resembles *F. varium* morphologically. Fructose fermentation will differentiate *F. ulcerans* (fructose-negative) from indole-negative strains of *F. varium.* We have found that *F. periodonticum* is biochemically indistinguishable from *F. nucleatum.* This finding was later confirmed by 16S rRNA sequencing (256). *F. russii* is indole-negative and a common finding in animal bite infections (330).

Capnocytophaga and *Leptotrichia* species (see Figure 6-3, Table 6-3) are morphologically similar to fusobacteria on Gram stain but are saccharolytic. Smaller cells with tapered ends (Figure 6-10) that grow in CO_2 and display yellowish-pinkish or bluish speckled colonies (Figure 6-11) are *Capnocytophaga* species, a part of the indigenous oral flora.

table 6-6. Characteristics c

						F
***Fusobacterium* sp.**	**Distinctive cellular morphology**	**Indole**	**Growth in 20% bile**	**Lipase**	**Gas in glucose agar**	**Glucose**
F. gonidiaformans	Gonidial forms	+	–	–	4^2	–
F. mortiferum[a]	Bizarre; round bodies	–	+	–	4	$+^w$
F. naviforme	Boat shape	+	–	–	$-^2$	w^-
F. necrophorum[b]	Large, pleomorphic	+	$-^+$	$+^-$	4^2	$-^w$
F. nucleatum[c]	Slender, pointed ends	+	–	–	$-^2$	$-^w$
F. russii	Large, rounded ends	–	–	–	2^-	–
F. varium	Large, rounded ends	$+^-$	+	$-^+$	4	w^+
F. ulcerans[d]	Large, rounded ends	–	+	–	2	+

[a] O-nitrophenyl-β-D-galactopyranoside (ONPG) -positive.

[b] Lipase-positive strains *F. necrophorum* subsp. *necrophorum* and lipase-negative strains *F. necrophorum* subsp. *funduliforme.*

[c] *F. periodonticum* shares the same characteristics and probably is *F. nucleatum.*

[d] Nitrate-positive.

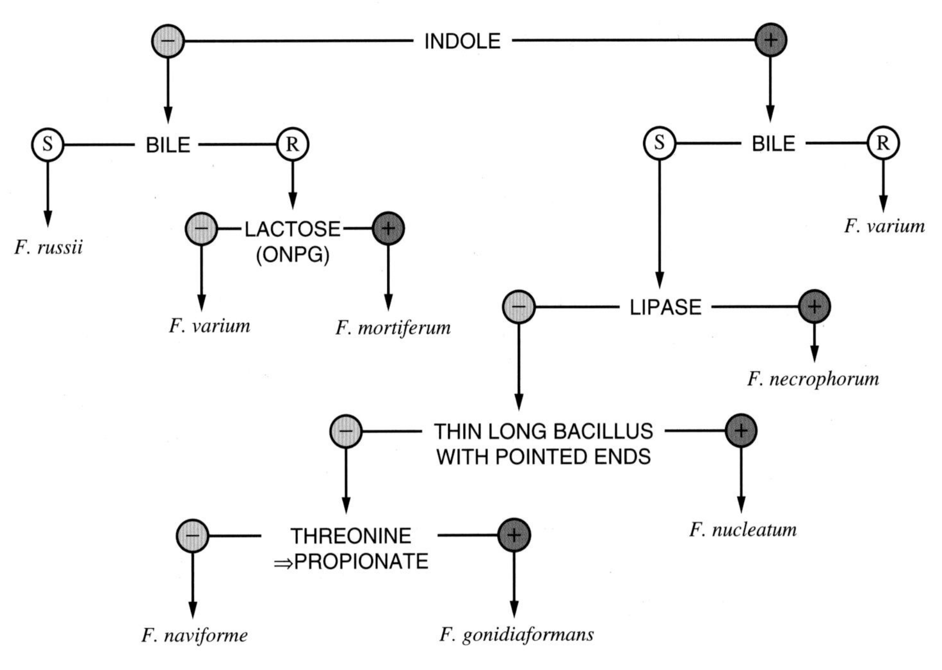

Figure 6-9. Flow chart illustrating scheme for identification of *Fusobacterium* sp.

usobacterium species

entation of						
ructose	**Lactose**	**Mannose**	**Esculin hydrolysis**	**Lactate converted to propionate**	**Threonine converted to propionate**	**Fatty acids from PYG**
	–	–	–	–	+	A,p,B(l,s)
w	+	$+^{w}$	$-^{+}$	–	+	a,p,B(v,l,s)
	–	–	–	–	–	a,B,L(p,s)
W	–	–	–	+	+	a,p,B(l,s)
v	–	–	–	–	+	a,p,B(L,s)
	–	–	–	–	–	a,B,L
+	–	$+^{w}$	–	–	+	a,p,B,L(s)
	–	$+^{-}$	–	–	+	a,p,B,l(s)

Very large fusiform rods isolated from the mouth or urogenital tract and with one pointed and one blunt end (Figure 6-12) and grey, relatively large, sometimes spreading, convoluted (brain texture) colonies (Figure 6-13) are suggestive of *Leptotrichia* species. Sucrose- and xylose-positive strains are *L. buccalis;* xylose-negative strains probably are *L. sanguinegens. Selenomonas* species and related *Centipeda periodontii* are motile curved rods (Figure 6-14) that do not grow in CO_2 and are often isolated from oral samples. Doughnut-shaped (*Selenomonas,* Figure 6-15) or swarming (*Centipeda*) colonies are helpful features

Figure 6-10. Gram stain of *Capnocytophaga* sp., a capnophilic organism that can be confused with *Fusobacterium nucleatum* due to similar cellular morphology.

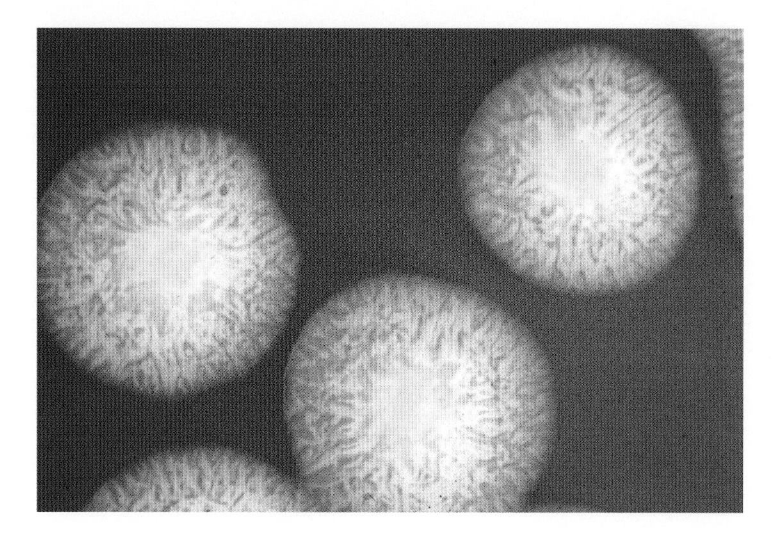

Figure 6-11. *Capnocytophaga* sp. colony; note speckled bluish-yellowish texture.

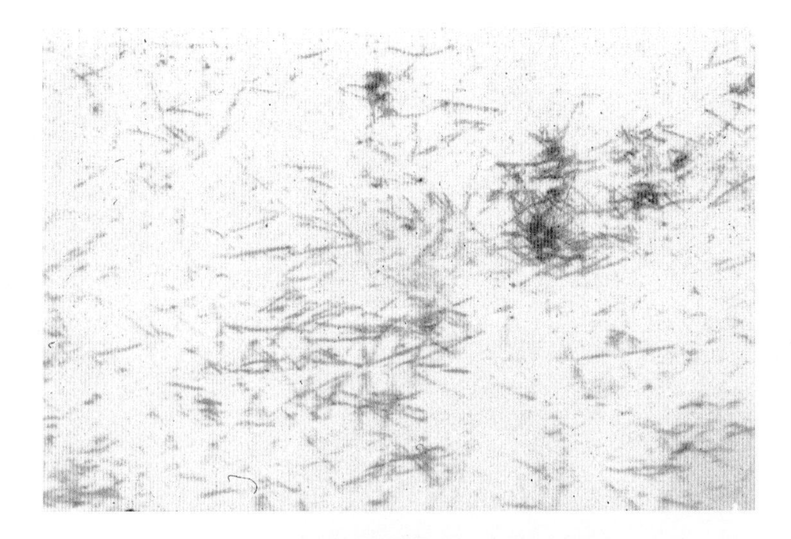

Figure 6-12. Gram stain of *Leptotrichia buccalis,* showing gram-negative bacillus with one square end and one pointed end.

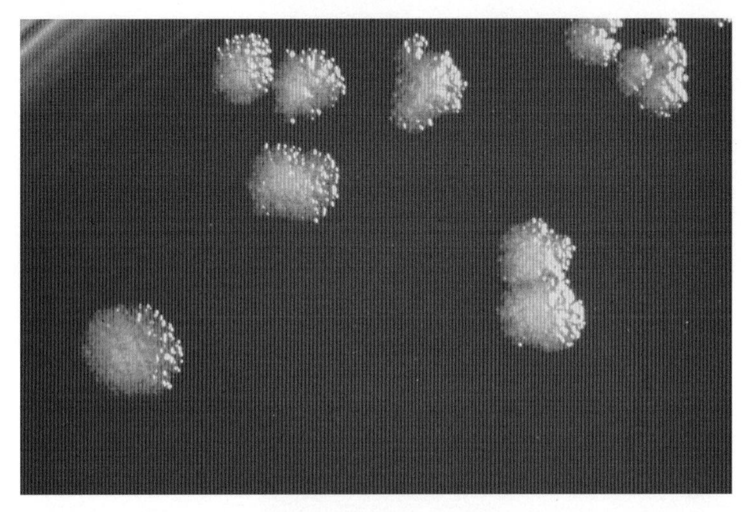

Figure 6-13. *Leptotrichia buccalis* colony; note convoluted (brain) texture.

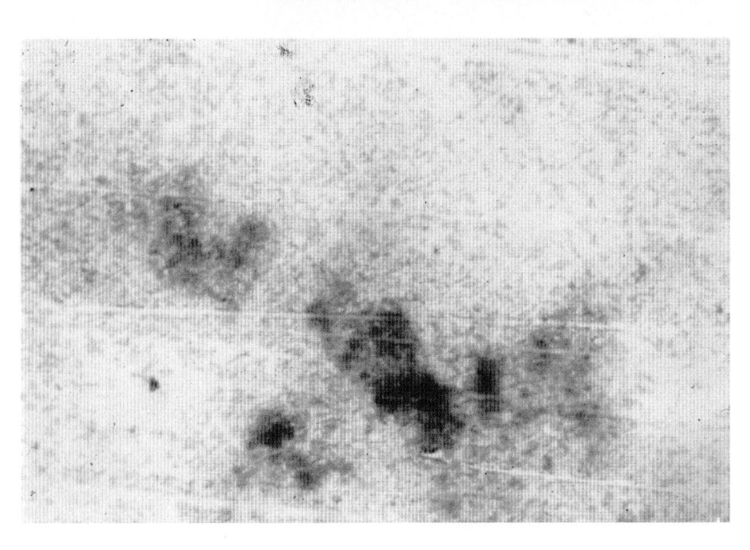

Figure 6-14. Gram stain of *Selenomonas* sp.; note gram-negative curved cells.

in identification. *Actinobacillus actinomycetemcomitans,* a capnophilic gram-negative rod, often isolated from TSBV medium (as are the former two), is the key organism in juvenile and rapidly progressing periodontitis. It is a nitrate- and catalase-positive tiny coccobacillus whose colonies have a distinctive starlike inner structure (Figure 6-16).

Figure 6-15. *Selenomonas* sp. colony; note donut-shaped colony with inner crater.

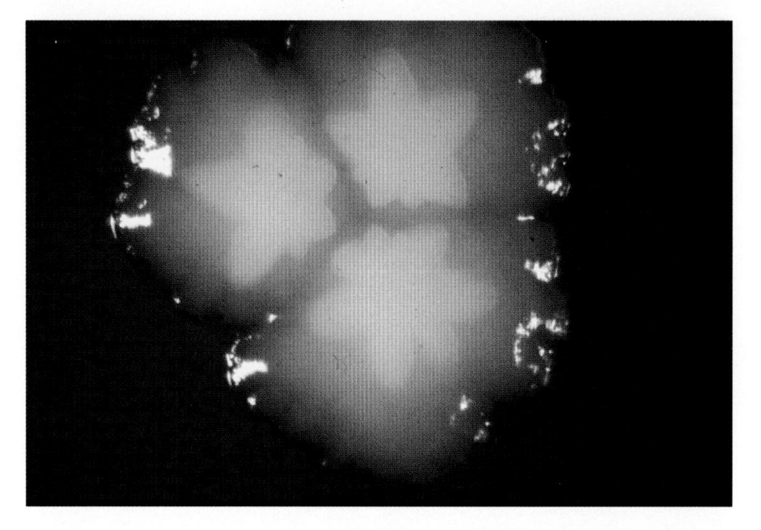

Figure 6-16. *Actinobacillus actinomycetemcomitans* on TSBV; note starlike inner structure.

Gram-Negative Cocci

Veillonella, Acidaminococcus, and *Megasphaera* comprise the genera of anaerobic gram-negative cocci. *Veillonella* species are part of the normal mouth, upper respiratory, and gastrointestinal tracts and the vaginal flora; *Megasphaera* and *Acidaminococcus* are part of the intestinal flora. *Veillonella* is isolated more frequently from clinical specimens than the other gram-negative cocci. Table 6-7 contains a key for differentiating these three genera.

Gram-Positive Cocci

The anaerobic gram-positive cocci are part of the normal flora of humans (see Table 1-1) and are frequently isolated from clinical specimens (see Table 1-3). Extensive taxonomic changes have occurred recently among the anaerobic gram-positive cocci (238;240). The genera of clinical importance are *Finegoldia, Micromonas, Peptostreptococcus,* and *Staphylococcus.* In addition, the microaerobic genera *Gemella, Streptococcus, Abiotrophia,* and *Granulicatella* are often isolated from anaerobic cultures only. Other anaerobic cocci, *Coprococcus, Peptococcus, Ruminococcus,* and *Sarcina* are rarely isolated from clinical specimens. Tables 6-8 to 6-9 and Figure 6-17 show differential characteristics for the gram-positive cocci.

Gas-liquid chromatography and biochemical tests are needed for genus- and species-level identification of most anaerobic gram-positive cocci except for *P. anaerobius* (SPS sensitive) and *M. micros,* which is SPS variable with cells <0.6 μm in diameter and usually forms short chains (see Figure 4-27). *M. micros* colonies are small and white with a halo of discoloration due to incomplete β-hemolysis surrounding them (see Figure 4-26). *P. asaccharolyticus* is indole-positive and has regular cell and colony morphology, whereas *P. harei* has irregular cell shape and size. *P. indolicus,* another asaccharolytic, indole-positive, gram-positive coccus, is rarely isolated from human specimens, and is nitrate- and coagulase-positive. Saccharolytic, indole-positive cocci can be differentiated by the PYR and ADH tests: *P. trisimilis* is PYR-positive and ADH-negative; *P. hydrogenalis* is PYR- and ADH-negative; and *P. vaginalis* is ADH-positive and PYR-negative. The indole-negative, butyric acid–producers *P. tetradius* and *P. prevotii* phenotypically closely resemble each other. The *P. tetradius/prevotii* group is saccharolytic, urease-positive, and lactose-negative, whereas *P. lactolyticus* is lactose-positive.

table 6-7. Characteristics of gram-negative Cocci

Organism	Nitrate reduction	Catalase	Glucose	Fatty Acids from PYG
Veillonella spp.	+	V	–	A,p
Acidaminococcus fermentans	–	–	–	A,B
Megasphaera elsdenii	–	–	+	a,ib,b,iv,v,C

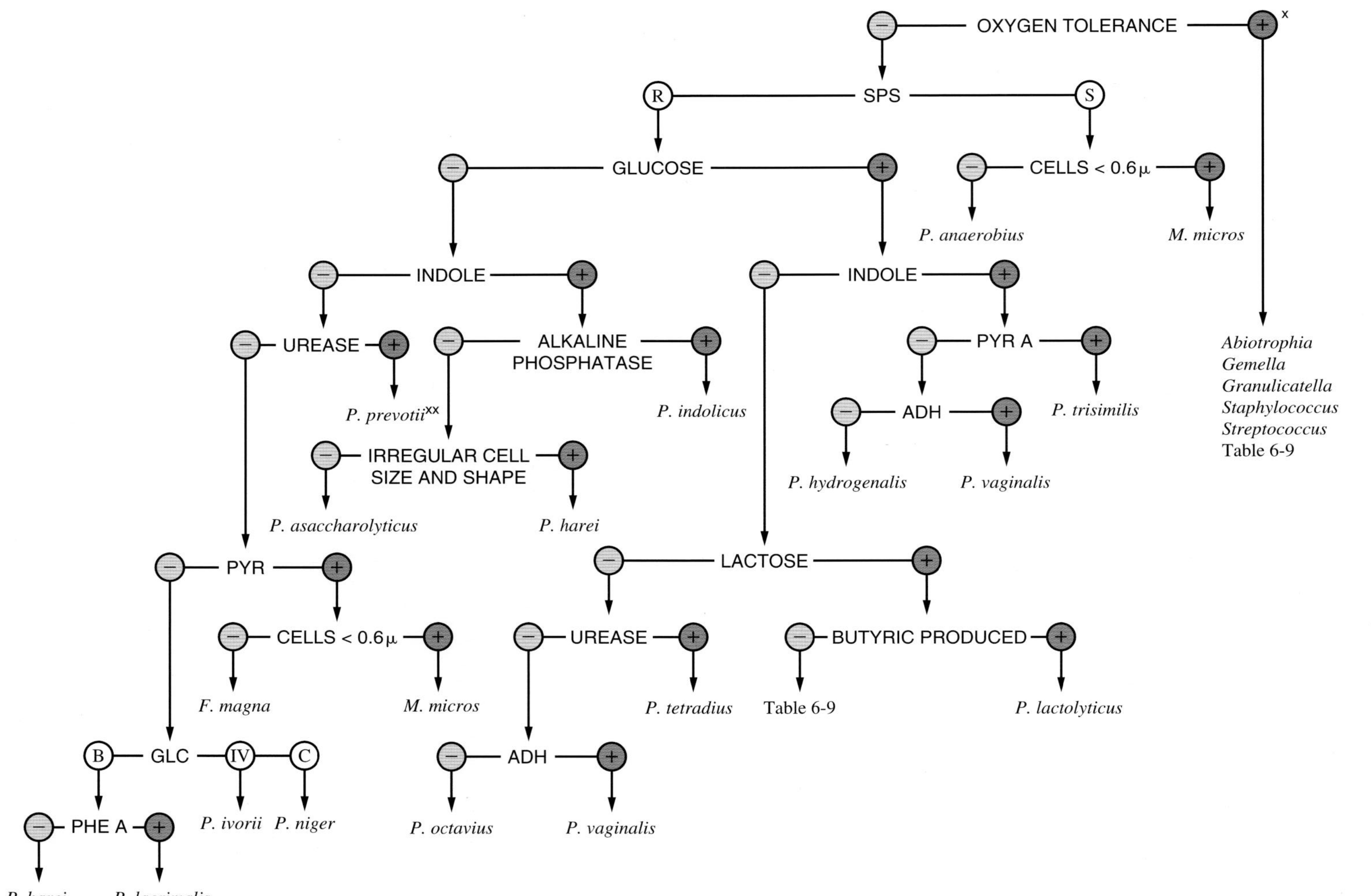

[x] These genera are often isolated on anaerobic primary culture; they become oxygen tolerant on successive passages.
[xx] Clinical strains of *P. prevotii*, usually urease–negative; see Table 6–8.

Figure 6-17. Flow chart illustrating scheme for identification of anaerobic cocci.

F. magna cells are >0.6 μm (see Figure 4-28) and are PYR-positive. The caproic acid–producing *P. octavius* and *P. niger* can be differentiated by carbohydrate fermentation; *P. octavius* is positive.

The three microaerophilic anginosus–group streptococci—*S. intermedius, S. constellatus* (two subspecies), and *S. anginosus*—*Abiotrophia* species, *Granulicatella* species, *G. morbillorum,* and *Staphylococcus saccharolyticus* may appear as anaerobic organisms on primary culture. The streptococci are strongly saccharolytic and produce lactic acid as their sole major endproduct. Cellular morphology varies in size and form (chains and pairs). Unlike true anaerobes, these organisms are resistant to metronidazole and usually become aerotolerant after subculturing. Anginosus-group streptococci have been associated with infections and abscess production. A recent study differentiates the species based on specialized enzymatic and biochemical tests (360). *Streptococcus anginosus* is unlikely to be identified using anaerobic methods, since most isolates will grow well on the CO_2 aerotolerance plate, but *S. intermedius* and *S. constellatus* (Figure 6-18) may require several subcultures to grow in a non-anaerobic atmosphere and thus would be identified initially as anaerobes (see Table 6-9). *S. saccharolyticus* is occasionally isolated from blood cultures (116).

The obligately anaerobic cocci (*S. pleomorphus, Ruminococcus* spp., and *Atopobium parvulum*) are rarely, if ever, isolated from clinical specimens. While the origin of *A. parvulum* is unknown (155), *S. pleomorphus* and *Ruminococcus* species have been isolated from human feces. Ribosomal RNA analysis of *S. pleomorphus* has shown closer similarity to the subphylum of *Clostridium* than to the genus *Streptococcus* (109;368). Further studies are required to establish the validity of this organism as an anaerobic streptococcus.

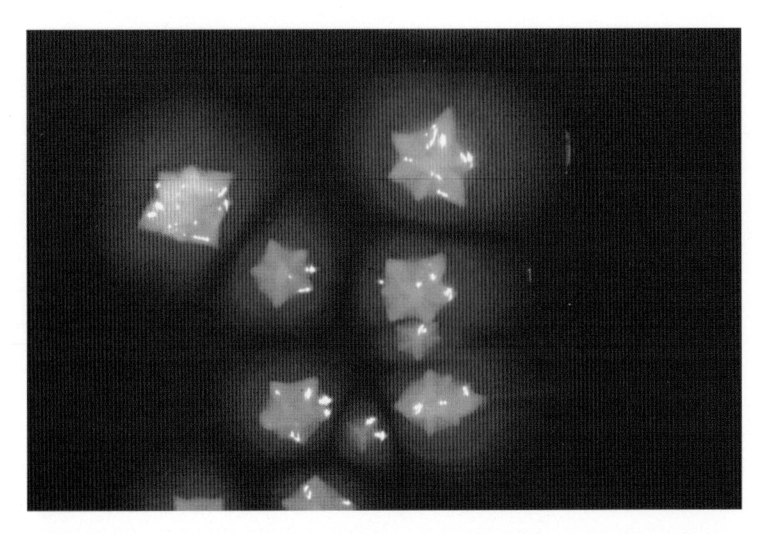

Figure 6-18. *Streptococcus constellatus* colony; note star-like structure.

table 6-8. Characteristics of *Finegoldia, Micromonas,*

	indole	urease	ALP	ADH	Fermentation of glucose	lactose	raffinose	ribose	mannose	α-gal	β-gal
F. magna	–	–	v	v	$-^{w}$	–	–	–	–	–	–
M. micros	–	–	+	–	–	–	–	–	–	–	–
P. asaccharolyticus	$+^{-}$	–	–	–	–	–	–	–	–	–	–
P. indolicus[c]	+	–	+	–	–	–	–	–	–	–	–
P. harei[d]	v	–	–	–	–	–	–	–	–	–	–
P. lacrimalis	–	–	–	–	–	–	–	–	–	–	–
"P. trisimilis"	+	–	v	–	+	w	$-^{w}$	$-^{w}$	+	–	v
P. hydrogenalis	+	v	$-^{w}$	–	+	+	+	–	+	–	–
P. prevotii	–	+[e]	–	–	–	–	+	+	+	+	–
P. tetradius	–	+	–	–	+	–	–	–	+	–	–
P. prevotii/tetradius	–	v	–	–	v	–	v	v	v	v	v
P. lactolyticus	–	+	–	–	+	+	–	–	+	–	+
P. vaginalis	v	–	$-^{w}$	+	+	–	–	–	v	–	–
β–GAL group	–	–	w	v	+	+	–	+	+	–	w^{+}
P. ivorii	–	–	–	–	–	–	–	–	–	–	–
P. anaerobius	–	–	–	–	+	–	–	–	w	–	–
P. octavius	–	–	–	–	+	–	–	+	+	–	–
P. niger	–	–	–	–	–	–	–	–	–	–	–

[a] For recent proposals of new classification of *Peptostreptococcus* spp. see Table 1-4 footnotes b–d.

[b] Adapted from Murdoch (Clin Microbiol Rev 1998:81-120); ALP, alkaline phosphatase; ADH, arginine dihydrolase; α-gal, α-galactosidase; β-gal, β-galactosidase; α-glu, α-glucosidase; β-gur, β-glucuronidase; ArgA, arginine arylamidase; ProA, proline arylamidase; PheA, phenylalanine arylamidase; LeuA, leucine arylamidase; PyrA, pyroglutamyl arylamidase; TyrA, tyrosine arylamidase; HisA, histidine arylamidase. (Enzyme reactions based on ATB 32A.)

[c] Of animal origin; coagulase-positive

[d] Differentiated from *P. asaccharolyticus* by irregular colony and cell morphology of *P. harei*.

[e] Clinical strains of *P. prevotii* are usually urease-negative.

Gram-Positive Bacilli

Clostridia

Clostridium species form a heterogenous group of anaerobic sporeforming gram-positive bacilli, some of which stain gram-variable and gram-negative. They are widely distributed in the environment and constitute part of the normal human intestinal flora. Most infections with these organisms are endogenous in origin, although a few noteworthy exogenous intoxications such as food poisoning (*C. perfringens*), botulism (*C. botulinum*), tetanus (*C. tetani*), and soft-tissue infections and toxic shock among injecting drug users (*C. novyi* type A) occur. Phylogenetically the genus *Clostridium* is extremely heteroge-

Peptostreptococcus spp.[a], and *Peptococcus niger*[b]

α-glu	β-gur	ArgA	ProA	PheA	LeuA	PyrA	TyrA	HisA	GLC
–	–	+	–	–	+	+	–[w]	–[w]	A(l,s)
–	–	+	+	+	+	+	+	+	A(l,s)
–	–	+	–	–	v	–	v	w	A,B,p
–	–	+	–	+	+	–	w	+	A,B,p(l,s)
–	–	+	–	–	–[w]	–	w	+	A,B(p,iv,v)
–	–	+	–	+	+	–	v	+	A,B,p,l
–	–	–	–	–	–	+	–	–	B
v	–	–	–	–	–	–	–	–	a,B,p,l
+	+	+	–	–	–	+	w	+	B(a,l,p,s)
+	+	+	–	w	+	w	w	w	B,L(a,p)
+	+	+	–	v	v	v	v	v	B
–	–	+	–	–	–	–	–	–	a,B,L
–	–	+	–	–	+	–	–	+	a,B,l
–[w]	–	+	–	–	+	w	–	–	B
–	–	–	+	–	–	–	–	–	B,IV(a,p,ib,v)
+	–	–	+	–	–	–	–	–	A,IC(ib,b,IV)
–	–	–	+	–	–	w	–	–	B,iv,C
–	–	–	–	–	–	–	–	–	A,ib,b,iv,C(p,l,s)

neous, with many species intermixed with other sporeforming and nonsporeforming genera (77). It is generally recognized that homology group I will form the basis of the genus *Clostridium;* the non–group I clostridia will require reclassification (77).

Including the lecithinase-positive clostridia discussed in Chapters 4 and 5, the *Clostridium* species of clinical importance may be subdivided into three groups based on having proteolytic (gelatin hydrolysis), saccharolytic (glucose fermentation), or both proteolytic and saccharolytic properties. The organisms listed in Table 6-10 represent the more commonly isolated pathogenic clostridia. Potentially clinically significant organisms such as *C. butyricum, C. clostridioforme, C. innocuum,* and *C. ramosum* can be resistant to clindamycin and multiple cephalosporins. Other species not listed may

table 6-9. Characteristics of other anaerobic and

	Oxygen tolerance	Esculin	Glucose	Cellobiose	Lactose	Raffinose
Ruminococcus hansenii	–	v	+	–	+	+
R. productus	–	+	+	+	+	+
Streptococcus pleomorphus	–	+	+	–	–	
Staphylococcus saccharolyticus	$-^{+}$	–	+	–	–	–
Gemella morbillorum	+	–	+	–	–	–
Streptococcus anginosus[a]	+	v	+		v	v
S. constellatus[a]	+	+	+	+	–	–
S. intermedius[a]	+	+	+	+	+	v
Abiotrophia spp.[b]	+					

[a] Anginosus group streptococci (formerly "Streptococcus milleri" group). VP and ADH positive.

[b] Satelliting behavior, formerly known as "nutritionally variant streptococci". Recently separated into two genera:

be identified using Bergey's Manual or the VPI Anaerobe Laboratory Manual and current literature (154;155).

Certain factors must be considered when identifying *Clostridium* species. First, it is important to note that some of the clostridia destain regularly and appear gram-negative, especially the RIC group (*C. ramosum, C. innocuum, C. clostridioforme*); however, the special-potency antibiotic disk pattern (colistin resistant and vancomycin sensitive) will demonstrate the gram-positivity of the isolate (except for *C. innocuum,* which is only moderately sensitive to vancomycin) (4). Second, the spores of some species are rarely detected microscopically, so that an ethanol or heat spore test may be necessary. Even by these means, however, it may sometimes be difficult to detect spore formation. Third, the colony morphology of pure cultures may be variable, so that the culture appears mixed. Subcultures of single colonies yield the same variable types. Fourth, the aerotolerant clostridia may be confused with *Bacillus* or *Lactobacillus* species. *Clostridium* species sporulate anaerobically only, grow much better anaerobically (larger colonies), and are almost always catalase-negative, whereas *Bacillus* species sporulate aerobically only, usually grow better aerobically, and are usually catalase-positive. Aerobically grown *C. tertium* shares similar colonial and cellular morphology with *Lactobacillus* species (Figure 6-19).

C. perfringens, C. sordellii, and *C. bifermentans* are discussed in Chapters 4 and 5. The rare urease-negative strains of *C. sordellii* are difficult to differentiate from *C. bifermentans. C. bifermentans* does not produce phenylacetic acid (PAA), its growth is enhanced by 1% mannose, and it is glutamic acid decarboxylase–negative (only a few strains have been tested); whereas *C. sordellii* may produce PAA, its growth is not enhanced by 1% mannose, and it is glutamic acid decarboxylase–positive. Enterotoxin-

microaerophilic cocci

N-acetyl-β-glucosaminidase	α-glucosidase	β-glucosidase	GLC
			L,a,s
			A,l,s
			L(a)
			A,l
			L(a)
–	–[+]	+	L(a)
–	+	–	L(a)
+	+	v	L(a)
			L

Granulicatella and *Abiotrophia* (76,178). PYR-positive.

producing *C. perfringens* has been implicated as an etiologic agent of persistent diarrhea in elderly patients at nursing homes and tertiary care institutions (28).The diagnosis of enterotoxigenic *C. perfringens* is based on quantitative culture and spore counts as well as demonstration of the toxin by a passive hemagglutination test (Unipath, England) or enzyme immunoassay (TechLab, Blacksburg,VA)(28).

C. novyi type A is lecithinase- and lipase-positive and may swarm. It is infrequently isolated; however, a recent report has implicated *C. novyi* type A as an etiologic agent in soft-tissue infections, toxic shock, and deaths among users who inject drugs intramuscularly or subcutaneously (1). Types B, C, and D vary in these properties and are unlikely to be isolated from human clinical material. *C. sporogenes* is lipase-positive, swarms, and has colonies that firmly adhere to the agar. It is phenotypically the same as *C. botulinum* types A, B, and F. Toxin neutralization in mice, polyacrylamide gel electrophoresis (PAGE) of soluble cellular proteins, or GLC of trimethylsilyl derivatives of whole-cell hydrolysates are necessary to differentiate *C. sporogenes* and *C. botulinum.* The different *C. botulinum* types vary in saccharolytic, proteolytic, and lipase activity. Suspected *C. botulinum* isolates or possible *C. botulinum*–containing material (e.g., suspected food in food poisoning cases) should be referred to the local or state public health laboratories. Even minute amounts of toxin can be fatal; great care should be exercised when handling suspicious material. An association between botulism and subcutaneous injection of Mexican black tar heroin has been reported (60). The swarmer *C. septicum* produces a Medusa's head–like colony on Brucella agar, but not on PEA, in 4–8 h (Figure 6-20) that becomes a heavy film of growth that covers the plate by 24–48 h. A Gram stain of *C. septicum* reveals large, gram-positive bacilli with oval and subterminal spores (Figure 6-21). A positive blood culture for *C. septicum* is strongly associated with neutropenic colitis and colonic malignancy and occasionally with

table 6-10. Characteristics of *Clostridium*

		Gelatin	Glucose	Lecithinase	Lipase	Indole	Esculin	Nittrate	Milk	Arabinose	Cellobiose	Fructose
Saccharolytic, proteolytic	*C. bifermentans*[a]	+	+	+	–	+	$+^{-}$	–	d	–	–	$-^{w}$
	C. botulinum[b]											
	types A, B, F	+	$+^{-}$	–	+	–	+	–	d	–	–	$-^{w}$
	types B, E, F[c]	+	+	–	+	–	–	–	c^{-}	w^{-}	–	a^{w}
	types C, D	+	+	$-^{+}$	+	$-^{+}$	–	–	d^{c}	–	–	v
	C. cadaveris	+	+	–	–	+	–	–	d^{c}	–	–	v
	C. difficile[d]	+	$+^{w}$	–	–	–	+	–	–	–	v	a^{w}
	C. novyi A	+	+	+	+	–	–	–	c^{d}	–	–	$-^{w}$
	C. perfringens	+	+	+	–	–	v	v	c^{d}	–	$-^{a}$	a
	C. putrificum	+	$+^{w}$	–	–	–	$-^{+}$	–	d	–	–	$-^{w}$
	C. septicum[e]	+	+	–	–	–	+	v	c^{d}	–	a^{w}	a
	C. sordellii[a]	+	+	+	–	+	$-^{+}$	–	d^{c}	–	–	v
	C. sporogenes[e]	+	$+^{-}$	$-^{+}$	+	–	+	–	d	–	$-^{w}$	–
Saccharolytic, non-proteolytic	*C. baratii*	–	+	+	–	–	+	$+^{-}$	c	–	a	a
	C. butyricum	–	+	–	–	–	+	–	c	a^{-}	a	a
	C. carnis[f]	–	+	–	–	–	+	–	a^{c-}	–	w^{a}	v
	C. clostridioforme[g]	–	+	–	–	$-^{+}$	+	$+^{-}$	c	v	v	a^{w}
	C. glycolicum	–	+	–	–	–	$-^{+}$	–	–	–	–	a^{w}
	C. indolis	–	+	–	–	+	+	$+^{-}$	c	w^{-}	a^{w}	w^{a}
	C. innocuum[d]	–	+	–	–	–	+	–	–	$-^{a}$	a	a
	C. paraputrificum	$-^{w}$	+	–	–	–	+	$-^{+}$	c	–	a	a^{w}
	C. ramosum	–	+	–	–	–	+	–	c	–	a	a
	C. sphenoides	–	+	–	–	+	+	$+^{-}$	c	$-^{w}$	a^{w}	a^{w}
	C. symbiosum	$-^{w}$	$+^{w}$	–	–	–	–	–	$-^{c}$	v	–	a
	C. tertium[f]	–	+	–	–	–	+	$+^{-}$	c	–	a^{w}	a
Asaccharolytic	*C. argentinense*	+	–	–	–	–	–	–	d	–	–	–
	C. hastiforme	+	–	–	–	–	–	$-^{+}$	$-^{d}$	–	–	–
	C. histolyticum[f]	+	–	–	–	–	–	–	d	–	–	–
	C. limosum	+	–	+	–	–	–	–	d	–	–	–
	C. subterminale	+	–	$-^{+}$	–	–	$-^{+}$	–	d	–	–	–
	C. tetani[e]	+	–	–	$-^{w}$	v	–	–	d^{-}	–	–	–
	C. malenominatum	$-^{w}$	–	–	–	+	–	$-^{+}$	–	–	–	–
	Filifactor villosus[h]	+	–	–	–	–	–	–	–	–	–	–

[a] *C. bifermentans* is urease-negative, *C. sordellii* is urease-positive. *C. bifermentans* usually forms chalk-white colonies on EYA.

[b] Toxin neutralization test required for identification. Send suspected isolates or *C. botulinum* containing material to the approppriate local or state public health agency.

[c] Non-proteolytic.

[d] Proline differentiates: *C. difficile* is l-proline-aminopeptidase (pro) -positive; *C. innocuum* pro-negative.

spp. of clinical significance

Lactose	Maltose	Mannitol	Mannose	Melibiose	Ribose	Salicin	Sucrose	Xylose	Spore location	GLC
–	w^{-}	–	$-^{w}$	–	–	–	–	–	ST	A(iv,ic,p,ib,b,l,s)
–	$-^{w}$	–	–	–	–	–	–	–	ST	A,B,IV,ib(ic,v,p)
–	a^{-}	$-^{w}$	a^{w}	$-^{w}$	v	–	a^{w}	–	ST	B,A(l)
–	v	–	v	$-^{w}$	v	–	–	–	ST	B,P,A(v,l,s)
–	–	–	$-^{w}$	–	–	–	–	–	T	B,A
–	–	w^{a}	v	–	$-^{w}$	$-^{w}$	–	$-^{w}$	ST^{T}	B,A,ic,iv,ib (v,l)
–	v	–	–	$-^{w}$	v	$-^{w}$	–	–	ST	A,B,P
a	a	–	a	v	v	$-^{a}$	a	–	ST	A,B,L(p,s)
–	$-^{w}$	–	–	–	–	–	–	–	T^{ST}	A,B,ib,iv(p,ic,v,l,s)
a	a	–	a	–	v	v	–	–	ST	B,A,(p,l)
–	w	–	$-^{w}$	–	$-^{w}$	–	–	–	ST	A(IC,p,ib,iv,l)
–	$-^{w}$	–	–	–	–	–	–	–	ST	A,B,iv,ib(p,ic,v,l,s)
w^{a}	w^{a}	–	a	$-^{w}$	w^{-}	a^{-}	a	–	ST	B,A,L(p,s)
a	a	$-^{w}$	a	a^{w}	a^{w}	a^{w}	a	a	ST	B,A(l,s)
v	w^{a}	–	w^{a}	–	–	w	w^{a}	–	ST	B,A,L(s)
v	a^{w}	–	a^{w}	v	v	v	a^{w}	a^{w}	ST	A(l,s)
–	v	–	–	–	–	–	–	a^{-}	ST	A,IV,IB(p,l,s)
w^{a}	w^{a}	$-^{w}$	$-^{w}$	$-^{w}$	$-^{w}$	w^{-}	v	v	T	A
$-^{w}$	–	a^{w}	a	–	v	a^{w}	a^{w}	$-^{w}$	T	B,L,a(s)
a	a	–	a	–	w^{-}	a	a	–	T^{ST}	B,A,L(s)
a	a	a^{-}	a	a^{-}	v	a	a	$-^{w}$	T	A,l(s)
w^{a}	a^{w}	w^{a}	a^{w}	v	$-^{w}$	v	w^{-}	v	ST^{T}	A(l,s)
$-^{a}$	–	–	v	–	–	–	–	–	ST	A,B,L
a	a	w^{a}	a^{w}	a^{w}	a^{w}	a^{w}	a	v	T	A,B,L
–	–	–	–	–	–	–	–	–	ST	A,b,ib,iv(l)
–	–	–	–	–	–	–	–	–	T	A,B,iv,ib(p,ic)
–	–	–	–	–	–	–	–	–	ST	A(l,s)
–	–	–	–	–	–	–	–	–	ST	A(l,s)
–	–	–	–	–	–	–	–	–	ST	A,B,IV,ib(p,ic,l,s)
–	–	–	–	–	–	–	–	–	T	A,B,p(l,s)
–	–	–	–	–	–	–	–	–	T^{ST}	A,B(p,l,s)
–	–	–	–	–	–	–	–	–	ST	B,a,iv,ib,l

[e] Swarming

[f] *C. tertium, C. carnis,* and most *C. histolyticum* grow aerobically.

[g] Cigar-shape.

[h] Yellowish rhizoid colonies adhere tightly to agar.

9

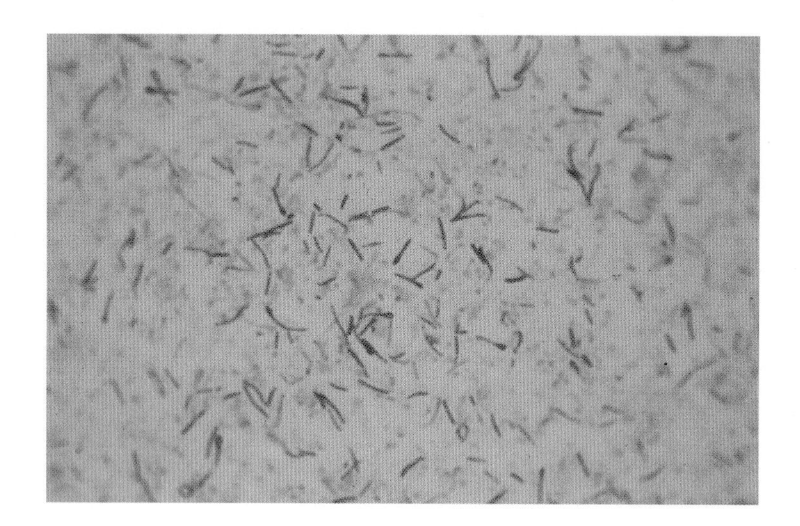

Figure 6-19a. Gram stain of *Clostridium tertium* grown anaerobically; note presence of spores.

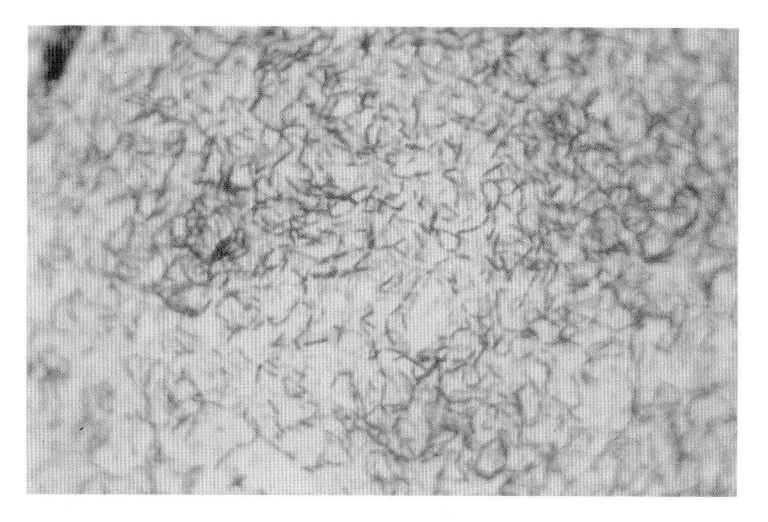

Figure 6-19b. Gram stain of *Clostridium tertium* grown aerobically; note absence of spores and lactobacillus-like morphology.

E. coli O157 infection with hemolytic uremic syndrome (HUS) after contact with farm animals (15). The isolation of *C. difficile* from stool samples of patients with suspected antimicrobial-associated diarrhea (AAD) or pseudomembranous colitis is discussed in Chapter 7. This organism produces a characteristic horse stable odor, forms pleomorphic colonies, fluoresces chartreuse, and produces a characteristic yellow ground-glass colony on cycloserine cefoxitin fructose agar (CCFA) (123). *C. innocuum* may grow on CCFA because of its resistance to cefoxitin, and even on KVLB medium because of its

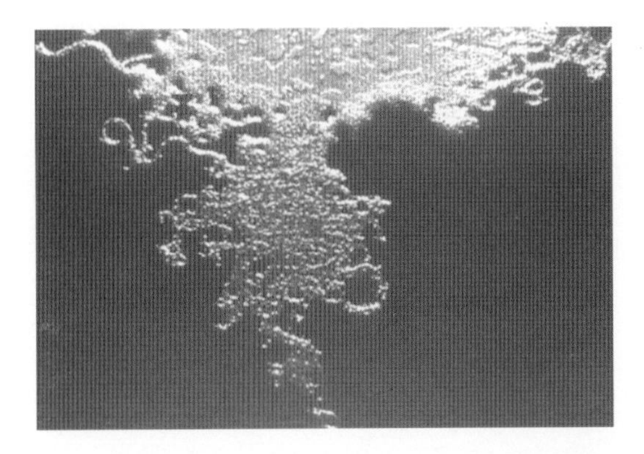

Figure 6-20. *Clostridium septicum* on blood agar after 8 hours; note Medusa-head colony morphology.

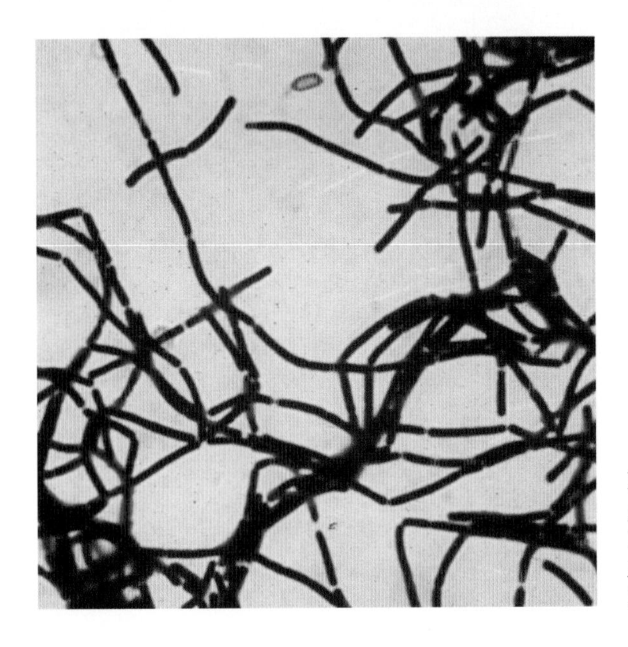

Figure 6-21. Gram stain of *C. septicum*. Long filamentous gram-positive bacilli with rare spores. (From Finegold, Baron, and Wexler, *A Clinical Guide to Anaerobic Infections*, 1992. Courtesy Princeton Scientific Productions, Inc. and Star Publishing Co.)

reduced susceptibility to vancomycin (4), although the colonies usually have an entire edge. An L-proline aminopeptidase test can be used to differentiate the species: *C. difficile* is proline-positive whereas *C. innocuum* is proline-negative (111). *C. innocuum* is an important finding in infections of immunocompromised hosts as it often is multiresistant to the drugs often used to treat anaerobic infections (4). The only other reported cause for AAD is *C. perfringens* in immunocompromised patients (28).

C. ramosum is a thin, gram-variable rod that has a small round or oval terminal spore, when present (Figure 6-22). Colonies that resemble those of gram-negative organisms on blood-containing agar turn the background brownish after air exposure. It is the second most commonly isolated *Clostridium* from clinical specimens including blood cultures, abdominal, and soft-tissue infections. It is one of the *Clostridium* species that may be resistant to clindamycin and multiple cephalosporins. *C. clostridioforme* is also often multiresistant and occasional strains produce β-lactamase. It is a relatively common isolate from blood cultures and other clinical specimens. Cells stain gram-negative and form characteristic elongated football shapes that rarely show spores (Figure 6-23). Colonies on blood agar turn the agar green on air exposure due to H_2O_2 production. It may be misidentified initially as a *Bacteroides* or *Fusobacterium* species. *C. oroticum* has a similar cell shape, but it often has spores and forms chains seen on the Gram stain smear. *C. symbiosum* has the same cell morphology as *C. oroticum* and may be differentiated by carbohydrate fermentation reactions. The performance of commercial kits in the identification of the RIC-group clostridia varies considerably, and usually is not reliable when used as the sole means of identification (4). *C. tertium* is aerotolerant (see Figure 6-19). It is often isolated from blood cultures of immunocompromised patients and has been reported as a cause of neutropenic enterocolitis (75;313).

C. tetani may form a thin film of growth over the entire plate, especially on moist media. This characteristically drumstick-shaped gram-positive bacillus is rarely isolated from clinical material of patients with tetanus. *C. tetani* also has been associated with tetanus in injecting drug users (61;332). In this context, it is important to emphasize a proper tetanus prevention program as tetanus is almost entirely preventable through vaccination, appropriate wound care, and/or tetanus immune globulin (TIG) (61;332).

Gram-Positive Nonsporeforming Bacilli

The nonsporeforming gram-positive bacilli consist of several genera that are differentiated from each other by their metabolic endproducts, as described in Table 6-11 and Figure 6-24 (see Appendix B). Major taxonomic revisions, especially within the genus *Eubacterium,* have introduced several new genera and species (see Table 1-4). The nonsporeforming gram-positive bacilli form part of the normal human flora at most sites (see Table 1-1) but occur infrequently in infections. Various *Actinomyces* species and *P. propionicus* are the major agents in classical actinomycosis. *Bifidobacterium dentium* has been isolated from a few serious pulmonary infections. *Propionibacterium* species are involved

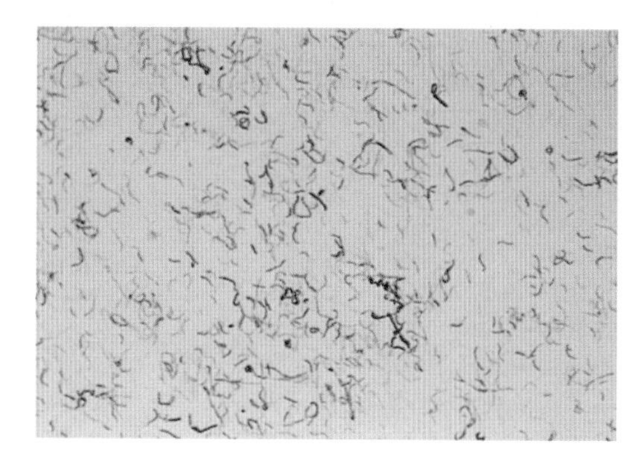

Figure 6-22. Gram stain of *C. ramosum;* note thin, gram-variable bacilli with distinct spores.

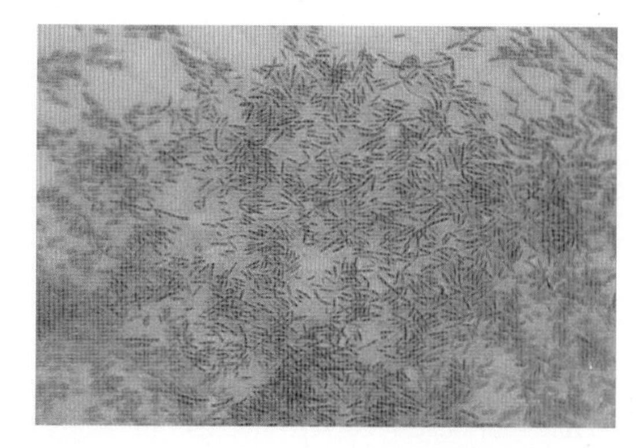

Figure 6-23. Gram stain of *C. clostridioforme;* note gram-negative staining, cigar-shaped cells.

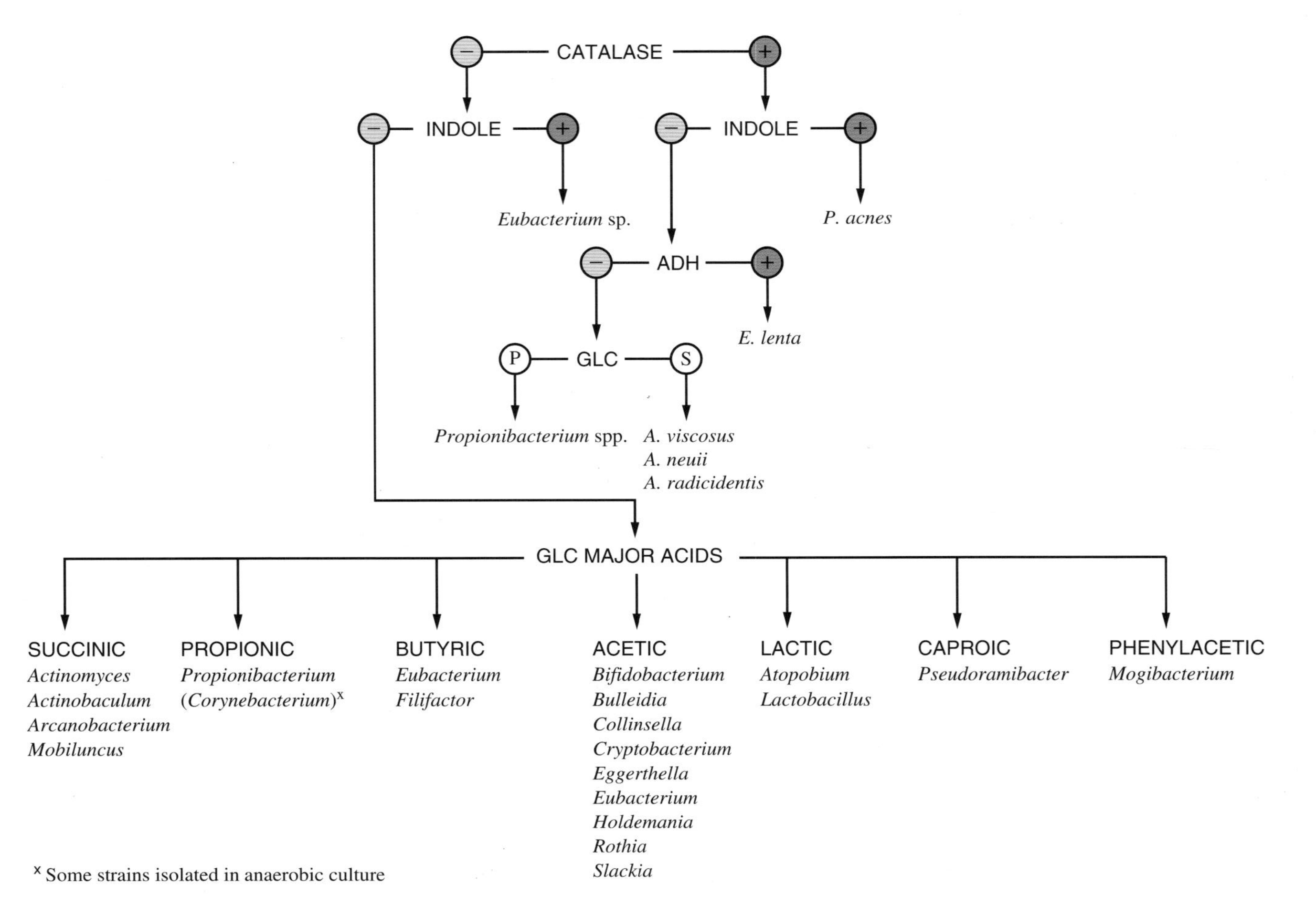

Figure 6-24. Flow chart illustrating scheme for identification of gram-positive nonspore forming bacilli.

table 6-11. Differentiation

	Strictly anaerobic	Metabolic end products	Catalase	Indole	Nitrate
Actinobaculum spp.	–	A,s	–	–	–
Actinomyces spp.	–	S,L,a	$-^{+}$	–	$+^{-}$
Arcanobacterium spp.	–	A,S,l	–	–	–
Atopobium spp.	$+^{-}$	L,a,(s)	–	–	–
Bifidobacterium spp.	$+^{-}$	A>L	–	–	–
Bulleidia sp.	+	a,L(s)	–	–	–
Collinsella sp.	+	A,L(s)	–	–	–
Corynebacterium matruchotii	–	L,S,p,(a)	+	–	+
Cryptobacterium sp.	+	-	–	–	–
Eggerthella sp.	+	(a,l,s)	+	–	+
Eubacterium spp.	+	See table 6-14	–	V	$-^{+}$
Filifactor spp.	+	B,a,(iv,ib,l)	–	–	–
Holdemania sp.	+	A,L,s	–	–	–
Lactobacillus spp.	V	L,(a,s)	–	–	$-^{+}$
Mobiluncus spp.	$+^{-}$	S,L,A	–	–	V
Mogibacterium spp.	+	pa	–	–	–
Propionibacterium spp.	V	A,P,l,s,iv	$+^{-}$	V	V
Pseudoramibacter sp.	+	C,a,b	–	–	–
Rothia dentocariosa	–	A,L,(s,p)	+	–	+
Slackia spp.	+	(a)	–	–	V

of the genera of gram-positive rods

Motility	Other Characteristics
–	16:0, 18:1w9c, 18:0 major CFA's Straight to curved rods, some branching
–	16:0, 18:1(9c,cycl), 18:0 major CFA's; 10:0, 12:0,14:0 CFA's present Irregular, branching
–	16:0, 18:1(9c,cycl), 18:0 major CFA's Coccobacilli predominate, no branching
–	Short rods or small elliptical cocci, no branching
–	16:0, 18:1cis9, 18:1cis9DMA major CFA's Irregular, branching
–	Short rods, no branching. Slow growth, saccharolytic, ADH positive.
–	Irregular rods, central swellings common. May appear as short pleomorphic cocci to short rods on original isolation.
–	16:0, 18:1, 18:0 major CFA's; 10:0, 12:0, 14:0 CFA's not present. "Whip-handle" cell morphology, branching. Colonies flat, spider-like, filamentous and adherent.
–	Very short rods, that easily decolorize in older cultures. ADH positive.
–	14:0i, 15:0ai major CFA's, 15:0ai DMA present. Pleomorphic bacilli, speckled colonies, catalase +, fluoresce red, H_2S+. Arginine stimulates growth.
V	Variable cellular morphologies, no branching (exception: *E. nodatum* exhibits branching)
–	Regular rods with rounded to tapered ends
–	18:1cis9, 18:1cis9DMA, and 16:1cis9 major CFA's. Pairs and short chains, some central to terminal swellings, no branching.
–	Variable cellular morphologies, chain formation common, no branching
+	16:0, 18:2, and 18:1 major CFA's. Gram-negative or gram-variable curved rods, no branching.
–	Shorts rods, no branching. Tiny translucent colonies. Asaccharolytic.
–	15:0, 15:0ai, 16:0 major CFA's Irregular, branching
–	Often in "flying birds" or Chinese character arrangements, no branching
–	16:0, 15:0ai, 17:0ai major CFA's. Irregular, branching. Colonies white, raised, convoluted, and adherent.
–	Arginine stimulates growth

occasionally in infections (usually endocarditis and infections related to implanted prosthethic devices such as ventriculoatrial shunts or artificial hip joints). Detailed species identification of most of the gram-positive nonsporeforming bacilli is beyond the scope of this manual and of clinical need, except for *Actinomyces* and certain other species.

Table 6-11 and Figure 6-24 show characteristics used in identifying the more commonly isolated or clinically important bacteria in this group. Several generalizations can be made using results from the nitrate, catalase, and indole reactions: (1) a catalase-positive organism is probably a *Propionibacterium* species, *A. viscosus, A. neuii,* or *Eggerthella lenta*; (2) a nitrate-positive organism is probably not a *Bifidobacterium* or *Lactobacillus*; and (3) an indole-positive organism is probably a *Propionibacterium,* or in rare cases, *Eubacterium.* In addition, many strains of most *Actinomyces* are microaerobic (CO_2 is required for maximum growth) and are slow growers, requiring more than 48 h incubation, even up to 5 d, for growth to appear in the primary culture. On the other hand, some of the recently described *Actinomyces* species are facultatively anaerobic bacteria.

Branching, beading, and diphtheroid-shaped gram-positive bacilli are seen on the Gram stain smear of *Actinomyces* species (Figure 6-25). *A. israelii* is noted for its molar tooth colony (Figure 6-26). The same morphology is shared by *A. graevenitzii,* a recently described *Actinomyces* species, but the colonies form pinkish pigment on RLB agar (289). Although it is reported that the red pigment production of *A. odontolyticus* is enhanced by exposure to air, these red colonies are also seen after anaerobic incubation (see Figure 4-6). *A. viscosus* and *A. neuii* are catalase-positive *Actinomyces* species. The CAMP test can be used to differentiate: *A. neuii* is CAMP-positive. *A. meyeri* is a small bacillus and is mostly a strict anaerobe. *A. turicensis,* formerly recognized as aerotolerant *A. meyeri*–like bacteria, is often misidentified by the Rap ID ANA II as *A. meyeri* with a code 020671 (29). Improvements in molecular identification methods have clarified the taxonomy of *Actinomyces* species and led to recognition of several new species (see Table 1-4). Table 6-12 lists the biochemical information necessary for the identification of *Actinomyces* species and related taxa (275;289).

Propionibacterium are pleomorphic bacilli that may appear to branch (see Figure 4-30). *P. acnes* is the most commonly isolated species and is often a contaminant in blood cultures. *Propionibacterium propionicus* (formerly *Arachnia propionica*) is found in human actinomycosis and lacrimal canaliculitis and resembles *A. israelii* (Figure 6-27) in colonial and microscopic morphology, but can be differentiated by propionic acid production. *Bifidobacterium* species, with bifurcated ends, may also appear to be branching; however, these bacilli are generally larger in diameter than *Actinomyces* or *Propionibacterium* (Figure 6-28). The biochemical characteristics of *Bifidobacterium* species and *Propionibacterium* species are listed in Tables 6-13 and 6-14, respectively. *Lactobacillus* species have straight sides and may occur in chains (Figures 6-29, 6–30). Many strains grow well in facultative or microaerobic conditions, produce greening of blood agar around colonies, and are resistant to vancomycin. A new genus, *Atopobium,* has been proposed to include anaerobic gram-positive elliptical cocci or rod-shaped organisms that produce major amounts of lactic acid (79) and is discussed here with *Lactobacillus* species (Table 6-15). *A. rimae* is a common isolate from the oral cavity (88). The characteristics of clinically significant *Eubacterium* and other related genera are listed in Table 6-16. *Eggerthella lenta* is the most commonly isolated species from

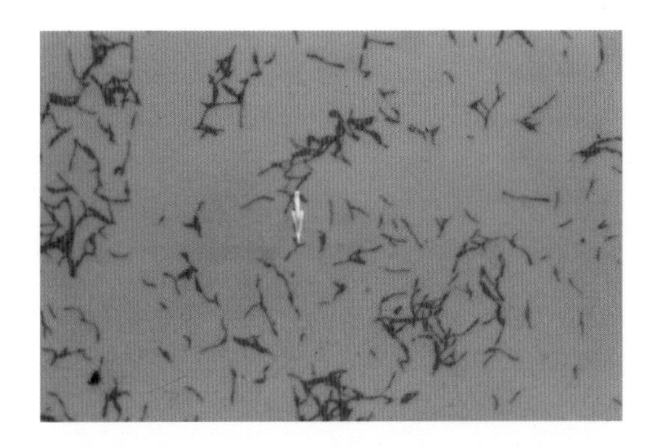

Figure 6-25. Gram stain of *Actinomyces viscosus*; note branching.

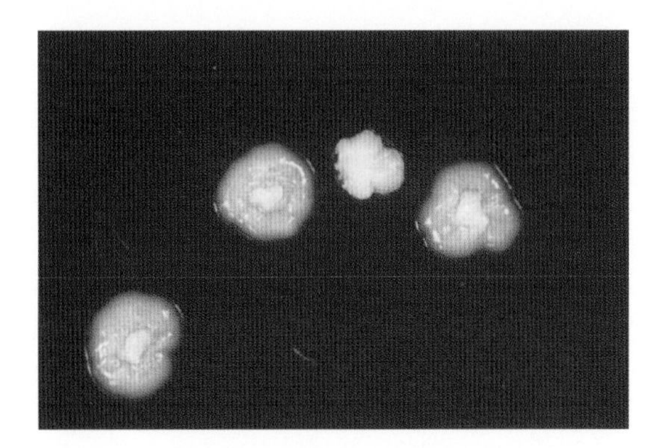

Figure 6-26. *Actinomyces israelii;* note molar-tooth colony.

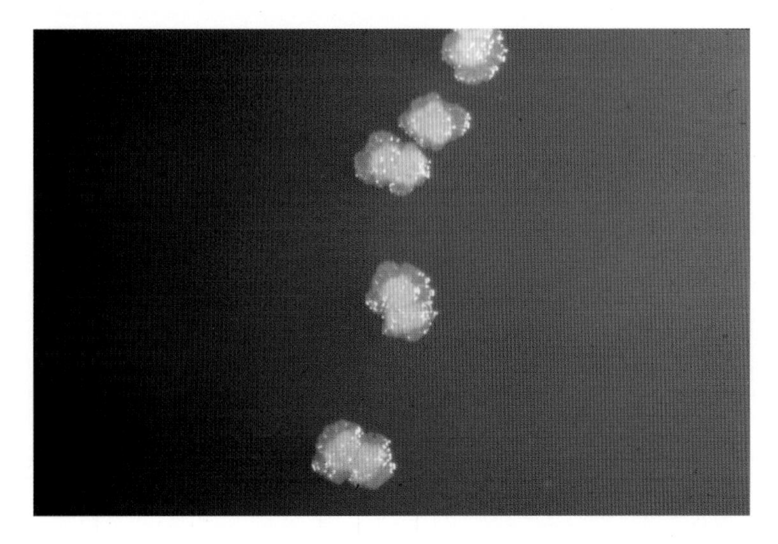

Figure 6-27. *Propionibacterium propionicus* colonies resembling *A. israelii.*

non-oral sources. The various genera of anaerobic gram-positive nonsporeforming bacilli of oral origin that were previously included in the genus *Eubacterium* or remained unidentified can be differentiated by GLC and Rapid ID 32A code patterns (88; 92) (see Table 6-16, footnote d).

table 6-12. Characteristics of *Actinomyces* spp., *Actinobaculum*

						Fermentation		
	Pigment	Catalase	Nitrate	Urease	Esculin	Arabinose	Mannitol	Raffinose
A. europaeus	–	–	+[b]	–	V	–	–	–
A. funkei (193a)	–	–	+	–	–	+	–	–
A. georgiae	–	–	$+^-$	–	$+^-$	$-^+$	V	–
A. gerensceriae	–	–	V	–	+	–	$+^-$	+[c]
A. graevenitzii	$-^+$[d]	–	$-^+$	–	–	–	–	$-^+$
A. israelii	–	–	$+^-$	–	+	+	$+^-$	+
A. meyeri	–	–	$-^+$	V	–	$-^+$	–	–
A. naeslundii	–	–	V	+	+	–	–	+
A. neuii ssp. *neuii*	–	+	+	–	–	V	+	V
A. neuii ssp. *anitratus*	–	+	–	–	+[f]	–	+	+
A. odontolyticus	+	–	$+^-$	–	V	–	–	–
A. radicidentis(79a)	$-^+$[d]	+	V	V	+	–	+	+
A. radingae	–	–	–	–	+	V	–	V
A. turicensis	–	–	–	–	–	V	–	V
A. urogenitalis(248a)	$-^+$[d]	–	+	–	+	V	V	$-^+$
A. viscosus	–	+	+	V	V	–	–	+
A. bernardiae	–	–	–	–	–	–	–	–
A. hemolyticum	–	–	–	–	–	–	–	–
A. pyogenes	–	$-^+$	–	–	–	V	V	–
A. schalii	–	–	–	–	–	V	–	–

[a] Adapted from Sarkonen et al. (289).

[b] The type strain is positive. In the original description (121) the type strain is nitrate, esculin, and xylose negative by API Coryne.

[c] In the original description 92% of strains positive (161). The type strain is negative.

[d] May produce a red pigment on RLB.

[e] The type strain is positive. Lawson et al. report a negative reaction (193) by API 2YM.

[f] In the original description esculin is negative (122) by API Coryne. The type strain is positive (+).

[g] The type strain is positive, clinical isolates positive. Original description reports a negative reaction by API ZYM (255).

[h] Most clinical strains negative; type strain is positive.

[i] Funke et al. (121) and Lawson et al. (193) reported negative reaction, method not stated; Hall *et al.* (138a) described β-N-Acetyl-glucosaminidase positive by Rosco.

NA; not available

Arcanobacterium bernardiae, A. hemolyticum, A. pyogenes, and *schalii*[a]

of								
Rhamnose	Sucrose	Trehalose	Xylose	CAMP	α-fucosidase	β-galactosidase	α-glucosidase	β-NAG
–	V	–	+[b]	–	–	+	+	–
–	+	–	$+^{-}$	+	+	+	+	+
+	+	+	+	–	–	+	+	–
$-^{+}$	+	V	+	–	–	+	+	–
–	+	V	–	–	–	+	–	+
$-^{+}$	+	+	+	–	–	+	+	–
–	+	–	+	+	–	+[e]	+	+[e]
–	+	+	$-^{+}$	–	–	V	V	–
V	+	–	+	+	–	+	+	V
–	+	+	+	+	–	+	+	V
$-^{+}$	+	–	V	–	$+^{-}$	+[g]	+[g]	–
–	+	+	–	–	–	+	+	–
–	+	–	+	+	+	+	+	+
–	+	V	+	–	–[h]	–	+	–
–	+	+	+	–	–	+	+	+
–	+	–	–	–	–	V	V	–
–	–	–	–	–	$+^{-}$	–	+	+[i]
–	V	–	–	+reverse	+	+	+	+
–	V	–	+	–	–	+	+	–
+	V	–	+	W	–	–	+	–

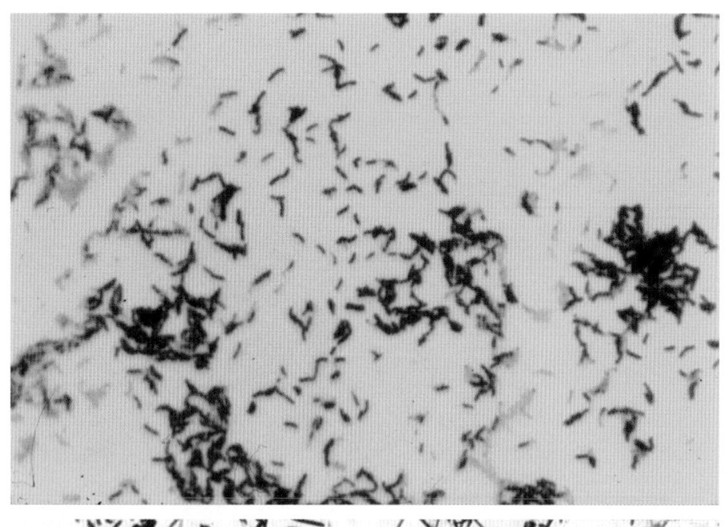

Figure 6-28. Gram stain of *Bifidobacterium* sp.; note pleomorphic bacilli with bifurcated ends.

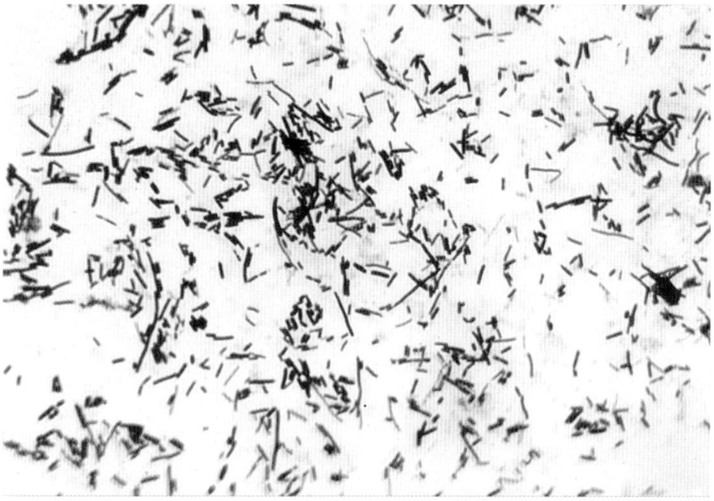

Figure 6-29. Gram stain of *Lactobacillus* sp.; note straight gram-positive rods.

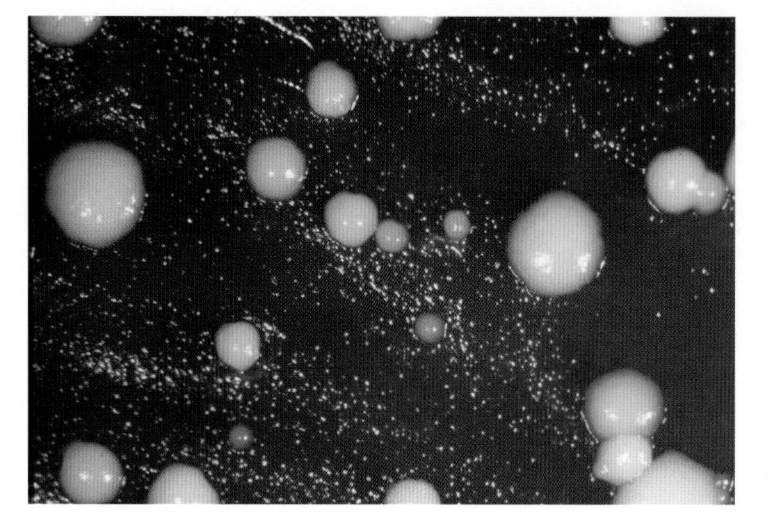

Figure 6-30. *Bifidobacterium* (tiny) and *Lactobacillus* (large) colonies on MRS (de Man, Rogosa, Sharpe) agar.

table 6-13. Characteristics of *Bifidobacterium* spp. of human origin[a]

	Fermentation of									Occurrence in humans		
	Arabinose	Cellobiose	Glycogen	Mannose	Melezitose	Salicin	Sorbitol	Starch	Sucrose	Intestinal	Dental	Clinical
B. adolescentis[b]	+	+	+	V		+	V	+	+	+		+
B. angulatum[c]	+		+	$-^{w}$	–	+	V	+	+	+		
B. bifidum	–	–	–	–	–	–	–	–	$-^{w}$	+		+
B. breve	–	$+^{-}$	V	+	V	+	V	V	+	+		+
B. catenulatum	$+^{-}$	+	V	–	–	+	+	–	+	+		
B. denticolens	V	+		–	–	+		+	+		+	
B. dentium	+	+	+	+	$+^{-}$	+	–	+	+		+	+
B. infantis[d]	$-^{+}$	$-^{+}$	–	V	–	–	–	–	+	+		
B. inopinatum	–	–		–	V	$+^{-}$		+	+		+	
B. longum[d]	+	$-^{+}$	V	V	$+^{-}$	–	–	–	+	+		+
B. pseudocatenulatum	+	V		+	–	+	V	+	+	+		

[a] All strains ferment glucose, lactose, and maltose.

[b] Sorbitol-negative strains are differentiated from *B. dentium* by transaldolase migration or PAGE.

[c] Cells V or palisades arrangement, no branching.

[d] Strains fermenting arabinose but not melezitose are separated by transaldolase migration (anodal: *B. infantis;* less anodal: *B. longum*).

table 6-14. Characteristics of *Propionibacterium* spp.

	Catalase	Nitrate	Indole	Esculin	Gelatin	GLC
P. acnes	$+^{-}$	$+^{-}$	$+^{-}$	–	+	A,P(iv,l,s)
P. avidum[a]	+	–	–	+	+	A,P(iv,l,s)
P. granulosum	+	–	–	–	$-^{+}$	A,P(iv,l,s)
P. propionicus[b]	–	+	–	–	v	A,P(l,s)

[a] β-hemolytic

[b] Morphologically similar to *A. israelii*, comparative pathogenicity to *Actinomyces* sp.

table 6-15. Characteristics of *Atopobium* spp.

	Strictly anaerobic	Fermentation of							
		Esculin	Arabinose	Cellobiose	Glycogen	Lactose	Maltose	Mannitol
Atopobium minutum	+	$-^{+}$	–	–	–	–	–	–
A. parvulum[a]	+	+	–	a	–	a	a	–
A. rimae	+	$+^{-}$	v	a^{-}	a^{-}	$-^{a}$	a^{-}	–
A. vaginae[b]	–	–						
Lactobacillus acidophilus	–	+	–	a	$-^{a}$	a	a	$-^{w}$
L. brevis	–	$-^{+}$	v	–	–	v	a^{w}	–
L. casei	–	+	–	a	–	a^{-}	a	a
L. catenaforme	+	+	–	a	a	v	v	–
L. crispatus	–	+	–	a	w^{a}	a	a	–
L. fermentum	–	–	v	–	–	a^{-}	a^{w}	–
L. gasseri	–	+	–	a		v	v	–
L. iners	–	–	–			–	v	–
L. jensenii	–	+	–	a^{w}	–	–	a^{w}	–
L. leichmannii	–	+	–	a	–	a	a^{w}	–
L. oris	–	v	a	$-^{a}$	–	a	a	–
L. paracasei subsp. *paracasei*	–	$+^{-}$	–	a^{-}	–	v	a	a
L. paraplantarum	–	+	v	a	–	a	a	a
L. plantarum	–	+	a^{w}	a	$-^{w}$	a	a	v
L. rhamnosus[c]	–	+	–	a	–	a^{-}	a^{w}	a
L. salivarius	–	–	–	–	–	a	a	a
L. uli[d]	+	$+^{-}$	–	$-^{a}$	a	v	a^{-}	–
L. vaginalis	–	$+^{-}$	–	–	–	a	a	–

[a] Exhibits coccal morphology.

[b] Fermentation reactions not available.

[c] *L. rhamnosus* GG is lactose-negative.

[d] *L. uli* was reclassified as *Olsonella uli* (Dewhirst et al., Int. J. Syst. Evol. Microbiol. 2001; 51: 1797–1804.)

6

and selected *Lactobacillus* spp.

Fermentation of						Occurrence in humans			
Melezitose	Raffinose	Rhamnose	Salicin	Sucrose	Xylose	Intestinal	Urogenital	Oral/ dental	Clinical
–	–	–	–	–	–	+			+
–	–	–	–[w]	a	–[a]			+	+
–	–	–	v	a	–			+	
							+		
–	v	–	a	a	–	+	+	+	+
–	–[a]	–	–[w]	v	v	+	+	+	+
a	–	–	a	a	–	+	+	+	+
–	–	–	a	a	–[w]	+		+	+
–	–	–	a	a	–	+	+	+	+
–	a[–]	–	–	a	v	+	+	+	+
–	v	–	a	a	–	+	+	+	+
–	–	–		–	–		+		+
–[a]	–	–	a[w]	a	–		+		+
–	v	–	a[w]	a	–[a]	+	+		+
–	a	–	v	a	–			+	
a[–]	–	–	a	a	–	+			+
a	v	v	a	a	–	+			
v	a[–]	–[a]	a	a	v	+		+	+
a[–]	–	a	a	a	–	+		+	+
–	a	–[a]	–	a	–	+	+	+	
–	–	–	a[–]	a	–			+	
–	a	–	–	a	–		+		

table 6-16. Characteristics of *Eubacterium* and related

	Glucose	Indole	Nitrate	Esculin	Gelatin
Collinsella aerofaciens	a	–	–	V	–
Eubacterium contortum	a	–	–	+	–
E. cylindroides	a	–	–	+	–
E. limosum	a	–	–	+	V
E. moniliforme[b]	a	–	V	–	–[w]
E. saburreum	a	+	–	+	–
E. rectale	a	–	–	+	–
E. tenue	w[–]	+	–	–	+
Pseudoramibacter alactolyticus	a	–	–	–	–
Eggerthella lenta	–	–	+	–	–
Eubacterium combesii	–	–	–	V	+
Eubacterium WAL1	–	–	+	–	–
E. nodatum	–	–	–	–	–
E. yurii[b,c]	–[w]	+	–	–	–[+]
Other *Eubacterium* and related species	–[a]	–	–	–	

[a] See Table 1-4 for recent taxonomic changes.

[b] Motile

[c] Three subspecies: *E. yurii* subsp. *margaretiae* (alkaline phosphatase-); *E. yurii* subsp.*yurii* (alkaline phosphatase+); *E. yurii* subsp. *schtitka* (alkaline phosphatase-).

[d] *Bulleidia extcructa:* a,l(s); *Cryptobacterium curtum:* (none); *E. brachy:* ib,iv,ic,(p,b,l,v); *E. infirmum:* a,b; *E. minutum:* B/b; *E. saphenum:* a,b; *E. sulci:* B,a(s); *M. timidum:* pa,(a,s); *Filifactor alocis:* b,a; *Slackia exigua:* (none, a,s) .Rapid ID32A (bioMerieux) codes can be used to further identify oral *Mogibacterium, Eubacterium,* and related spp. (63,88,92).

genera[a] of clinical significance

Metabolic end products	Other distinguishing characteristics
A,L(s)	Irregular rods, central swellings common. May appear as short pleomorphic cocci to short rods on original isolation.
A(l,s)	Ovoid cells, short or long twisted chains.
B,L(a,s)	Cells occur as single, pairs, and long chains. Central swellings common. Decolorize easily. Does not ferment maltose and arabinose.
A,B,L(s,ib,iv)	Pleomorphic, mottled, "slimy" colonies on Brucella agar. Cells often with swollen ends and bifurcations.
A,B,L(p,s)	Straight rods, often palisades arrangement.
A,B,L(s)	Stains mostly gram-negative, filamentous, often curving and with swellings. Flat colonies with rhizoid edges and granular, raised centers.
b,L(a,p,s)	Cells occur as single, in short chains or lumps. Central or terminal swellings may occur. Ferments maltose and arabinose.
A(p,ib,iv,ic)	Large rods often with blunt or widened ends. Flat colonies with lobate to diffuse edges, mottled appearance.
C,a,b	Irregular rods, often arranged in "flying birds," clump or chinese characters.
a(l,s)	Pleomorphic bacilli, speckled colonies, catalase +, fluoresce red, H_2S+. Arginine stimulates growth.
A,B,iv,l,ib(pf)	Cells occur as single, pairs, chains, and palisades. Often beta-hemolytic.
(a,l,s)	Short cocco-bacilli, translucent colonies, catalase-, fluorescence-, H_2S-.
a,B(l,s)	Exhibits *Actinomyces*-like morphology: cells appear branched, club-shaped or filamentous, colonies heaped, berry-shaped, or molar-tooth. Slow growth.
A,B,p	Gram-variable to gram-negative straight rods, form brush-like aggregates. Colonies speckled, maybe spreading.
See footnote[d]	

table 6-17. PRAS and preformed enzyme tests for advanced

B. fragilis group	Pigmented gram-negative bacillus	Gram-negative bacillus	Fusobacterium	Gram-negative coccus
PY	PY	PY	PY	PY
PYG	PYG	PYG	PYG	PYG
ARA	CEL	ARA	ESC	IND-NIT
CEL	ESC	CEL	FRUC	
ESC	LAC	ESC	LAC	
RHA	SUC	LAC	MANO	
SAL	IND-NIT	(RHA)	TH	
SUC		SAL	LT	
TRE	α-fuc	SUC	IND-NIT	
IND-NIT	α-gal	(TRE)	BILE	
XYLAN	β-gal	XYLAN		
XYL	β-NAG	XYL	ONPG	
BILE	(chymotrypsin)	GEL		
α-fuc	(trypsin)	IND-NIT		
l-ara		BILE		
PGUA				
		α-fuc		
		ONPG		
		β-NAG		
		β-xyl		
		urease		

identification of anaerobic isolates

Gram-positive coccus	Bifidobacterium	Clostridium	Actinomyces/ Propionibacterium	Gram-positive bacillus; unknown genus
PY	PY	PY	PY	PY
PYG	PYG	PYG	PYG	PYG
CEL	ARA	ARA	ARA	ARA
ESC	CEL	CEL	ES	CEL
LAC	GLYC	ESC		ESC
MANO	MANO	FRUC	MANIT	FRUC
RAF	MELZ	LAC	RAF	GLYC
RI	SAL	MAL	RHA	LAC
TW	SORB	MANIT	SUC	MAL
IND-NIT	ST	MANO	TRE	MANIT
Rapid ID 32 or	SUC	MEB	XYL	MANO
		RI	GEL	MEB
other enzyme		SAL	IND-NIT	MELZ
system		SUC		RAF
		XYL	α-fuc	RHA
		GEL	β-gal	RI
		MILK	α-glu	SAL
		IND-NIT	β-NAG	SORB
			urease	SUC
				ST
				TRE
				XYL
				GEL
				MILK
				IND-NIT
				urease

Identification Systems

Several identification methods are currently available, including methods utilizing PRAS biochemicals, many rapid biochemical kits, the Presumpto plate system, and a method based on whole-cell fatty acid analysis. The PRAS biochemical system is still considered the standard method.

PRAS biochemicals are available commercially (Anaerobe Systems) or they can be prepared according to the methods published in the VPI Anaerobe Laboratory Manual (154). CDC utilizes aerobically prepared thioglycolate-based media with bromthymol blue indicator (87; 102a). Table 6-17 lists the biochemicals needed to identify different groups of organisms. Inoculation is done under anaerobic conditions in an anaerobic chamber, or using a special apparatus that allows continuous gas flow into the tubes during manipulation. Another technique utilizes rubber stoppers and the inoculation can be done using a needle and syringe; this has been further modified by replacing the rubber stopper with a Hungate-type stopper and screwcap. Inoculation and interpretation procedures are outlined in Appendix C. A short PRAS biochemical scheme for identification of clinical isolates of bile-resistant *Bacteroides* species was described by Citron and colleagues (69). For further guidelines on identification of particular organisms, refer to the VPI Manual (154;230), Bergey's Manual (155), and the CDC manual (87; 102a). It should be noted that definitive identification with PRAS biochemicals often requires additional tests such as gas-liquid chromatography and other biochemical tests including catalase, urease, and chromogenic enzyme tests.

The PRAS system is expensive, labor intensive, and not problem free. The major problem is interpretation of biochemical results. The following three points may aid in reducing potential errors.

Box 6-1. Growth supplements

Growth supplement	Concentration %	Organism
Arginine	0.5	*E. lenta*
Formate - fumarate	0.3–0.3	*Campylobacter* spp./ *B. ureolyticus*
Hemin	0.001	*B. fragilis* group, *Prevotella* spp., and *Porphyromonas* spp.
N-acetylmuramic acid	0.001	*B. forsythus*
Pyruvate	1	*Bilophila* sp. and *Veillonella* spp.
Pyruvate-$MgSO_4$	1–0.25	*Desulfomonas* sp. and *Desulfovibrio* spp.
Serum	1	gram-negative and gram-positive anaerobes
Sodium bicarbonate	0.1	*Fusobacterium* spp., other gram-negative bacilli
Tween-80	0.5	gram-positive anaerobes

One, the inoculum must be a young culture of at least 2+ turbidity. An old culture contains predominantly nonviable organisms and may lead to unreliable results. Since some anaerobes grow poorly in thioglycolate or peptone-yeast broth, the addition of supplements may be necessary to enhance growth. If growth in the thioglycolate broth is less than 2+, perform growth stimulation tests (see Appendix C) and add an appropriate supplement to each PRAS tube. Tween-80 and hemin can be added directly to the inoculum broth.

Two, the anaerobe tested must be in pure culture and viable in PRAS biochemical tubes. Do not read biochemical reactions of mixed cultures; re-isolate the organism and repeat the biochemical inoculation. If the isolate fails to grow on the purity and viability plate, but turbidity appears in the biochemical tubes, then subculture the PYG to determine that the turbidity is growth of the isolate and not merely inoculum or a contaminant. If the organism grows on the plate, but not in the biochemical tubes after 2–3 d of incubation, then perform growth stimulation tests and re-inoculate. Do not incubate nonturbid biochemicals for >5 d, as growth rarely occurs beyond that time.

Three, some anaerobes characteristically grow with heavy turbidity while others do not. Any inconsistencies in growth throughout the series of tubes should be investigated. Unusually heavy growth in one or two tubes suggests either the presence of a growth factor or contamination. Gram stain or subculture an aliquot of material from the questionable tubes to rule out contamination. Poor or no growth in one or two tubes of a series showing good growth may be due to an oxidized tube (pink color due to resazurin indicator change) or a tube skipped during inoculation. *Propionibacterium acnes* and *Staphylococcus* species are the usual contaminants. If the indole and nitrate tests are both positive, suspect that *P. acnes* is present in the tubes. An aerotolerance test performed along with the purity and the viability check will help to determine the presence of an aerobic contaminant.

The API 20A (bioMerieux, Marcy l'Etoile, France) is a microtube biochemical system. The API 20A strip consists of 16 cupules containing carbohydrates and 4 cupules containing tests for indole, urease, gelatin and esculin hydrolysis, and catalase. After 24–48 h of incubation, the color reactions of the API 20A system is read manually and scored. A profile number is generated and interpreted from the code book or computerized data base provided by the manufacturer. The color reactions are not always clear-cut, as shades of brown or no color can make interpretation of the test results difficult. API 20A is best suited for identification of saccharolytic, rapidly growing organisms, such as the *B. fragilis* group and many clostridia. Most asaccharolytic organisms cannot be identified and some fastidious organisms fail to grow in the systems (e.g., some *Prevotella* species). Even for many saccharolytic organisms, supplemental tests, including bile and GLC, are often required for definitive identification.

Rapid (2–4 h) identification systems are based on detection of preformed (constitutive) enzymes by use of chomogenic or fluorogenic substrates or a combination of both. These systems include RapID ANA II (Remel), Rapid ID 32A (bioMerieux), ANI Card (Vitek Systems), API ZYM (bioMerieux), MicroScan (American MicroScan), and BBLCrystal Anaerobe (ANR) Identification (ID) system (Becton-Dickinson). Overall performance of these systems has varied from moderate to good; 60–90% of the isolates are identified to the species level (4;56;59;89;180;290;319). The performance of all these systems is, of course, affected by the source and nature of the isolates; the accuracy can be further increased by certain simple supplemental tests and by GLC. The API ZYM system, which allows the determination of 19 preformed enzymes in 4 h, does not have a data base. It can be used with compiled data from different publications and is a useful supplement for identification of clinically encountered anaerobic bacteria, especially the *Porphyromonas* species (see Table 6-5) (99;135;348).

Individual single, dual, or triple enzyme substrate–containing tablets are available commercially (Rosco Diagnostic Tablets, Rosco, Taastrup, Denmark; WEE-TAB system, Key Scientific Products, Round Rock, TX). They are much cheaper than commercial kits, can be applied in a number of situations, and allow flexibility in tailoring the set to best suit special needs (Chapter 5) (99;156). The use of 4-methylumbelliferone derivatives of many substrates (Sigma, St. Louis, MO) permits rapid and inexpensive spot tests based on fluorescence (210;216;227;264; 289). It should be noted, however, that reactions obtained by fluorogenic and the different chromogenic test applications may not completely coincide owing to divergent substrate concentrations, affinities, and buffering conditions in the different systems. Therefore, it is important to report the system used when reporting enzyme reactions (289).

The Centers for Disease Control (CDC) has developed three four-quadrant plates (Presumpto plates I, II, III) that incorporate test substrates in Lombard-Dowell agar medium (204). These plates contain tests for determination of growth in 20% bile; lecithinase and lipase production; esculin, starch, casein, and gelatin hydrolysis; production of H_2S and indole; DNase activity; and carbohydrate fermentation (glucose, mannitol, lactose, and rhamnose). Presumpto I plates are available from Remel Laboratories; Anaerobe Systems sells egg yolk agar and will make the other media as monoplates on request.

One method for identifying or assisting in identification of anaerobes is based on analysis of whole-cell fatty acids by capillary column GLC. The method is discussed in Appendix B.

The MicrobeLynx System (Micromass UK Limited) applies biopolymer mass spectrometry techniques (MALDI, matrix-assisted laser desorption ionization) to the analysis of intact bacteria. This method allows the unique population of macromolecules expressed on the surface of bacteria to be rapidly sampled and characterized by molecular weight. The resulting mass spectrum provides a unique physiochemical fingerprint for the species tested. The system is currently for research use only.

chapter 7

Laboratory Tests for Diagnosis of *Clostridium Difficile* Enteric Disease

General Principles

Antimicrobial-associated colitis (AAC) is usually caused by overgrowth of *Clostridium difficile* in the bowel. Compared to colitis, *C. difficile* is less often isolated in antimicrobial-associated diarrhea (AAD) without colitis. Previously it was thought that only *C. difficile* strains producing both the enterotoxin (toxin A; TcdA) and the cytotoxin (toxin B; TcdB) cause clinical disease. However, it has been recently reported that strains producing only the cytotoxin ($TcdA^-$ / $TcdB^+$) may be implicated in nosocomial infections and infections in pediatric patients and in the immunocompromised host (2;174;199). Toxin A and toxin B are encoded by genes *cdtA* and *cdtB*, respectively, that together with genes *tcdC, tcdD,* and *tcdE,* whose functions are yet to be determined, form a 19 kb pathogenicity locus (PaLoc) (283). Strains producing both toxins ($TcdA^+$/ $TcdB^+$) are cytotoxic and responsible for the majority of clinical disease in humans and animals. Nontoxigenic strains ($TcdA^-$ / $TcdB^-$) do not produce either toxin, are not cytotoxic nor virulent, but may possess parts of toxin genes as demonstrated by PCR methods. Toxin-variant strains lack parts of the toxin genes and may produce only cytotoxin ($TcdA^-$ / $TcdB^+$), both toxins, or no toxin depending on the extent of deleted areas on PaLoc. At present, 15 different toxinotypes are recognized, and some of the toxin-variable strains produce a binary toxin that is an actin-specific ADP-ribosyltransferase (283;314). Importantly, strains that form only cytotoxin ($TcdA^-$ / $TcdB^+$) cannot be detected by commercial EIA methods designed for the detection of toxin A only (208). As such strains may comprise around 10% of clinical *C. difficile* isolates (6;176) and even >40% in pediatric patients (174), laboratories relying only on single-toxin detection should reevaluate their choice of diagnostic tests. Testing to detect toxin alone compared to culture and toxin determination of the isolates (called "toxigenic culture") may miss up to 30% of *C. difficile*–associated diarrhea (CDAD) cases (261).

Definitive diagnosis of pseudomembranous colitis requires visualization of pseudomembrane via endoscopy, relevant patient data, and documentation of the microbial cause. Although none are diagnostic by themselves, laboratory tests can contribute evidence to substantiate a clinical diagnosis in primary cases. During epidemics and outbreaks in institutions, culture provides isolates for more accurate toxin detection, molecular typing, and tracing the clonal spread of virulent strains (261).

Tests currently available to the routine laboratory include (1) toxigenic culture; (2) enzyme immunoassays (EIA) for enterotoxin A only (from several manufacturers, see (9)), for both enterotoxin A and cytotoxin B (Meridian Diagnostics Inc., Rohm Pharma,

TechLab Inc.), or for both glutamate dehydrogenase and toxin A (Biosite Diagnostics, San Diego, CA); (3) cell culture assay for cytotoxin using commercial systems (Advanced Clinical Diagnostics, Bartels Inc., Biowhittaker Inc., TechLab Inc.) or conventional method; (4) noncommercial PCR applications including detection of TcdA$^-$/ TcdB$^+$ strains (176); and (5) EIA detecting glutamate dehydrogenase (Meridian Diagnostics Inc.) as a screening test (does not detect toxins and is therefore inadequate as a sole detection method). All methods test for different characteristics, so test results cannot be compared directly. A combination of methods increases accuracy and timely reporting.

Because some patient populations may harbor the organism in stool in the absence of symptoms, it is important that the laboratory define specimen selection to increase the specificity of the test results. Children less than 1 year old without symptoms have a relatively high carriage rate, even with toxigenic strains. Strains from pediatric patients, especially those from asymptomatic carriers, often belong to serogroup F, in which most of the TcdA$^-$ / TcdB$^+$ strains cluster (176). Their stools should not be routinely tested, unless the presence of both toxins is determined and clinical symptoms support the microbiologic diagnosis. In addition, we recommend that only unformed stools (stools assuming the container shape) be tested routinely for *C. difficile*. Of course, extenuating circumstances identified by physicians should allow for flexibility in testing of certain specimens.

Cultures alone are the least specific and should not be used without toxin determinations. However, toxigenic culture is both the most specific and most sensitive method to detect possible disease related to *C. difficile*. However, studies have shown that as many as 20% or more of patients in hospitals where this organism is endemic become colonized asymptomatically with the organism, particularly if they are receiving certain antimicrobial agents (190;237). Recent data support the view that patients who become asymptomatic carriers have significantly higher levels of IgG antibody against toxin A compared to patients who develop symptomatic disease (190). Culture results are not usually available for 48 h after specimen receipt. Thus, in urgent cases, an EIA method detecting both toxins or a screening EIA combined with toxigenic culture is indicated.

Alternative broth culture screening procedures using gas-liquid chromatography for detection of *p*-cresol or other endproducts (such as isocaproic acid) have been used for epidemiologic purposes (162). The addition of 0.1% sodium taurocholate or cholic acid to broth or to solid agar media after alcohol shock for spore selection may enhance spore germination and increase culture yield (31). Add cefoxitin (16 mg/L) and sodium taurocholate or cholic acid as a sodium salt (0.1%) to commercially purchased PRAS broths (PY fructose or chopped meat) to prepare a suitable enrichment broth for epidemiologic studies. Cycloserine cefoxitin fructose agar (CCFA), a selective agar developed by George and colleagues (123), is recommended for plate cultures. An alternative commercial medium with half the selective supplements and cholic acid, CCEY medium, is available from Lab M, Bury, England (31) and is especially supportive for specimens that have been pretreated with alcohol shock spore selection. Additional tests are in the development stage. The polymerase chain reaction (PCR) has been used to amplify the enterotoxin A gene of *Clostridium difficile* and also for the deletions of *tcdA* of TcdA$^-$ / TcdB$^+$ strains and may prove useful for direct detection in specimens (176;177). However, because the organism and its toxins may be present in asymptomatic individuals, laboratory tests for its presence will never constitute the sole diagnostic modality.

LABORATORY PROCEDURES

Specimen Collection for All Procedures

Collect stool in a leakproof container. Do not contaminate with urine. Deliver promptly to the laboratory, refrigerate if unable to process within 2 h, up to 2 d, or freeze at −20 °C or colder if unable to process within 2 d of collection. Proteolytic inactivation of toxin has been noted in up to 20% of *C. difficile* culture–positive fecal samples that were left in room temperature (31).

Culture Procedures

Principle

Stool specimens from patients with diarrhea can be tested for the presence of *Clostridium difficile* by culturing on a selective medium, CCFA. The antibiotics cycloserine (500 μg/ml) and cefoxitin (16 μg/ml) inhibit the growth of most bacteria other than *Clostridium difficile*. Fructose is metabolized by *C. difficile* more actively than is glucose, and neutral red acts as an indicator to detect proteolysis of the medium. Colonies of *C. difficile* break down proteins in the medium, which results in production of alkaline endproducts that turn neutral red a yellow color. *C. innocuum* is resistant to cefoxitin and may grow on CCFA, thus posing a problem in differentiation unless further tests are performed. Many laboratories advocate alcohol shock pretreatment of stool samples to encourage spore formation and simultaneous elimination of competing fecal flora before plating on selective or nonselective media. The medium often used after spore selection is CCEY agar (Lab M), which contains half the amount of selective supplements in CCFA, and also (1) cholic acid to promote germination of spores after plating; (2) *p*-hydroxyphenylacetic acid to promote formation of *p*-cresol, responsible for the typical colony odor; and (3) blood to promote growth. Phenyl red is omitted as it autofluoresces under UV light (31). The original and other formulations of the selective media are available from different manufacturers (Becton-Dickinson Microbiology Systems, Dimed, Lab M, Oxoid, Remel).

Note that not all isolates of *Clostridium difficile* produce toxin, and that even the presence of toxin-producing *Clostridium difficile* may not be contributing to disease in particular hosts such as small children and asymptomatic carriers.

Specimen evaluation

Stool specimens less than 2 h old, refrigerated for <48 h, or frozen at –20 °C are acceptable for culture. Rectal swabs are not recommended except for broth enrichment epidemiological studies.

CCFA or CCEY medium

Prepare the medium according to the directions in Appendix C. Store the plates in plastic bags at 2–8 °C for up to 8 weeks (CCEY for 1 week). CCFA plates are available commercially, but as for most anaerobic media, freshly prepared or anaerobically prepared and stored (PRAS) plates are best. Peterson and colleagues found that the original formulation of CCFA agar reduced in an anaerobic atmosphere for at least 4 h best promoted isolation and better growth of *C. difficile* from fecal specimens from patients with CDAD (261).

Specimen processing

(1) Mix the stool specimen thoroughly and save a 2 ml aliquot for toxin testing. If the toxin test will not be performed from a fresh sample or within 1 d from the refrigerated sample, freeze the aliquot at −20 °C or colder.

(2a) Perform quantitative cultures for *C. difficile* in an anaerobic chamber. Dilute 1 g or 1 ml of stool in 9 ml of diluent (e.g., 0.05% yeast extract) and vortex in the chamber in tubes containing glass beads until the suspension appears homogeneous (approximately 60 s). Make serial 10-fold dilutions in diluent and plate duplicate 0.1 ml samples from dilutions 10^{-1}, 10^{-3}, and 10^{-5} onto individual CCFA plates using a rotator-pipet method. After 24–48 h anaerobic incubation at 37 °C, count the number of characteristic colonies (see below) on those plates yielding 30–300 CFU.

Note: If plates are incubated in an anaerobic chamber or anaerobic pouch, they may be inspected after 24 h; if colonies are large enough to count at this stage, the incubation can be concluded. Jars, on the other hand, must not be opened before 48 h.

Calculate the numbers of CFU per ml stool by multiplying the average CFU on the plates chosen times all dilution factors (e.g., an average of 30 colonies on the two plates that received 0.1 ml from the 10^{-3} dilution means a count of 3×10^5 CFU/ml stool).

(2b) Perform semiquantitative cultures by inoculating approximately 0.1 g or 2–3 drops of liquid stool (use a Pasteur pipet) onto the first quadrant of a CCFA plate and streak for isolation. Use the same grading system as described in the text for clinical specimens (1+ to 4+); this roughly correlates with CFU/ml of *C. difficile* in stool (4+ corresponds to 10^5 CFU/ml).

(2c) Perform an alcohol shock to select spores by adding equal volumes of stool and industrial grade or absolute alcohol in a tube, mix well, and let stand for 30–60 min (mix a few times or use a gentle mixer) and plate 0.1 ml on a selective medium (preferably CCEY), plus nonselective medium as an option if the original formulation of CCFA medium with higher concentrations of selective antibiotics is used.

Isolation and identification

(1) The distinctive colonies of *C. difficile* on CCFA are approximately 4 mm in diameter (if well isolated), yellow, ground-glass in appearance, circular with a slightly filamentous edge, and low umbonate to flat in profile (see Figure 3-10). The initial orange-pink color of the medium is often changed to yellow for 2–3 mm around the colonies. *C. difficile* produces a horse stable odor that is detectable from the primary culture plate, but is easier to detect from a pure culture plate. The colonies on CCEY are greyish, later with whitish centers, fluoresce bright yellow-green under long-wave UV light, and have a strong odor due to production of *p*-cresol. Other organisms, especially *C. innocuum,* may grow on CCFA, and must be differentiated by tests described below. Organisms in Gram stain are gram-positive straight rods, with subterminal oval spores (not usually seen from selective media, but typical from subcultures on the nonselective media).

(2a) If isolated colonies are available on the selective plate, test for L-proline aminopeptidase activity (PRO) by smearing the colony on a wet PRO-KIT (Remel), wait 2 min, add one drop of *p*-dimethylaminocinnamaldehyde reagent, and let stand for 1 min. Development of dark pink to red color of inoculum is a positive reaction (111). Alternatively, make a suspension in 0.2–0.3 ml of water or saline, add 1 tablet or disk (Rosco, WEE-TABS), incubate 2–4 h, add reagent, and interpret as above. *C. difficile* is positive; *C. innocuum* is negative. Other bacteria including clostridia (*C. sordellii, C. bifermentans, C. clostridioforme, C. sporogenes,* and *C. glycolicum*) may test positive if colonies are picked from a nonselective medium and sometimes even from CCFA-based media (111). However, the two former are lecithinase-positive, *C. sporogenes* is lipase-positive, and those growing on CCFA have different morphologies than that of *C. difficile* (111).

(2b) Alternatively, pick a typical suspect colony from CCFA or CCEY and test by Latex agglutination (Becton-Dickinson Microbiology Systems, Mercia Diagnostics) or by the EIA screening test. All detect glutamate dehydrogenase. *C. difficile* is positive; *C. innocuum* is negative. The same clostridia mentioned that interfere with the PRO reaction may also cause cross-reactions with the glutamate dehydrogenase–detecting tests. Use the features mentioned above to differentiate species.

(2c) With the blunt end of a Pasteur pipette, cut an agar plug of a typical colony and add it to 0.5 ml of PY or other suitable broth. Shake well, acidify, and test by gas-liquid chromatography (GLC) for volatile fatty acids; a positive peak for isocaproic acid (IC) is diagnostic for *C. difficile*. Alternatively, a colony may be suspended in PY or thioglycolate broth, incubated overnight, and tested as above.

(2d) If enough colony material is present on the selective medium, test agar plugs of typical colonies for toxin A/B combination or each toxin separately (see below).

(3) Issue a presumptive preliminary report when preliminary identification results are ready.

(4) If preliminary identification tests give indeterminate results and there is not enough colony material for toxin determination, subculture *C. difficile*–like colonies to blood agar or another CCFA or CCEY plate for purity and to chocolate agar for an aerotolerance test (incubate in 5–7% CO_2). Perform a Gram stain to detect the presence of spores and verify the morphology. *C. difficile* sporulates readily on blood agar but rarely on CCFA. Inoculate well-isolated colonies into PY or thioglycolate broth for subsequent endproduct analysis by GLC and to chopped meat broth, incubate 2 d, and test filtrate for toxin (261).

(5) Examine the pure culture plate for colony morphology, fluorescence, and odor. After the thioglycolate broth has incubated for 24 h, it should be processed by GLC as described in Appendix B.

(6) If colonies were not tested for toxins from the selective medium, do it from the pure culture.

Reporting results

(1) Report culture positive for *C. difficile* if an organism is isolated that is an anaerobic gram-positive spore-forming bacillus with typical colony morphology that produces a horse stable odor, fluoresces chartreuse under long-wave UV light, is

PRO- or glutamate dehydrogenase–positive when colonies growing on the selective media are tested, and/or produces isocaproic acid among a number of endproducts. If determined, report the CFU/ml or CFU/g dry weight of stool. Alternatively, report the semiquantitative culture result.

(2) If the above criteria are not met, report the culture as negative for *C. difficile*.

(3) Inoculate equivocal isolates into PRAS media for definitive identification.

Quality control

Maintain a viable culture of *Clostridium difficile* (ATCC 9689 and ATCC 17858) in chopped meat broth. Subculture a few drops to each new batch of medium and weekly thereafter and observe for characteristic colonies. Note that a laboratory-acclimatized strain may yield better growth than fresh clinical isolates; thus its growth is only a rough approximation of media acceptability. *E. coli* ATCC 25922 must be inhibited by CCFA.

Cytotoxin Testing Using Commercial Systems

Principle

Production of cytotoxin by fecal isolates of *Clostridium difficile* increases the possibility of involvement in clinical disease. Certain mammalian cells (including human foreskin fibroblasts, MRC-5 cells, Chinese hamster ovary K-1 cells) display a characteristic cytopathic effect (CPE) in the presence of *C. difficile* toxin that is neutralized by specific antitoxin. Stool from the majority of patients with *C. difficile* enteric disease will contain detectable cytotoxin in the filtered supernatant. However, cytotoxin may be inactivated if fecal samples are held at ambient temperatures (31). Also, the sensitivity of different tissue culture cell lines may vary and contribute to insensitivity compared to toxigenic culture (261). A number of commercial products are available at this time (Advanced Clinical Diagnostics, Bartels Inc., Biowhittaker Inc., TechLab Inc.) in which patient stool filtrates are tested concurrently with filtrates neutralized with antitoxin. A cytopathic effect that is inhibited by antitoxin is assumed to have been produced by *C. difficile* cytotoxin.

Note that the cytotoxin is not thought to play as strong a role in disease pathogenesis as is the enterotoxin with the few exceptions mentioned above. Enterotoxin, however, is produced in much smaller amounts than cytotoxin and is thus more difficult to detect in a bioassay system. About 10% of toxigenic strains do not produce toxin A, as noted before. Due to the restrictions mentioned above, a negative toxin B assay result must be confirmed by other methods in cases clinically compatible with CDAD or colitis.

Procedure

Detailed procedures of each commercial system are given in package inserts.

(1) Follow manufacturers' instructions for preparing fresh or thawed stool. Briefly, centrifuge stool and remove supernatant.

Note: Use aerosol-protective containers in the centrifuge. If possible, place the centrifuge in a biological safety cabinet or fume hood.

(2) Filter sterilize supernatant through a membrane filter.

Note: Wear gloves and face shield and perform the procedure in a biological safety cabinet. The back pressure on the filter/syringe assembly has been known to cause stool to be expelled with force, creating aerosols that may contain infectious hepatitis, HIV, other viruses, or pathogenic bacteria.

(3) Dilute in diluent to produce final concentration indicated. Alternatively, to avoid nonspecific toxicity, perform tests at a final concentration of 1:100 stool filtrate to diluent. Patients with clinically significant amounts of cytotoxin typically will be positive at this dilution, but not all nonspecific CPEs are diluted out even at this dilution (261).

(4) Continue as directed in individual test inserts.

Interpretation of results

Specimen well	Neutralized well	Interpretation
CPE positive	Normal monolayer	Cytotoxin present
Normal monolayer	Normal monolayer	Cytotoxin negative
CPE positive	CPE positive	Nonspecific cytotoxicity. Must perform test again with sample diluted (1:10, 1:20, 1:40, 1:80). If able to dilute out CPE in neutralized wells, then cytotoxin test is positive.

Reporting results

(1) Report all positive results and preliminary negative results after 24 h.

(2) Reincubate the plate for an additional 24 h to detect late-appearing cytotoxicity. Report all final negative results at 48 h.

(3) For nonspecific toxicity, report results as highest titer at which nonspecific toxicity still persists and titer at which specific cytotoxicity is detected. If unable to dilute out nonspecific toxicity, report "Nonspecific toxicity; unable to interpret results."

Note: Freezing and thawing the stool filtrate several times may remove nonspecific toxic factors (and reduce the toxin titer).

Quality control

(1) As each new microtiter plate is received, examine all wells to ensure that the monolayers are confluent and intact-looking. Before performing each test, again observe the monolayers to be used and verify their suitability.

(2) Each time the test is performed, include a positive control (diluted 1:10 in diluent), a neutralized positive control, an antitoxin control, and a diluent control. The positive control should exhibit CPE throughout the well, which is absent in the neutralized well. No CPE should be visualized in wells containing antitoxin alone and diluent alone.

Enzyme Immunoassay Tests for Enterotoxin and/or Cytotoxin and/or Cellular Antigen

Principle

Both enterotoxin A and cytotoxin B of *C. difficile* are thought to be important for development of symptomatic disease; however, toxin A is a potent enterotoxin in the intestine and toxin B less so. EIA tests (Meridian Diagnostics Inc., Rohm Pharma, TechLab Inc., Biosite) may use slightly different formats. The TechLab TOX A/B test has been recently evaluated against cytotoxin and enterotoxin testing from fecal samples and found to correlate highly (208). In this EIA test, microwells are coated with affinity-purified goat polyclonal antibody against toxins A and B. The detecting antibody consists of monoclonal antibody against toxin A and affinity-purified goat polyclonal antibody against toxin B, each conjugated to horseradish peroxidase (208). If either toxin is present in stool supernatant, it will bind to the polyclonal antibodies attached to the inner surface of wells in the microtiter plate. Addition of the monoclonal/polyclonal conjugates and their substrate allows detection of the bound toxin antigens. Meridian's A/B EIA toxin test is somewhat similar in format.

The membrane EIA systems for toxin A (Wampole Laboratories, Meridian Diagnostics) are easier to perform singly, require less expertise, and yield faster results. The Biosite product detects both enterotoxin and a cell-associated antigen, glutamate dehydrogenase, allowing detecting of toxin A–negative strains.

Another format available for testing is the Vidas automated enzyme-linked fluorescent antibody assay for toxin A (bioMerieux Vitek, Hazelwood, MO). A recent abstract showed several EIA tests to be less sensitive but equally specific as cytotoxin tests (57).

Procedure for Microwell EIA

Follow the package inserts of individual manufacturers.

(1) Before testing, be certain that all reagents are at room temperature.
(2) Prepare buffer as described in manufacturer's instructions.
(3) Add sample diluent (supplied) to a small test tube for each specimen to be tested.
(4) Prepare stool for test. Some manufacturers require centrifugation and filtration; others use a simple dilution.

Note: Do this in a biological safety cabinet to avoid creating aerosols.

(5) Remove sufficient microdilution well strips from the package to test all specimens and controls.
(6) Follow specific manufacturer's instructions for handling, inoculating, incubating, washing, and reading plate.
(7) Read spectrophotometrically, if at all possible, or visually if no other method is available.

Note: Check manufacturer's instructions since only certain test systems can be interpreted visually. If an EIA reader is required by the manufacturer, do not try to read tests visually.

7

Interpretation of results

For visual interpretation, any change in color signifies presence of enterotoxin A or cytotoxin B in the stool supernatant. If the color is so faint that there is a question about its presence, then the result is interpreted as negative. Spectrophotometrically, positive wells must read >0.10 OD at 450 nm or as stated in the product insert.

Quality control

Each time the test is performed, a positive and a negative control (supplied) must be tested along with patient samples.

chapter 8

Susceptibility Testing of Anaerobic Bacteria

Resistance Mechanisms in Anaerobic Bacteria

Significant resistance to β-lactams is due to the production of β-lactamases from genes that can be transferred between cells (286), and β-lactamases capable of hydrolyzing "stable" agents have been reported (269). Carbapenems are generally stable to most β-lactamases although imipenem resistance has been described in a number of *B. fragilis* isolates in Japan and elsewhere (14;380). The resistance is most often attributable to an imipenem-hydrolyzing β-lactamase but there may be other mechanisms as well. The carbapenem-hydrolyzing β-lactamase is generally a metalloenzyme (coded for by the *cfiA* or *ccrA* genes) that hydrolyzes imipenem; the CfiA and CcrA enzymes are identical (14;268;380). Non-metalloenzymes have also been associated with imipenem resistance (101). Resistance to macrolides (MLS) is usually due to rRNA methylases that modify the 23S component of the ribosome and has been found in both gram-negative (*Bacteroides, Campylobacter, Prevotella*) and gram-positive anaerobic rods (*Clostridium*, nonsporing rods) (274). Clindamycin resistance has been described on transferable plasmids (301). Quinolone resistance has been found due to both *gyrA* and *parC* mutations and resistance to DNA gyrase has been demonstrated in *B. fragilis* (253). Metronidazole resistance is coded for by the *nim* gene (342), and has been found on both the chromosome (273) and on plasmids (341); most of both the plasmid-borne and chromosomal markers are transferable (343). Chloramphenicol resistance is due to the production of chloramphenicol acetyltransferase.

Overview of Current Susceptibility Patterns

Summary of Data Collected at the Wadsworth Anaerobe Laboratory

The most active agents against the *Bacteroides fragilis* group are the β-lactam/β-lactamase inhibitor combinations, the carbapenems (imipenem, meropenem), the newer fluoroquinolones (trovafloxacin, moxifloxacin, gemifloxacin, gatifloxacin, sitafloxacin), and metronidazole. Some resistant strains are seen in all of these agent categories. We found that clinafloxacin, gatifloxacin, and moxifloxacin were active against >95% of *B. fragilis*–group strains. Gemifloxacin was only tested against the species *B. fragilis*, but was active against 85–95% of these strains. For *F. nucleatum*, 85–95% were susceptible to gemifloxacin, compared with 70–85% of strains of *F. mortiferum* and *F. varium*. Cefoxitin is active against most strains of *B. fragilis* but less active against other members of the group. Both *Prevotella* and *Porphyromonas* species exhibit some resistance to the macrolides, and ~30–40% of strains have tetracycline minimum inhibitory concentrations (MICs) of >16 μg/ml (many had MICs of 64 μg/ml). Among the gram-positive anaerobes, clostridia other than *C. perfringens* show considerable resistance to the macrolides and to many of the cephalosporin compounds. Agents with excellent activity include β-lactam/β-lactamase inhibitor combinations, carbapenems, and metronidazole. *Peptostreptococcus* species differ as to their resistance patterns. *P. anaerobius* and *P. asaccharolyticus* are more resistant to several agents (e.g., amoxicillin/clavulanate, clindamycin, penicillin G) than are the other species tested. Table 8-1 summarizes current patterns of susceptibility of anaerobic bacteria tested at the Wadsworth Anaerobe Laboratories. In an effort to simplify interpretation, antimicrobials with similar efficacies against groups of anaerobes have been grouped and the range of percent susceptible strains has been used to define the group. The ranking within a group does not reflect the degree of activity. This presentation also minimizes the significance of differences of <10%. At times an antimicrobial was borderline between two groups, and the decision about its grouping was necessarily arbitrary.

Summary of Data from Other Laboratories

Information from other studies generally parallels our findings. Snydman and coworkers tested *B. fragilis* group isolates in a multicenter survey of eight medical centers (including the northeast, midwest, southeast, southern, and western regions of the U.S.) between 1995 and 1996 as well as during an earlier 7-year period (1990 to 1996) (this study also includes data from our laboratory) (305). For *B. fragilis,* decreases in MICs were seen for many β-lactam agents (piperacillin-tazobactam, ticarcillin-clavulanate, piperacillin, ticarcillin, ceftizoxime, cefotetan, and cefmetazole); a significant increase in resistance rates was observed for clindamycin and cefoxitin. For the non–*B. fragilis* species, a significant decrease in the geometric mean MICs was observed for meropenem, ampicillin-sulbactam, ticarcillin-clavulanate, piperacillin, ticarcillin, ceftizoxime, and cefmetazole; a significant increase was observed for cefoxitin. Significant increases in resistance rates were observed within the *B. fragilis* strains for ticarcillin and ceftizoxime and within the non–*B. fragilis* isolates for cefotetan. Significant increases in percent resistance rates among all *B. fragilis*–group species were observed for clindamycin, while imipenem showed no significant change in resistance trends. No chloramphenicol or metronidazole resistance was seen. The data demonstrate that resistance among the *B. fragilis*–group species has decreased in the past several years, the major exception being clindamycin (305). In a study of 911 clinical strains of the *Bacteroides fragilis* group isolated from 1992 to 1997 in Canada, rates of resistance to metronidazole, imipenem, piperacillin-tazobactam, ticarcillin-clavulanic acid, penicillin, piperacillin, and cefoxitin remained essentially unchanged, but there was a significant increase in the rates of resistance to clindamycin (8.2% in 1992 to 19.7% in 1997 ($P < 0.0004$)) (191).

In a study of 463 anaerobes in five hospitals in France in 1996 (235), none of the 209 *B. fragilis*–group strains showed resistance to imipenem or ticarcillin-clavulanic acid. High resistance rates (29%) were observed for cefotetan and clindamycin. Clindamycin resistance had increased within the *B. fragilis* group (from 14% in 1992 to 29% in 1996) and also among strains of clostridia (32%), *P. acnes* (18%), and *Peptostreptococcus*

(28%). Only 5% clindamycin resistance was seen among *B. fragilis*–group isolates in Cape Town, South Africa (181).

A European study group (including 15 laboratories from 13 countries) studied non–*Bacteroides fragilis* group anaerobic gram-negative bacilli (179). They found that all isolates were susceptible to the carbapenems and all but three isolates of *Bilophila wadsworthia* were susceptible to amoxicillin-clavulanate. Most isolates (except *Bilophila*) were susceptible to cefoxitin. Most strains of fusobacteria were inherently resistant to the macrolides. While most strains of *F. varium* were clindamycin resistant, resistance to clindamycin among other species of *Fusobacterium* was rare. Three isolates (*Bacteroides ureolyticus, Sutterella wadsworthensis,* and an unnamed species) were metronidazole resistant. The majority of *Prevotella* species were susceptible to telithromycin; three isolates that were resistant were also resistant to clindamycin and to the macrolides.

While the earlier fluoroquinolones had limited activity and poor clinical efficacy for anaerobic infections, the more recently developed quinolones are much more active (144). Goldstein and group found marked differences in the activity of gemifloxacin for the various species of the *B. fragilis* group while *Prevotella* and *Porphyromonas* strains were generally susceptible (128). In another study, they noted that *Prevotella bivia* and *P. melaninogenica* were relatively resistant to gemifloxacin and other quinolones (127). In our laboratory, we found that 48% of *Prevotella* strains were resistant to >4 μg/ml of gemifloxacin (unpublished data). Gatifloxacin was active against *Prevotella, Porphyromonas,* and *Peptostreptococcus* isolates from bite wounds, but not against *Fusobacterium* species tested (including *F. nucleatum*) (126).

Resistance has been detected to virtually all of the drugs traditionally considered to have excellent activity against anaerobes. Resistance in carbapenems is due to metalloenzymes that hydrolyze the drug. The lack of inhibition of these enzymes by β-lactamase inhibitors means that strains producing this enzyme are also resistant to β-lactam/β-lactamase inhibitor combination agents. Although the percentage of strains producing high levels of this enzyme is still low, 2.5–4% of *B. fragilis* isolates harbor the gene for this enzyme. In these strains, the gene is expressed at very low levels and does not result in clinical resistance. However, conversion from low-level to high-level resistance has been observed in a patient undergoing therapy with imipenem (344). Thus the possibility of increasing resistance in the future must be considered. Metronidazole resistance detected by the Anaerobe Reference Unit in the United Kingdom was 7.5% in 1998, compared to 3.8% in 1997 and 1.9% in 1995 (36). The authors comment that the trend toward increasing resistance may be the result of a selection bias, as some isolates are referred to the unit in order to confirm resistance to a 5 μg metronidazole disk, or it may truly reflect an increase in resistance.

GENERAL CONSIDERATIONS

Methods that are used to determine minimum inhibitory concentrations (MICs) of antimicrobials for anaerobic bacteria include agar dilution and microbroth dilution. Broth disk elution and agar disk diffusion have also been used in the past but are not currently approved by the National Committee for Clinical Laboratory Standards (NCCLS). Techniques using a drug concentration gradient, such as the spiral gradient system or

table 8-1. Susceptibility of anaerobic Anaerobe Wadsworth Laboratory, VAMC

B. fragilis	Other B. fragilis group[3]	Other Bacteroides	Prevotella	Porphyromonas
Piperacillin Amoxicillin + clavulanate Ampicillin +sulbactam Cefoperazone +sulbactam Piperacillin + tazobactam Ticarcillin + clavulanate Cefoxitin Biapenem Imipenem Meropenem Chloramphenicol Clinafloxacin Gatifloxacin Sitafloxacin Levofloxacin Moxifloxacin Ofloxacin Trovafloxacin Metronidazole	Ampicillin + sulbactam Cefoperazone + sulbactam Piperacillin + tazobactam Ticarcillin + clavulanate Biapenem Imipenem Meropenem Chloramphenicol Clinafloxacin Gatifloxacin Moxifloxacin Sitafloxacin Trovafloxacin Metronidazole Minocycline	Piperacillin Amoxicillin + clavulanate Ampicillin + sulbactam Ticarcillin + clavulanate Cefoperazone Cefoperazone + sulbactam Cefotaxime Cefoxitin Cefizoxime Biapenem Imipenem Chloramphenicol Clinafloxacin Gatifloxacin Sitafloxacin Levofloxacin Moxifloxacin Trovafloxacin Metronidazole Clindamycin	Piperacillin Amoxicillin + clavulanate Ampicillin + sulbactam Piperacillin + tazobactam Ticarcillin + clavulanate Cefoxitin Ceftizoxime Biapenem Imipenem Meropenem Chloramphenicol Clinafloxacin Moxifloxacin Sitafloxacin Trovafloxacin Metronidazole Clindamycin Telithromycin	Piperacillin Amoxicillin + clavulanate Ampicillin + sulbactam Cefoxitin Ceftizoxime Ceftriaxone Biapenem Imipenem Meropenem Chloramphenicol Clinafloxacin Gatifloxacin Gemifloxacin Sitafloxacin Sparfloxacin Trovafloxacin Metronidazole Azithromycin Telithromycin Minocycline
Cefotetan Ceftizoxime Clindamycin Minocycline Gemifloxacin	Amoxicillin + clavulanate Piperacillin Cefoxitin Ceftizoxime	Cefotetan Ceftazidime Ceftriaxone Clarithromycin Erythromycin Roxithromycin Telithromycin Minocycline	Ceftriaxone Azithromycin Clarithromycin Erythromycin Roxithromycin	Ciprofloxacin Clarithromycin Clindamycin Erythromycin Roxithromycin
Moxalactam Ceftriaxone Clarithromycin	Levofloxacin Clarithromycin Clindamycin	Moxalactam Ofloxacin Sparfloxacin Azithromycin	Ciprofloxacin Ofloxacin Sparfloxacin Minocycline	Tetracycline
Cefoperazone Cefotaxime Ceftazidime Sparfloxacin	Cefoperazone Cefotetan Moxalactam Ofloxacin Sparfloxacin Telithromycin	Ciprofloxacin Tetracycline	Tetracycline Gemifloxacin	Fleroxacin Lomefloxacin
Penicillin G Ciprofloxacin Fleroxacin Lomefloxacin Azithromycin Erythromycin Roxithromycin Telithromycin Tetracycline	Penicillin G Cefotaxime Ceftazidime Ceftriaxone Ciprofloxacin Fleroxacin Lomefloxacin Azithromycin Erythromycin Roxithromycin	Penicillin G Fleroxacin Lomefloxacin	Penicillin G Fleroxacin Lomefloxacin	

PERCENT OF STRAINS

>95% susceptible	85–95% susceptible

The ranking within a color group does not reflect

bacteria from recent studies at the West Los Angeles

	Campylobacter gracilis	*Sutterella wadsworthensis*	*F. nucleatum*	*F. mortiferum* and *F. varium*	Other *Fusobacterium*	*Bilophila wadsworthia*
	Piperacillin Amoxicillin + clavulanate Piperacillin + tazobactam Ticarcillin + clavulanate Cefoxitin Ceftizoxime Ceftriaxone Biapenem Imipenem Meropenem Chloramphenicol Ciprofloxacin Clinafloxacin Sitafloxacin Fleroxacin Lomefloxacin Sparfloxacin Trovafloxacin Metronidazole Azithromycin Clindamycin Erythromycin Roxithromycin Minocycline Tetracycline	Amoxicillin + clavulanate Ticarcillin + clavulanate Cefoxitin Ceftriaxone Imipenem Meropenem Ciprofloxacin Fleroxacin	Piperacillin Amoxicillin + clavulanate Ampicillin + sulbactam Piperacillin + tazobactam Ticarcillin + clavulanate Cefoxitin Ceftizoxime Ceftriaxone Biapenem Imipenem Meropenem Chloramphenicol Clinafloxacin Gatifloxacin Levofloxacin Moxifloxacin Ofloxacin Sitafloxacin Sparfloxacin Trovafloxacin Clindamycin Metronidazole Minocycline Tetracycline	Piperacillin Ampicillin + sulbactam Piperacillin + tazobactam Ticarcillin + clavulanate Cefoxitin Biapenem Imipenem Meropenem Chloramphenicol Clinafloxacin Sitafloxacin Trovafloxacin Metronidazole Minocycline	Penicillin G Ampicillin + sulbactam Piperacillin + tazobactam Cefoxitin Biapenem Imipenem Meropenem Chloramphenicol Clinafloxacin Gatifloxacin Moxifloxacin Sitafloxacin Metronidazole Clindamycin Minocycline Tetracycline	Piperacillin Ticarcillin Amoxicillin + clavulanate Ampicillin + sulbactam Cefotetan Cefoxitin Ceftizoxime Imipenem Chloramphenicol Ciprofloxacin Sitafloxacin Fleroxacin Lomefloxacin Sparfloxacin Trovafloxacin Metronidazole Minocycline Tetracycline
70–84% susceptible	Telithromycin	Piperacillin Piperacillin + tazobactam Ceftizoxime Trovafloxacin Azithromycin Clarithromycin Erythromycin Roxithromycin Telithromycin	Ciprofloxacin Gemifloxacin Azithromycin	Amoxicillin + clavulanate Ceftizoxime Ceftriaxone	Piperacillin Amoxicillin + clavulanate Ticarcillin + clavulanate Cefoperazone + sulbactam Cefotaxime Cefotetan Ceftizoxime Ceftriaxone	Clindamycin
50–69% susceptible		Metronidazole Clindamycin	Penicillin G Telithromycin	Clindamycin Tetracycline Gemifloxacin Ciprofloxacin Sparfloxacin	Ceftazidime Moxalactam Ciprofloxacin Sparfloxacin Azithromycin	Telithromycin
<50% susceptible			Fleroxacin Lomefloxacin Clarithromycin Erythromycin Roxithromycin	Fleroxacin Lomefloxacin Azithromycin Clarithromycin Erythromycin Roxithromycin Telithromycin	Fleroxacin Lomefloxacin Clarithromycin Erythromycin Roxithromycin Telithromycin	Amoxicillin Ampicillin Penicillin G Azithromycin Clarithromycin Erythromycin Roxithromycin

SUSCEPTIBLE AT BREAKPOINT

70–84% susceptible   50–69% susceptible   <50% susceptible

(continued)

degree of activity; drugs are arranged by classes.

table 8-1. *Continued*

Gram-positive cocci	*C. difficile*[1]	*C. ramosum*	*C. perfringens*	Other *Clostridium* spp.	NSF-GPR[2]
Penicillin G Piperacillin Amoxicillin + clavulanate Ampicillin + sulbactam Piperacillin + tazobactam Ticarcillin + clavulanate Cefoperazone Cefoperazone + sulbactam Cefotetan Cefoxitin Ceftazidime Ceftizoxime Ceftriaxone Biapenem Imipenem Meropenem Chloramphenicol Clinafloxacin Gatifloxacin Moxifloxacin Sitafloxacin Sparfloxacin Trovafloxacin Metronidazole Telithromycin	Ampicillin Piperacillin Ticarcillin Amoxicillin + clavulanate Ampicillin + sulbactam Piperacillin + tazobactam Ticarcillin + clavulanate Cefotetan Imipenem Meropenem Clinafloxacin Sitafloxacin Trovafloxacin Metronidazole	Amoxicillin + clavulanate Piperacillin + tazobactam Ticarcillin + clavulanate Ceftizoxime Imipenem Clinafloxacin Sitafloxacin Metronidazole	Ampicillin Penicillin G Piperacillin Ticarcillin Ampicillin + sulbactam Amoxicillin + clavulanate Piperacillin + tazobactam Ticarcillin + clavulanate Cefotetan Ceftizoxime Biapenem Imipenem Chloramphenicol Ciprofloxacin Clinafloxacin Fleroxacin Gatifloxacin Moxifloxacin Sitafloxacin Sparfloxacin Trovafloxacin Metronidazole Azithromycin Clarithromycin Erythromycin Roxithromycin Telithromycin	Amoxicillin Ampicillin Carbenicillin Penicillin G Piperacillin Ticarcillin Ampicillin + sulbactam Amoxicillin + clavulanate Biapenem Imipenem Chloramphenicol Clinafloxacin Sitafloxacin Trovafloxacin Metronidazole Minocycline	Penicillin G Piperacillin Amoxicillin + clavulanate Ampicillin + sulbactam Piperacillin + tazobactam Ticarcillin + clavulanate Cefotaxime Ceftizoxime Biapenem Imipenem Meropenem Chloramphenicol Clindamycin Clinafloxacin Gatifloxacin Levofloxacin Moxifloxacin Sitafloxacin Minocycline
Gemifloxacin Levofloxacin Clindamycin Minocycline	Ceftriaxone Biapenem Chloramphenicol	Ampicillin Piperacillin Ampicillin + sulbactam Chloramphenicol Trovafloxacin Clindamycin	Lomefloxacin Clindamycin	Moxalactam	Penicillin G Cefotetan Cefoxitin Ceftriaxone Cefoperazone + sulbactam Trovafloxacin Azithromycin Clarithromycin Erythromycin Roxithromycin Telithromycin
Ciprofloxacin Ofloxacin Azithromycin Clarithromycin Erythromycin Penicillin G	Clindamycin Minocycline Tetracycline Azithromycin Clarithromycin Erythromycin Roxithromycin Telithromycin	Cefoxitin	Minocycline	Gatifloxacin Levofloxacin Moxifloxacin Ofloxacin Sparfloxacin Clindamycin Tetracycline	Cefoperazone Moxalactam Sparfloxacin Tetracycline
Fleroxacin Tetracycline Roxithromycin	Cefoxitin Ceftizoxime Ciprofloxacin Fleroxacin Lomefloxacin Sparfloxacin	Sparfloxacin Minocycline Tetracycline	Tetracycline	Cefoperazone Cefotaxime Cefoxitin Ceftizoxime Ceftriaxone Ciprofloxacin Azithromycin Clarithromycin Erythromycin Roxithromycin Telithromycin	Ciprofloxacin Ofloxacin Metronidazole
Lomefloxacin		Ciprofloxacin Fleroxacin Lomefloxacin Azithromycin Clarithromycin Erythromycin Roxithromycin Telithromycin		Ceftazidime Fleroxacin Lomefloxacin	Fleroxacin Lomefloxacin

[1] Breakpoint is used only as a reference point. *C. difficile* is primarily of interest in relation to antimicrobial induced pseudomembranous colitis. These data must be interpreted in the context of level of drug achieved in the colon and impact of agent on indigenous colonic flora.

[2] Nonsporeforming gram-positive rods. In various studies we included the following: *Actinomyces, Bifidobacterium, Eubacterium, Lactobacillus* and *Propionibacterium.*

PERCENT OF STRAINS SUSCEPTIBLE AT BREAKPOINT

>95% susceptible	85–95% susceptible	70–84% susceptible	50–69% susceptible	<50% susceptible

The ranking within a color group does not reflect degree of activity; drugs are arranged by classes.

Etest, have also been evaluated by many laboratories. The NCCLS standard does not address commercial susceptibility testing systems but states that clearance by the FDA indicates equivalence to results generated using the NCCLS reference methods as described in the package insert. In the fourth edition of the approved standard for susceptibility testing for anaerobic bacteria (246), Brucella agar supplemented with laked blood and vitamin K_1 (Wadsworth method) became the recommended reference agar, replacing Wilkins-Chalgren agar. The change in the reference standard was instituted because studies indicated that Brucella blood agar supplemented with hemin and vitamin K_1 was superior to both Wilkins-Chalgren agar and to Wilkins-Chalgren agar plus blood for growth of fastidious organisms. Values for ATCC reference strains were statistically similar among the three media.

The following caveats should be kept in mind for any method:

- Unless adequate growth of the organism is achieved, the MIC cannot be reliably determined.
- MICs for control organisms should fall within the acceptable range with each antimicrobial agent. MICs of quality-control organisms in some broth microdilution techniques are often one two-fold dilution lower than in an agar dilution technique. See Table 8-7 for the acceptable ranges for broth microdilution for some antimicrobial agents.
- Modifications of the methods (e.g., higher inoculum, longer incubation time) and media (e.g., added supplements) may permit the growth of most fastidious organisms, but such modifications should be undertaken only when necessary and with appropriate quality control. Reports should state what modifications of the reference method were used to obtain the result, especially if the data are to be published.
- Results will vary depending on the method used. Care should be taken when comparing results generated by use of different methods. As stated above, broth microdilution methods may produce MICs slightly lower than MICs obtained by agar dilution methods.
- The MICs of certain antimicrobial agents for *Bacteroides* cluster near the susceptibility breakpoint concentrations for those agents. Since the acceptable error for the dilution methods described is one two-fold dilution, **many strains may be designated as resistant on one occasion and susceptible on another, within the allowable error of the technique.**
- Determination of endpoint may be difficult and is probably one of the greatest sources of error in interpretation of susceptibility tests (Figure 8-1). The problem exists with some gram-negative anaerobic bacteria and problems of trailing endpoints occur in both agar (agar dilution, Etest, spiral gradient) and broth (microdilution) techniques. It may be particularly troublesome with those organisms that grow poorly, so that the growth control itself is fairly light. In agar dilution, evidence of light growth (or multiple tiny colonies) may persist in the presence of very high concentrations of drug, despite a sharp drop-off in growth at a much lower drug level. In the case of *Fusobacterium,* this has been shown to be due to the persistence of cell wall–deficient forms of the bacteria. Recent experiments have indicated that the haze seen with *Bilophila wadsworthia* is also due to spheroplast formation and, indeed, this may be the cause of the haze seen with many gram-negative organisms and β-lactam agents, all of which affect bacterial cell wall

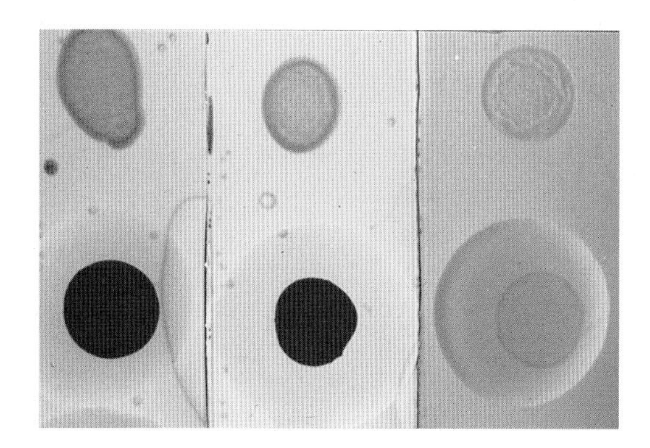

Figure 8-1. Growth of *Bilophila wadsworthia* on agar dilution plates containing imipenem. Top row: *B. wadsworthia* on agar dilution plates. Note that determination of an endpoint would be very difficult. Bottom row: *B. wadsworthia* on agar plates overlayed with triphenyltetrazolium chloride solution. Viable organisms produce a red color. Note how endpoint determination is facilitated.

formation. In these cases, the point at which the growth drops off sharply should be read as the MIC, and the persistence of haze noted. Some organisms (e.g., fusobacteria, *Campylobacter gracilis,* and *B. wadsworthia*) may show no sharp drop-off of growth, and the endpoint may be very difficult to determine. Determination of endpoint in some cases is necessarily arbitrary, and undue importance should not be attributed to these results. The NCCLS procedure instructs that the endpoint should be read at the point at which growth drops off sharply. In contrast, the Etest procedure instructs that the endpoint be read at the point where there are no more isolated colonies. In cases of trailing endpoints, these endpoints may be very different when determined by the agar and Etest procedures.

When susceptibility testing is performed for the purpose of evaluation of new antimicrobial agents or monitoring antibiograms, a few additional cautions are necessary:

- Recent clinical isolates should be used to determine current antibiograms. Patterns of antimicrobial susceptibility may change in an institution over time.
- Adequate numbers of strains (at least 10 of each species) should be tested if inferences will be made regarding the susceptibility of a particular group or species.
- Adequate representation of all clinically relevant species should be ensured. *B. fragilis*, for example, is much more susceptible to many agents than are the other members of the group, yet the other species account for at least half of the clinical isolates. The different species of the *B. fragilis* group should be reported separately.

PROCEDURES FOR TESTING SUSCEPTIBILITY

Indications for Susceptibility Testing

Economic realities necessitate that careful consideration be given to performing anaerobic susceptibility tests. Anaerobic susceptibility testing need not be performed routinely on all isolates in the clinical laboratory. Tables 8-2 and 8-3 list indications for anaerobic susceptibility testing and isolates that should be tested. Otherwise, determination of empiric therapy can be based on published antibiograms, but since patterns

table 8-2. Indications for susceptibility testing for anaerobes

A. Specific infections from which isolates should be considered for susceptibility testing
   - Refractory or recurrent bacteremia
   - Central nervous system infections
   - Endocarditis
   - Osteomyelitis
   - Joint infection
   - Prosthetic device infection
   - Organism isolated from any normally sterile site
   - Infections not responsive to empiric therapy
   - Infections which require long-term therapy

B. To determine patterns of susceptibility in a particular hospital or geographic area; these should be done at intervals of four to six months

C. Evaluation of new agents

table 8-3. Isolates to consider for testing

Isolates
Bacteroides fragilis group isolates
Prevotella spp.
Fusobacterium nucleatum
Fusobacterium mortiferum/varium
Clostridium perfringens
C. ramosum, C. innocuum, C. clostridioforme
Bilophila wadsworthia *
Sutterella wadsworthensis *

* Supplements recommended.

may vary within geographic regions and even within hospitals, it is strongly recommended that each hospital center periodically test their isolates to determine local patterns and detect any changes in resistance. This resistance may have important implications for clinical outcome. Several factors play a role in the difficulty of obtaining data that correlate susceptibility results with clinical outcome: many anaerobic infections are mixed and elimination of all the organisms isolated may not be necessary; the effects of drainage and/or debridement will affect the outcome, even if a resistant organism is involved; and the general health of the patient, always an important factor, may be especially significant in cases involving anaerobes.

However, recent studies show a correlation between susceptibility results and clinical outcomes in serious infections (279); one example is a study in which the MIC of cefoxitin, the dose, and the duration of therapy were predictors of the outcome in a retrospective analysis of 19 patients with *B. fragilis*–group infections (304). In a study of the outcome of cases of anaerobic bacteremia in Finland (285), only about half the patients received appropriate empirical therapy. For 18 patients, initially ineffective treatment was changed on the basis of bacteriologic results, and for 11 patients no change was made. The mortality in these patient groups was 18%, 17%, and 55%, respectively. The authors conclude that failure to pay attention to the results of anaerobic blood cultures may have serious consequences for the patients. Another report describes a case of *B. fragilis* bacteremia and infected aortic aneurysm that was diagnosed as fever of unknown origin, in part because routine anaerobic blood cultures were not obtained (251). These reports underscore the importance of timely identification and susceptibility results in serious anaerobic infections.

When therapeutic decisions are based, in part, on published susceptibility data, clinicians should be aware that methodological differences may sometimes result in extreme differences in certain test results, as with ceftizoxime, which shows excellent activity against anaerobes by broth microdilution tests, and poorer activity with agar dilution tests. Susceptibility testing of anaerobes may be undertaken in a clinical setting with individual isolates, or as a large study to evaluate antimicrobial agents and report current resistance patterns. For monitoring of patterns, the NCCLS agar dilution technique is the most efficient method that will support the growth of a variety of organisms. Broth microdilution may also be used for certain combinations of organisms and drug; further evaluation of these techniques for use in anaerobic testing is currently underway by the NCCLS. For individual isolates and selected agents, the Etest is a viable alternative (see below).

Agents to be Tested

8

The hospital formulary should serve as a guide for choosing the agents for testing. If susceptibility testing is not being performed routinely for anaerobes, it should be undertaken when a patient is nonresponsive to empiric therapy or when there is a reason to suspect resistance to one of these agents; the laboratory should include the agent being used therapeutically. Imipenem, chloramphenicol, metronidazole and the β-lactam/β-lactamase inhibitor combinations are presently active against most anaerobes in the United States. Of course, these agents must be monitored periodically in large centers in order to detect any developing resistance. Note that metronidazole resistance is transferable and care should be taken to detect developing resistance. Many cephalosporins, penicillins, and clindamycin are variably active against anaerobes, and one cannot predict the susceptibility of a particular strain based on published patterns. Norfloxacin, ofloxacin, and ciprofloxacin have relatively poor activity against anaerobes; the newer fluoroquinolones, including gatifloxacin, gemifloxacin, levofloxacin, moxifloxacin, sitafloxacin, and trovafloxacin, have a much improved *in vitro* spectrum against most anaerobes although there are reports of resistant strains.

Preparation of Inoculum

A McFarland turbidity standard should be used for preparation of inocula for susceptibility tests. Instructions for making McFarland standards are found in Appendix C. The 0.5 McFarland standard should correlate with approximately 1.5×10^8 CFU/ml, but will vary widely for organisms of different cell sizes. Both Remel and Hardy Diagnostics sell McFarland standards using latex beads; these may be stored and used indefinitely.

There are two acceptable methods for preparing the inoculum for susceptibility testing.

(1) Direct colony suspension: Use a sterile cotton swab to suspend colonies from a 24–72 h blood agar plate culture directly into a clear broth (such as Brucella) to achieve turbidity equal to a 0.5 McFarland standard (against a black stripe background).
(2) Growth broth suspension: Inoculate five or more colonies into enriched fluid thioglycolate medium (or other appropriate broth) and incubate 4–6 h (or overnight for slow-growing organisms). Dilute to the density of a 0.5 McFarland standard with a clear broth as above.

Preparation of Antimicrobial Stock Solutions and Dilutions

Stock solutions of antimicrobial agents generally may be prepared in advance and frozen in aliquots at −70 °C. After thawing they should not be refrozen. Then NCCLS recommends that stock solutions of most agents can be stored for 6 months without significant loss of activity, but cautions that directions provided by the manufacturer should be considered. Further, any deterioration should be reflected in the results of susceptibility testing using the quality control strains. The solvents and diluents suitable for many antibiotics are indicated in Table 8-4. Contamination is very rare, and solutions generally need not be sterilized, although sterile water and buffers should be used for making the solutions. If desired, solutions may be sterilized through a nitrocellulose membrane filter. Other filters such as paper, asbestos, or sintered glass may adsorb appreciable amounts of some agents and should not be used. Generally, unless the manufacturer indicates otherwise, cephalosporins and cephamycins may be dissolved in 0.1 M phosphate buffer (PB), pH 6.0, and further diluted in water. The stock solutions may be dispensed into sterile glass, polypropylene, polystyrene, or polyetheylene vials and stored at −70 °C.

Either of the following formulae may be used to determine the amount of powder or diluent needed for a standard solution:

$$\text{Weight (mg)} = \frac{\text{Volume (ml)} \times \text{Concentration needed } (\mu\text{g/ml})}{\text{Assay potency } (\mu\text{g/mg})}$$

or

$$\text{Volume (ml)} = \frac{\text{Weight (mg)} \times \text{Assay potency } (\mu\text{g/mg})}{\text{Concentration needed } (\mu\text{g/ml})}$$

Prepare the dilutions according to the scheme in Table 8-5. Avoid serial two-fold dilutions since any error made toward the beginning of the dilution scheme will be compounded. The dilution scheme shown in Table 8-5 will minimize the chance of error carry-over.

Addition of Growth Supplements to Antimicrobial Susceptibility Test Media

Unless good growth of the organism is achieved, an accurate MIC cannot be determined. We recommend that formate/fumarate (0.3/0.3%) routinely be added to the media (agar or broth) for testing of *Campylobacter gracilis, Sutterella wadsworthensis,* and *Bacteroides ureolyticus*–group strains. The light, nearly transparent growth of these

table 8-4. Solvents and diluents for antimicrobial agents

Antimicrobial Agent	Solvent	Diluent
Amoxicillin, Ticarcillin	0.1 M PB[2], pH 6	0.1 M PB, pH 6
Ampicillin, Cefoperazone, Sulbactam	0.1 M PB, pH 8	0.1 M PB, pH 6
Azlocillin, Carbenicillin	Water	Water
Cefamandole, Cefmetazole, Cefotaxime	Water	Water
Cefotetan	DMSO[1] or DMF[1]	Water
Cefoxitin, Ceftizoxime	Water	Water
Cephalothin	0.1 M PB, pH 6	Water
Chloramphenicol	95% ethanol[3]	Water
Clavulanic acid	0.1 M PB, pH 6	0.1 M PB, pH 6
Clindamycin	Water	Water
Erythromycin	95% ethanol[3]	Water
Imipenem[4]	0.01 M PB, pH 7.2	0.01 M PB, pH 7.2
Metronidazole	DMSO	Water
Mezlocillin	Water	Water
Moxalactam (diammonium salt)[5]	0.04 N HCl	0.1 M PB, pH 6
Penicillin G, Piperacillin	Water	Water
Sulbactam	0.1 M PB, pH 8	0.1 M PB, pH 6
Tetracycline	Water	Water
Vancomycin	0.05 N HCl[6]	Water

[1] Dissolve powder in minimum amount of solvent needed to solubilize agent and bring to volume with diluent. DMSO = dimethyl sulfoxide, DMF = dimethyl formamide

[2] Phosphate buffer

[3] Agent is dissolved in 1/10 final volume 95% ethanol, then brought to the proper volume with water.

[4] Imipenem is not very stable and should be freshly prepared.

[5] Let stand for 1.5 to 2 hours to allow the solution to come to an equilibrium of the two types of isomers that are present in solutions for clinical use.

[6] Stock solution is made with 0.05 N HCl, and further dilutions are made with water.

organisms in conventional media results in erroneously high MICs for certain strains. For testing of *Bilophila wadsworthia,* 1% pyruvate is added. If difficulty is encountered growing some gram-positive cocci or other gram-positive organisms, 0.1% Tween-80 may be helpful. Since supplements may affect MIC results, it is imperative that the MICs for the quality control organisms remain within their accepted range.

Quality Control Standards

The use of quality control strains is necessary to monitor the quality of the media and the stability of the antimicrobial preparations. Appropriate control tests should be performed whenever new media, reagents, or antimicrobials are prepared. **Include at least two quality control strains with each agar dilution test run.** When testing with other methods (e.g., broth microdilution or Etest), test one quality control strain with the test strain.

table 8-5. Scheme for preparing dilutions of antimicrobial agents to be used in agar dilution susceptibility tests

Antimicrobial Solution						
Step	**Conc.**	**Source**	**Vol +**	**Distilled Water =**	**Intermediate Concentration =**	**Final Conc. at 1:10 Dil. in Agar**
	5120 μg/ml (mg/l)	Stock	–	–	5120 μg/ml (mg/l)	512 μg/ml (mg/l)
1	5120	Stock	2 ml	2 ml	2560	256
2	5120	Stock	1	3	1280	128
3	5120	Stock	1	7	640	64
4	640	Step 3	2	2	320	32
5	640	Step 3	1	3	160	16
6	640	Step 3	1	7	80	8
7	80	Step 6	2	2	40	4
8	80	Step 6	1	3	20	2
9	80	Step 6	1	7	10	1
10	10	Step 9	2	2	5	0.5
11	10	Step 9	1	3	2.5	0.25
12	10	Step 9	1	7	1.25	0.125

The NCCLS-approved quality control (QC) strains and the acceptable MIC ranges (obtained by the agar dilution technique) for these strains are listed in Table 8-6. The NCCLS has recommended QC values for some antimicrobials with broth microdilution testing (Table 8-7).

Reference Agar Dilution Test Procedure

In Advance (Up to 1 Month)

Prepare the Brucella base supplemented with vitamin K_1 and hemin in the amount needed for a test run and autoclave; keep refrigerated, and do not store longer than 1 month. If media is to be prepared in advance and refrigerated, add 17 ml of agar to tubes; laked blood is added the day of the test (1 ml to each tube).

Preparation of Antimicrobial Plates (Up to 24 h Preceding Test)

(1) If media is prepared in advance, melt and cool the medium to 48 °C in a water bath and add laked blood to 5% final volume.

table 8-6. Acceptable ranges of MICs for control strains using NCCLS agar dilution reference method*

Antimicrobial Agent	*Bacteroides fragilis*[1] ATCC 25285	*Bacteroides thetaiotaomicron*[1] ATCC 29741	*Eggerthella lenta*[1] ATCC 43055	*Finegoldia magna*[2] ATCC 29328
Amoxicillin/clavulanate	0.25–1	0.5–2		
Ampicillin	16–64	16–64		0.125–0.25
Ampicillin/sulbactam	0.5–2	0.5–2	0.25–2	
Cefmetazole	8–32	32–128	4–16	
Cefoperazone	32–128	32–128	32–128	
Cefotaxime	8–32	16–64	64–256	
Cefotetan	4–16	32–128	32–128	
Cefoxitin	4–16	8–32	4–16	0.125–0.5
Ceftizoxime	NR[3]	4–16	16–64	
Ceftriaxone	32–128	64–256	NR[3]	
Chloramphenicol	2–8	4–16		2–4
Clinafloxacin	0.03–0.12	0.06–0.5	0.03–0.12	
Clindamycin	0.5–2	2–8	0.06–0.25	0.5–1
Ertapenem	0.06–0.25	0.25–1	0.5–2	
Imipenem	0.03–0.12	0.125–0.5	0.125–0.5	
Meropenem	0.06–0.25	0.125–0.5	0.125–0.5	
Metronidazole	0.25–1	0.5–2		0.25–2
Mezlocillin	16–64	8–32	8–32	
Penicillin G[4]	8–32 (16-64)	8–32 (16-64)		(0.063–0.125)
Piperacillin	2–8	8–32	8–32	
Piperacillin/tazobactam	0.12/4–0.5/4	4/4–16/4	4/4–16/4	
Tetracycline	0.12–0.5	8–32		0.25–1
Ticarcillin	16–64	16–64	16–64	
Ticarcillin/clavulanate	NR[3]	0.5/2–2/2	16/2–64/2	
Trovafloxacin	0.12–0.5	0.25–1	0.25–1	

* Permission to use excerpts of data from NCCLS publication M11-A5 (Methods for Antimicrobial Susceptibility Testing of Anaerobic Bacteria, Approved Standard - Fifth Edition) Tables 1-5. The interpretive data are valid only if the methodology in M11-A5 is followed. NCCLS frequently updates the interpretive tables through new additions of the standard and supplement. Users should refer to the most recent addition. The current standard may be obtained from NCCLS, 940 West Valley Road, Suite 1400, Wayne, PA 19087, U.S.A.

[1] NCCLS-approved control strains.

[2] *Finegoldia* (*Peptostreptococcus*) *magna* is not an NCCLS-approved quality control strain. These figures were obtained from data at the Wadsworth Anaerobe Laboratory using the Wadsworth method and may be useful for laboratories needing quality control at low concentrations.

[3] NR—no MIC range recommended.

[4] Penicillin values are in µg/ml (units/ml in parentheses).

table 8-7. Acceptable ranges of Minimum Inhibitory Concentrations (MICs) (μg/mL) for control strains for broth microdilution testing of anaerobes*

Antimicrobial Agent	*Bacteroides fragilis* ATCC 25285	*Bacteroides thetaiotaomicron* ATCC 29741	*Eggerthella lenta* ATCC 43055
Ampicillin/sulbactam (2:1)	0.5/0.25–2/1	0.5/0.25–2/1	0.5/0.25–2/1
Cefotetan	1–8	16–128	16–64
Cefoxitin	2–8	8–64	2–16
Ceftizoxime	NR[a]	NR[a]	8–32
Clindamycin	0.5–2	2–8	0.06–0.25
Ertapenem	0.06–0.5	0.5–2	0.5–4
Imipenem	0.03–0.25	0.25–1	0.25–2
Metronidazole	0.25–2	0.5–4	0.125–0.5
Piperacillin	4–16	8–64	8–32
Piperacillin/tazobactam	0.03–0.25	2/4–16/4	8/4–32/4
Ticarcillin/clavulanate	0.06–0.5	0.5/2–2/2	8/2–32/2
Trovafloxacin	0.125–0.5	0.5–2	0.25–2

* Permission to use excerpts of data from NCCLS publication M11-A5 (Methods for Antimicrobial Susceptibility Testing of Anaerobic Bacteria; Approved Standard—Fifth Edition) has been granted by NCCLS. The interpretive data are valid only if the methodology in M11-A5 is followed. NCCLS frequently updates the interpretive tables through new editions of the standard and supplements. Users should refer to the most recent edition. The current standard may be obtained from NCCLS, 940 West Valley Road, Suite 1400, Wayne, PA 19087, U.S.A.

NOTE: For four-dilution ranges, results at the extremes of the acceptable range(s) should be suspect. Verify control validity with data from other control strains.

[a] NR indicates that no MIC range is recommended with this organism/antibiotic combination.

8

(2) If media is prepared on the test day, autoclave and cool to 48 °C in a water bath and add laked blood to 5% final volume.

(3) Prepare dilutions of antimicrobial agents (see Table 8-5).

(4) Incorporate dilutions into supplemented Brucella laked blood agar. Add 2.0 ml of each antimicrobial dilution to a tube containing 18 ml of molten agar with laked blood added (~48 °C) to make a 1:10 dilution. Mix thoroughly by gently inverting the tube six times and pour into a Petri dish.

(5) After plates have solidified, place them in a 35–37° incubator with the top of the plates slightly ajar for 30–45 min to allow evaporation of excess moisture. Plates may be refrigerated if the test is to be performed the next day (except plates containing imipenem or clavulanate).

Performing the Test Run

(1) If plates are prepared the previous day, remove the plates from the refrigerator and allow to warm to room temperature.

(2) Add the individual inocula (diluted to a 0.5 McFarland density) to the wells of a replicator device (such as a Steers replicator manufactured by CMI Promex, Inc., Pedricktown, NJ).
(3) Stamp the inocula on the antimicrobial-containing plates, beginning with the lowest concentration of antimicrobial agent and proceeding to the highest. Stamp the bacteroiostatic antimicrobial agents first (e.g., macrolides, chloramphenicol).

Note: If several agents known to have potent antianaerobe activity are to be tested, it is best to stamp from different inoculum blocks to avoid any antimicrobial carry-over effect.

(4) At the beginning and end of each set of antibiotic plates, stamp two plates with no antibiotic; incubate one anaerobically (growth control) and one aerobically (aerobic contaminant control).
(5) Leave the plates at room temperature until the inoculum is absorbed (approximately 10 min).
(6) Stack the test plates and anaerobic growth control and incubate upside down (to prevent condensate from falling on the inoculated spots) in an anaerobic atmosphere at 35–37° for 48 h. Incubate the aerobic contaminant control aerobically at 35–37° for 48 h.
(7) Determine the MIC (see below) and record results.

Reading Endpoints

The MIC is defined as the lowest concentration of drug yielding no growth, a haze, one discrete colony, or multiple tiny colonies (Figure 8-2). See "General Considerations," above, for a detailed discussion of problems related to endpoint determination. In large research evaluations of antimicrobial agents, difficult endpoint readings should be especially noted, since they may cause considerable discrepancies in results. Breakpoints approved by the National Committee for Clinical Laboratory Standards (NCCLS) are listed in Table 8-8.

With some organism and antibiotic combinations, endpoints trail and are difficult to read. In recent studies, we have investigated the use of tetrazolium chloride (TTC, a dye that is reduced by many viable bacteria to a red compound, formazan) as an aid in endpoint determination. We found that endpoints determined with the aid of TTC correlated with the viability endpoints measured by determining the concentration of

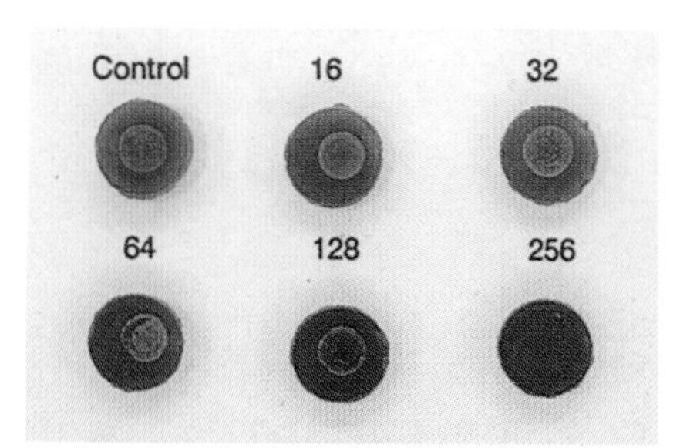

Figure 8-2. Illustration of an endpoint that is difficult to determine. *B. distasonis* on agar dilution plates containing concentrations of ceftizoxime (µg/ml) as indicated. Note that photography exaggerates the appearance of the "ahaze" seen at the higher concentrations.

table 8-8. Interpretive categories and correlative Minimum Inhibitory Concentrations (MICS)[a]

	MIC (μg/ml)		
Antimicrobial Agent	**Susceptible**	**Intermediate[a]**	**Resistant**
Amoxicillin/clavulanate[b]	≤ 4/2	8/4	≥ 16/8
Ampicillin[c]	≤ 0.5[c]	1[c]	≥ 2[c]
Ampicillin/sulbactam[b, d]	≤ 8/4	16/8	≥ 32/16
Cefmetazole	≤ 16	32	≥ 64
Cefoperazone	≤ 16	32	≥ 64
Cefotaxime	≤ 16	32	≥ 64
Cefotetan[b]	≤ 16	32	≥ 64
Cefoxitin[b, d]	≤ 16	32	≥ 64
Ceftizoxime[b]	≤ 32	64	≥ 128
Ceftriaxone	≤ 16	32	≥ 64
Chloramphenicol	≤ 8	16	≥ 32
Clindamycin[d]	≤ 2	4	≥ 8
Imipenem[b]	≤ 4	8	≥ 16
Metronidazole[d]	≤ 8	16	≥ 32
Meropenem	≤ 4	8	≥ 16
Mezlocillin	≤ 32	64	≥ 128
Penicillin	≤ 0.5[c]	1[c]	≥ 2[c]
Piperacillin[b, d]	≤ 32	64	≥ 128
Piperacillin/tazobactam[b]	≤ 32/4	64/4	≥ 128/4
Tetracycline	≤ 4	8	≥ 16
Ticarcillin	≤ 32	64	≥ 128
Ticarcillin/clavulanate[b]	≤ 32/2	64/2	≥ 128/2
Trovafloxacin[b, d]	≤ 2	4	≥ 8

[a] The intermediate range was established because of the difficulty in reading end points and the clustering of MICs at breakpoint concentrations. Where data are available, the interpretive guidelines are based on pharmacokinetic data, population distribution of MICs, and studies of clinical efficacy. To achieve the best possible levels of a drug in abscesses and necrotic or poorly perfused tissues, which are encountered commonly in these infections, maximum approved dosages of antimicrobial agents are recommended for therapy of anaerobic infections. With such dosages, it is believed that organisms with susceptible or intermediate end points are generally amenable to therapy. Ancillary therapy, such as drainage procedures and debridement, are obviously of great importance for the proper management of anaerobic infections.

[b] MIC values using either *Brucella* blood agar or Wilkins-Chalgren agar (former reference medium) are considered equivalent, based upon published *in vitro* literature and a multicenter collaborative trial for these antibiotics.

[c] Members of the *Bacteroides fragilis* group are presumed to be resistant. Other gram-negative anaerobes should be screened for β-lactamase activity with a chromogenic cephalosporin, and, if positive, reported as resistant to penicillin and ampicillin. Higher blood levels are achievable; infection with non-β-lactamase producing organisms with higher MICs might be treatable.

[d] MIC values for agar or broth microdilutions are considered equivalent.

antibiotic that permitted no net growth of organisms on Steer's replicator "spots" over a 48 h period. The TTC was applied by flooding the plates with a 0.05% TTC in 1% molten agar solution (320) . We are continuing to investigate the use of viability indicator dyes in aiding reading of endpoints and determination of MIC (Figure 8-3).

Broth Microdilution Test

A number of manual, semiautomated, and automated devices are available that allow rapid and economical preparation of large numbers of plates containing two-fold dilution series of antimicrobial agents. Plastic (microtiter) trays containing 8 by 12 rows of small wells are used. The recommended medium is Brucella broth supplemented with hemin (5 μg/ml), vitamin K_1 (1 μg/ml), sodium bicarbonate (1mg/ml), and lysed horse blood (5%). There are some commercial sources for frozen or lyophilized trays. Microtech Medical Systems sell frozen trays for use with anaerobes. PML Microbiologicals will supply laboratories with custom-poured MIC panels. When using commercial trays, follow the manufacturer's directions for test performance and quality control, while ensuring that the basic principles mentioned above are maintained (adequate inoculum and final volume).

In Advance

(1) Prepare antimicrobial agent dilutions in at least 10 ml of Brucella broth with 5% lysed horse blood.

(2) Dispense 0.1 ml of the antimicrobial solutions into the microtiter plates using either a manual or automated dispensing device.

Note: The trays should not contain less than 0.1 ml per well. Volumes of less than 0.1 ml are not recommended for anaerobes because of evaporation, which can result in concentration of antimicrobial agents, and because of the marked effect of the higher inoculum size in smaller volumes.

(3) Seal the prepared trays in plastic bags and freeze at −70 °C until needed. Such trays usually remain stable for 4–6 months. **Freezers with self-defrosting units should not be used.** Appropriate quality control should be performed to monitor shelf life.

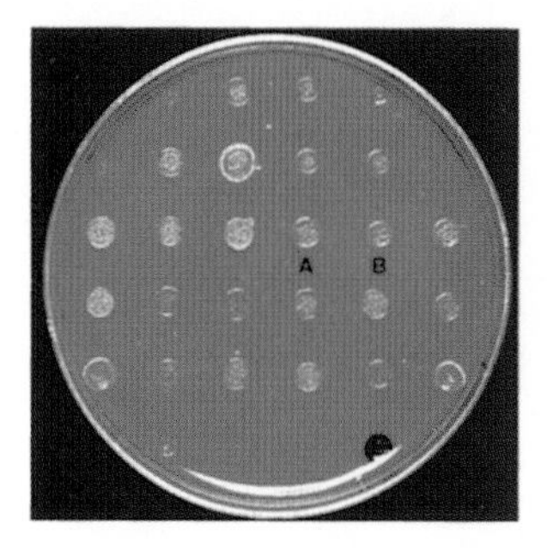

Plate 1. Growth Control.

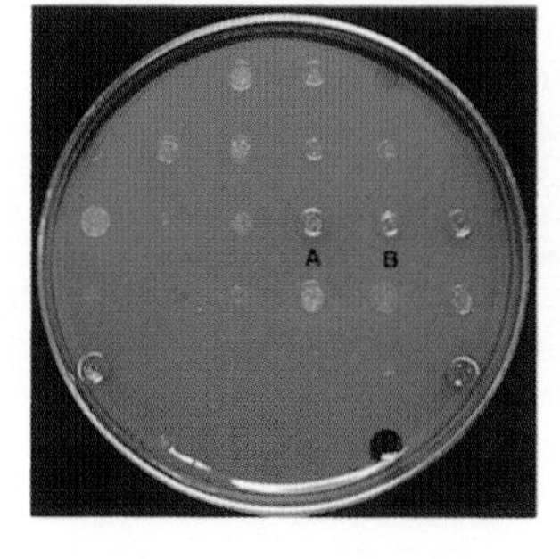

Plate 2. Antimicrobial Agent (16 μg/ml).

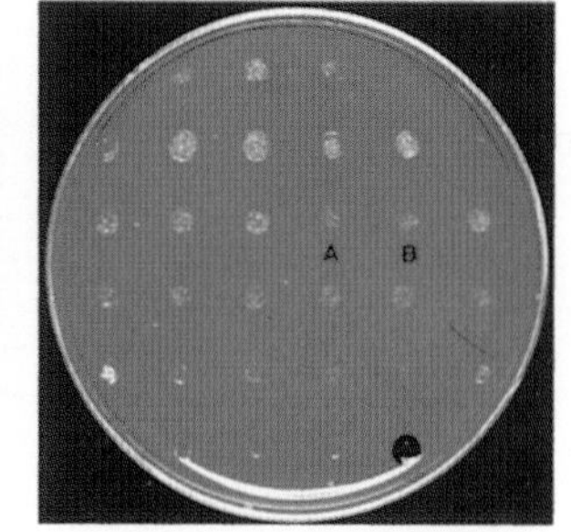

Plate 3. Antimicrobial Agent (32 μg/ml).

Figure 8-3. Agar dilution susceptibility testing. (From Finegold, Baron, and Wexler, *A Clinical Guide to Anaerobic Infections*, 1992. Courtesy Princeton Scientific Productions, Inc. and Star Publishing Co.)

8

Alternatively, frozen or lyophilized plates containing diluted antimicrobial agents to which inoculum is added may be purchased (see below). At least one well should contain broth with no drug to serve as a growth control. Another well containing broth and no drug, but not inoculated, will serve both as a sterility check and as a negative control for visual comparison with wells containing growth. A sterility check may also be performed by incubating an uninoculated tray.

Performing the Test Run

(1) Remove the trays from the freezer and allow them to thaw.

(2) Prepare the inoculum as described earlier. The final inoculum in each well should be 1×10^5 CFU (1×10^6 CFU/ml).

Note: The volume of the inoculum is generally 0.01 ml (10 μl); an inoculum adjusted to the turbidity of a 0.5 McFarland (~1.5×10^8 CFU/ml) and diluted 1:15 (i.e., ~1×10^7 CFU/ml) will result in a final inoculum of ~1×10^5 CFU/well. If diluted inoculum (usually 0.1 ml) is used to reconstitute lyophilized trays, the broth must contain 1×10^6 CFU/ml (resulting in a final inoculum of ~1×10^5 CFU/well).

(3) Inoculate trays within 15 min of inoculum preparation. If resistance to metronidazole in a gram-negative organism is observed, the plate should be reduced 2–4 h and the test repeated (the antimicrobial activity of metronidazole depends upon the formation of an active intermediate which demands a reduced atmosphere). Manufacturer's guidelines should be followed. Different commercial suppliers may give different guidelines as to whether plates should be inoculated in the chamber or on the bench.

(4) Stack trays in multiples of four or less, to allow even incubation temperature. The top tray may be covered with a plastic lid to prevent evaporation. **Cellophane tape should not be used unless the plates were inoculated inside an anaerobic chamber.**

(5) Incubate the trays in an anaerobic atmosphere for 48 h.

(6) Plates should be examined with a viewing device, such as a stand with mirrors.

(7) Read the MIC as the concentration where the most significant reduction of growth is observed. This may be a complete inhibition of growth or a tiny, gradually diminishing button of growth (Figure 8-4).

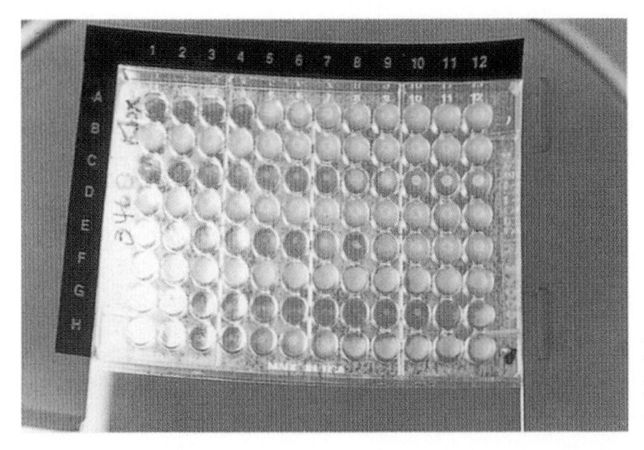

Figure 8-4. Microdilution plate. The MIC is defined as the concentration in the first well showing no button of growth.

MBC Determination

Although some workers have attempted to determine minimum bactericidal concentrations (MBCs) using microbroth trays, this cannot be done reliably and we do not recommend it.

Broth Macrodilution Tests

Broth dilution tests are not used for routine susceptibility testing. However, in selected instances this test is useful if MBCs must be determined (which may be indicated in a clinical setting such as endocarditis). It may also be used for testing organisms with swarming growth (such as some *Clostridium* spp.); these strains may also be tested with the microdilution technique. Serial two-fold dilutions of antimicrobial stock solutions are prepared in 2.5 ml of Brucella broth containing hemin (5 μg/ml), $NaHCO_3$ (1 mg/ml) and vitamin K_1 (1 μg/ml). Tubes should be prepared within 3 h of use, or else frozen at −70 °C.

The inoculum is a 1:200 dilution of a 0.5 McFarland in the supplemented Brucella broth. A final inoculum volume of 2.5 ml is added to the broth containing the drug (the final inoculum will be ~3×10^5 CFU/ml). Incubate the tubes in an anaerobic atmosphere at 37 °C for 48 h. Include an inoculated broth containing no antimicrobial agent as a growth control for each strain tested. Include a tube of uninoculated broth with each day's tests to serve as a negative growth control. Also include tests on control strains, results of which should fall within appropriate reference values. Values for broth tests will often be one dilution lower than for agar dilution tests. The MIC is read as the lowest concentration showing no visible growth.

MBCs may be determined by streaking 0.1 ml from each tube that shows no visible growth (after allowing 48 h for growth and recording MICs) to a blood agar plate. Plates are incubated anaerobically for 48 h and the MBC is read as the lowest concentration of drug resulting in fewer than 30 colonies (99.9% killing rate).

Other Methods of Susceptibility Testing

Etest

This test produced by AB Biodisk uses a plastic strip coated with an antibiotic gradient on one side and an MIC interpretive scale on the other. The strip is applied to the surface of an agar plate that has been inoculated with a 0.5 McFarland suspension of the organism, and the plate is incubated anaerobically for 24–48 h depending on the growth rate of the organism. Six Etest strips may be placed in a radial fashion on a 150 mm plate. The point at which the teardrop-shaped zone of inhibition intersects the interpretive scale is read as the MIC (Figure 8-5). These strips have been evaluated by numerous investigators and one of us has studied their use for anaerobic susceptibility testing of 105 strains with 98% categorical agreement with the agar dilution reference method (71). The Etest has also been used satisfactorily to determine the susceptibility of *Prevotella intermedia* to metronidazole (262) as well as organisms associated with bacterial vaginosis (82); results from one study with *C. difficile* did not show satisfactory correlation with standard techniques (376). Others have reported generally good results, but caution that some organism and antimicrobial combinations are prone to a significant number of very major errors (280). This test is particularly suited for studying individual patient isolates against a few se-

Figure 8-5. Example of the Etest for determining the susceptibility of an anaerobic bacillus to six antimicrobial agents. The MIC is read from the point of the strip at which the teardrop-shaped zone of inhibition ends.

lected agents; testing large numbers of isolates or a wide range of antimicrobials would not be accomplished efficiently with this method. The method is expensive.

Spiral Gradient Endpoint System

The spiral streaker (Spiral Systems Instruments) deposits the antimicrobial stock solution in a spiral pattern on an agar plate, resulting in a radially decreasing concentration gradient. The isolates are deposited on the plate using an automated inoculator in a radial pattern. MICs are determined by measuring the distance from the center of the plate to the point where growth stops (Figure 8-6). A computer program translates this number into the MIC. Details of the procedure may be found in the manufacturer's guidelines. A multilaboratory collaborative evaluation found good correlation with the agar dilution technique, as we have in our own tests (358). The spiral gradient system is more precise than the two-fold serial agar dilution susceptibility test. This method is especially suited for batch testing, and would not be particularly efficient for testing single isolates against a variety of antimicrobials.

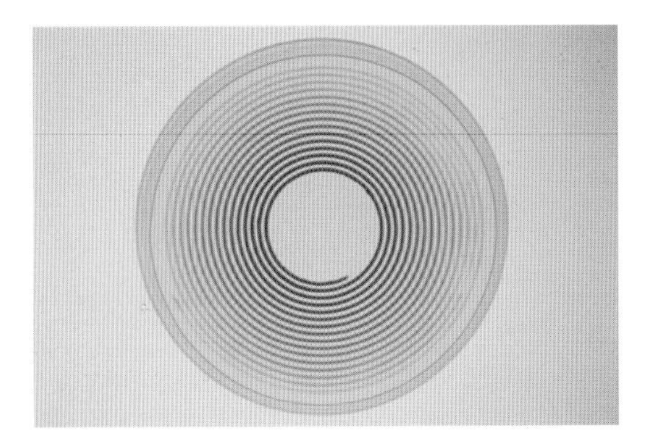

Figure 8-6a. A spiral gradient on the surface of a Petri plate formed with crystal violet dye to illustrate the antibiotic gradient formed by a spiral plater (photograph courtesy of Spiral Systems Instruments, Bethesda, MD).

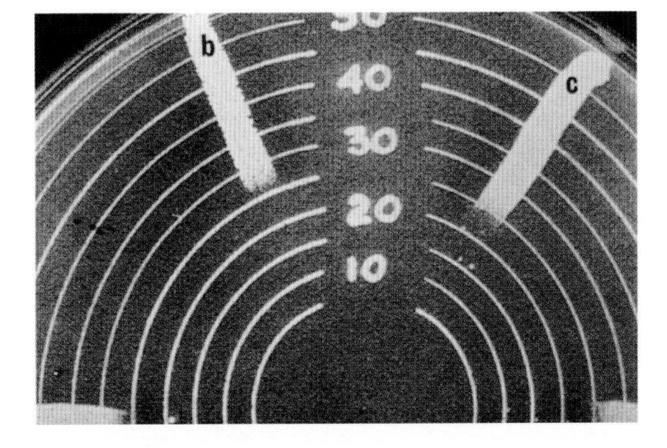

Figure 8-6b. Organisms inoculated in radial lines on an agar plate containing an antibiotic gradient created by the spiral plater. The plate is viewed through the Spiral Gradient Endpoint (SGE) template for measurement of the endpoint location. Spiral gradient endpoints are measured as the distance along the radius from the center of the plate to the point of growth inhibition (photograph courtesy of Spiral Systems Instruments).

table 8-9. β-lactamase producing anaerobes

Bacteroides fragilis group organisms
Prevotella species
Bilophila wadsworthia
Fusobacterium species
Clostridium species

β-Lactamase Testing

The β-lactamase test is simple and rapid and may be used as a supplement to conventional susceptibility tests. If a strain is β-lactamase-positive, it will probably be resistant to the β-lactamase-labile agents (e.g., many penicillins and cephalosporins) and susceptible to the β-lactam agents when combined with β-lactamase inhibitors such as sulbactam, tazobactam, or clavulanic acid. However, if a strain does not produce β-lactamase, one cannot assume that it will be susceptible to the β-lactam agents. Many strains of *Bacteroides distasonis* do not produce β-lactamase, yet are quite resistant to β-lactam agents. β-lactamase-producing anaerobes are listed in Table 8-9.

The most reliable method for detecting β-lactamases (specifically cephalosporinases) in anaerobes utilizes nitrocefin disks (Cefinase; BBL). The disk should be moistened with sterile water, and then several well-isolated colonies should be smeared on the disk with a sterile loop or stick. A positive reaction is indicated by a color change from yellow to red. Most reactions occur within 5–10 min but there are some β-lactamase-positive strains that react more slowly (up to 30 min). Tests should not be incubated beyond 30 min because nonspecific degradation of substrate may result in a false positive result. *Bilophila wadsworthia* requires the addition of pyruvate to the growth medium for reliable β-lactamase testing results (taurine is not adequate, even though it supports luxuriant growth of *B. wadsworthia*) (317).

appendix A

Special Procedures for Specimen Collection and Research Studies

Laboratory Handling of Oral Site Specimens to Avoid Normal Oral Flora

Principle

Saliva and normal mucosal surfaces of the oral cavity contain anaerobic bacteria, streptococci, and other normal flora in large numbers. Certain of these bacteria contribute to pathologic processes when they are able to colonize a different environmental niche aided by local disease, trauma, poor oral hygiene, or other infectious processes. The clinical significance of any bacteria recovered from specific oral sites can be assessed only when contamination by the normal oral flora has been avoided.

Numerous strategies for collection of site-specific specimens from within the oral cavity have been developed (120;203;233;272). Transcutaneous aspiration through iodine-disinfected, uninvolved skin is still the most desirable collection method for obtaining specimens from oral infections pointing toward the skin surface. Also, infected tissue taken during surgery and transported in a loosely capped container in an anaerobic pouch provides a proper specimen (272). In severe complications of oral-associated infections, such as deep-seated head and neck and mediastinal suppurations and Lemierre's syndrome (postanginal sepsis syndrome), blood cultures must be taken along with focal sites. Several collection methods are presented in this section along with a general scheme for processing oral site specimens for quantitative anaerobic bacteriology. It should be remembered that oral infections originating from periodontal pathology often contain the key pathogenic bacteria *Porphyromonas gingivalis, Actinobacillus actinomycetemcomitans*, and *B. forsythus* along with other oral bacteria. Periapical and mucosal abscesses secondary to infected root canals and pulpal pathology often include gram-positive, nonsporeforming bacteria such as the newly described *Eubacterium*-like species (88) and *M. micros* together with the *S. anginosus*–group streptococci (189;272). In many cases, culture severely underestimates the presence of fastidious organisms, such as *P. endodontalis* and *P. dentalis*, and naturally nonculturable phylotypes that are demonstrated only by 16S rRNA sequencing (352). After mechanical stress, infection is the second major cause of dental implant failures. The bacterial findings of periimplantitis often resemble those of oral abscesses or periodontitis, depending on the pre- or postimplantation health status and indigenous microbial composition of the oral cavity and periodontium (169;225;333). Timely diagnosis and therapeutical intervention may prevent failures that not only are expensive but cause major masticatory disability and patient discomfort for long periods.

Organisms in gingival sites are present as microcolonies on the solid surfaces, often held together by complex polysaccharide matrices (glucans, etc.) forming a biofilm; these aggregates must be dispersed to obtain reliable quantitative results. Sonication or physical disruption using a Vortex mixer and glass beads (anaerobically) are usually employed. The sensitivity to sonication of oral bacteria vary; the viability of some fastidious species may drop. For transport of all oral specimens, Hungate-stoppered tubes or vials containing prereduced, sterile anaerobic diluent (e.g., one-quarter strength Ringer's solution with 0.1% sodium metaphosphate) in 0.5–2.0 ml volumes with six to eight glass beads (1.0–1.5 mm in diameter) may be used. Another transport system, VMGA III (183;236), utilizes sterile semisolid anaerobic transport medium (2 ml screwcap tube + glass beads) filled almost to the brim of the transport tube (Figure A-1). All specimens should be transported as soon as possible after collection. VMGA III will allow overnight transportation from distant centers, but longer transit will permit overgrowth by enteric rods (8).

Saliva Collection

(1) Instruct the patient to pass saliva into the sterile tube, either after paraffin stimulation or without stimulation. A minimum of 2.0 ml is usually collected (under timer control to determine the salivary flow rate, which relates to quantity of findings). From small children, collect the drool or saliva sublingually by a blunt-tipped plastic pipette.

(2) Immediately draw the saliva into a syringe or a blunt-tipped plastic Pasteur pipette.

(3) Inject the saliva sample or expel the pipette contents into an anaerobic transport tube.

(4) Deliver to the laboratory for immediate processing. If not able to process immediately, collect in cryotubes and freeze at −70° C or lower until processed.

Supragingival Plaque Specimen Collection

(1) Isolate the area to be sampled with cotton rolls and dry the sample site with sterile cotton swabs.

(2) Collect a sample of supragingival or coronal plaque with the curette or scaler.

(3) Open the cap of the transport tube and immediately insert the scaler or curette tip into the transport medium and swirl a few times to detach the plaque. **Do not open transport tube until specimen is ready to be inserted. Minimize exposure to air!** Close the tube immediately. If a gassing cannula is available, maintain continuous flow of oxygen-free gas into the tube.

(4) Transport the tube to the laboratory for processing as soon as possible.

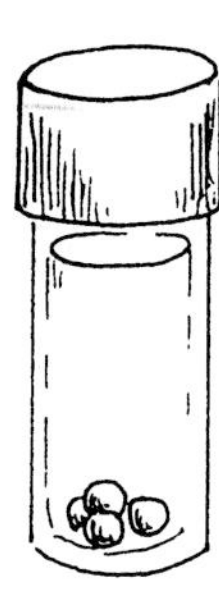

Figure A-1. VMGA III transport vial for dental specimens. Note glass beads and medium filled almost to the top of the vial.

A

Subgingival Plaque Specimen Collection

(1) Isolate the sample site with cotton rolls and dry the area with sterile cotton swabs.
(2) Remove supragingival plaque with a sterile scaler.

Using the Sterile Scaler

(3) Obtain subgingival plaque by inserting scaler or curette in the pocket and gently rubbing against the root surface.
(4) Immediately transfer the scaler tip into the anaerobic transport vial and swirl a few times to detach the plaque. If a gassing cannula is available, maintain continuous flow of oxygen-free gas into the tube when the top is open.
(5) Tighten the cap and deliver to the laboratory for processing. VMGA III medium is used if subgingival samples are sent to a distant oral microbiology diagnostic laboratory for screening for periodontitis-associated pathogens. Viability of these organisms is good if delivered within 24 h (8).

Using the Paper Point Method

(3) Retract the gingival margin and insert three paper points to the depth of the sulcus and hold for 10–60 s.
(4) Without delay, transfer the paper points into the anaerobic transport vial. If a gassing cannula is available, maintain continuous flow of oxygen-free gas into the tube when the top is open.
(5) Tighten the cap and deliver to the laboratory for processing as above.

Note: Significant differences in the sampling efficiency of paper points by different manufacturers or with different size strips may occur (142).

Using a Barbed Broach

(3) Retract the gingival margin and insert the needle to the depth of the sulcus.
(4) Pass the barbed broach through the needle.
(5) Obtain specimen and retract the broach back into the needle.
(6) Without delay, insert the needle containing the broach through the septum of the transport vial and extend the broach into the medium.
(7) Remove the needle, leaving the broach inside the vial.
(8) Deliver to the laboratory for processing.

Periimplant Plaque Specimen Collection (169)

Failed Implants

(1) Dismantle the superficial prosthetic structures and crown.
(2) Minimize the salivary contamination by chlorhexidine mouth rinse.
(3) Isolate the sample site with cotton rolls and dry the area with sterile cotton swabs and prevent salivary contamination by frequent suction.
(4) Explant the fixture (screw).

(5) Place the fixture in VMGA III medium and transport to the laboratory.

Note: The fixture is a 100 to 1,000× more effective sampling device compared to scoop samples from the implant socket since the threads of the fixture retain granulation tissue material.

Suspected Periimplantitis

If using the paper point method, proceed as with a subgingival sample. If using the scaler method (**nonabrasive, such as plastic, scalers only**), proceed as with a subgingival sample.

Processing Oral Specimens in the Laboratory

Specimen Processing

(1) Perform all manipulations of the specimen in an anaerobic chamber if possible. Quantitative studies of oral flora cannot be performed adequately in room air.

(2) If the specimen is to be sonicated (VMG III transport system), place the entire transport vial into the sonicator for 30 s before manipulating the sample in the chamber.

(3) If the transport medium contains glass beads for dispersal of the specimen, mix the vial or tube vigorously on a Vortex mixer for a minimum of 20 s before further manipulations in the chamber.

(4) Make serial 10-fold dilutions in PRAS Ringer's solution or other appropriate broth to 10^{-6}. These dilutions should allow enumeration of all isolates on at least one set of plates. Metaphosphate (0.1%) may be added to the first dilution tube to prevent reaggregation of the organisms.

(5) Spread 0.1 ml of inoculum from each dilution onto the surface of each agar medium being used for the study. Table A-1 contains suggested media. Many selective agars for isolation of dental organisms are available from Anaerobe Systems. Inoculum can be spread using a pipette to run the suspension slowly onto the surface as the plate is rotated mechanically, or manually using a sterile wire or glass spreader, or with the aid of the spiral streaker instrument.

(6) Prepare a Gram stain, semiquantitate, and describe all morphologies.

(7) Prepare a wet mount from warm specimen (some advocate separate sample in dilute 1% gelatin) and examine under darkfield microscopy; describe all bacterial morphologies and types of motility and count the relative share of major morphological groups.

Incubation and Examination

Table A-1 lists suggested incubation conditions and periods for different media used. It also lists the organisms isolated on different selective media. Descriptions of typical colonies on selective agars are found at the end of this appendix. For identification of anaerobic organisms refer to Chapters 4, 5, and 6; the facultative and aerobic organisms may be identified as described in the Manual of Clinical Microbiology (242) and Diagnostic Microbiology (17) update.

table A-1. Suggested media for oral flora studies[1]

Medium	Incubation atmosphere	Dilutions plated [2]	Days of incubation	Purpose
Brucella PRAS blood agar	Anaerobic	10^{-3}, 10^{-4}, 10^{-5}, 10^{-6}	4–7	Total counts and predominant flora, *Porphyromonas gingivalis*
TSBV agar	Air + 5–7% CO_2 & Anaerobic	Undiluted, 10^{-1}, 10^{-2}	2–7	*Actinobacillus actinomycetemcomitans*
KVLB–2 agar	Anaerobic	Undiluted, 10^{-1}, 10^{-2}	4–7	*Bacteroides,* Pigmented *Prevotella,* some *Porphyromonas* spp.
JVN or NV agar	Anaerobic	Undiluted, 10^{-1}, 10^{-2}	4–7	*Fusobacterium* and *Leptotrichia*
Lactobacillus selective medium or MRS agar	Anaerobic	10^{-1}, 10^{-2}, 10^{-4}	2–5	*Lactobacillus Bifidobacterium*
Veillonella neomycin agar	Anaerobic	10^{-1}, 10^{-2}, 10^{-4}	2–7	*Veillonella* and other gram-negative cocci
CFAT agar	Air + 5–7% CO_2 & Anaerobic	10^{-1}, 10^{-2}, 10^{-4}	4–7	*Actinomyces*
Chocolate bacitracin agar	Air + 5–7% CO_2	10^{-1}, 10^{-2}, 10^{-4}	2–5	*Haemophilus, Campylobacter,* and other gram-negative capnophiles
Mitis-Salivarius agar	Air + 5–7% CO_2	10^{-2}, 10^{-4}, 10^{-6}	2–5	Alpha- and nonhemolytic streptococci, mutans streptococci
Brucella or trypticase soy blood agar	Air + 5–7% CO_2	10^{-2}, 10^{-4}, 10^{-6}	1–2	Total counts and predominant capnophilic flora

[1] See end of Appendix A for interpretive information on media.

[2] 0.1 ml of each dilution is placed on each medium used.

ANAEROBES IN BITE WOUND INFECTIONS

Animal and human bite wounds provide a challenge for the microbiology laboratory. Such wounds typically harbor oral flora of the biter which is complex and frequently contains anaerobic bacteria. It is important to isolate and report the presence of anaerobes in these infections because the choice of antimicrobial therapy should be based on the susceptibilities of the organisms and inadequate therapy can lead to serious sequelae.

Dog bites may include a crush injury (if the dog is large and has powerful jaws), tearing of tissues, or puncture wounds. Coupled with the introduction of dog oral flora, more than half of such wound infections harbor anaerobic bacteria. Cat bites are frequently deep puncture wounds or scratches, and two-thirds contain anaerobes (330). These infections can be very serious if they occur on the hands or fingers in the proximity of tendons and bones, and if not adequately treated, may result in permanent disability. Unusual anaerobic species are typically cultured from these wounds (330), including *Prevotella heparinolytica, Bacteroides pyogenes, Bacteroides tectus,* and animal species of *Porphyromonas.* Since these organisms are not included in the data bases of the rapid identification systems used in most laboratories, the microbiologist should suspect their presence and use other methods for identification (5;70;119;156). *B. pyogenes* and *Bacteroides tectus* resemble *P. bivia* in the RapID ANA II system, while *P. heparinolytica* resembles *B. uniformis.* Most of the animal strains of *Porphyromonas* are identified as *P. gingivalis* because of a negative test for α-fucosidase. The characteristics of the animal strains of *Porphyromonas* and *P. heparinolytica* are included in Chapters 4, 5, and 6.

Human bites include direct bites and also clenched fist injuries sustained by the hitter during a fight. If the clenched fist contacts the mouth and teeth of the victim, oral flora is introduced into the wound sustained to the knuckles, fingers, or other parts of the hand. The wound infections frequently contain organisms such as *Eikenella corrodens* and a plethora of anaerobes. *Prevotella* species isolated from human bites are more likely to be β-lactamase producers than strains isolated from dog and cat bite wounds, but the opposite holds true for *Porphyromonas* species. Table A-2 summarizes the organisms isolated from human, cat, and dog bite wounds.

RESPIRATORY TRACT SPECIMENS SUITABLE FOR ANAEROBIC CULTURE

Principle

Anaerobic bacteria are often involved in infections of the lung and pleural space, particularly in patients who have aspirated oral or gastric secretions or who have significant periodontal disease or xerostomia (113;202;338). Because normal oral secretions contain large numbers of the same organisms that may become pathogens when they reach the normally sterile alveoli, specimens for definitive bacteriologic evaluation must be collected so as to bypass oral flora. Only two methods were well established previously: percutaneous transtracheal aspiration and the use of protected bronchial brush in a double-lumen catheter (53;62;113). However, it is likely that bronchoalveolar lavage (BAL) specimens are also suitable. Transtracheal aspiration is seldom done presently and many of the specimens obtained, in fact, have been contaminated by oropharyngeal flora (73). Performed correctly, protected bronchial brush catheter (PBC) culture collection is well tolerated and should take less than 2 min (53). Both PBC and BAL specimens must be cultured quantitatively.

Respiratory specimens collected via the upper airway and thus contaminated with saliva and other oral secretions (such as sputum and endotracheal suction specimens) may also be evaluated for the presence of nonanaerobic organisms known to be pathogens: *Staphylococcus aureus, Streptococcus pneumoniae, Haemophilus influenzae, Pseudomonas, Moraxella (Branhamella) catarrhalis,* and *Enterobacteriaceae,* **all of which may simply be colonizing the respiratory tract and not be the etiologic agent of pneumonia.** Accordingly, a culture report indicating the recovery of any of these organisms does not establish the diagnosis. In cases of suspected legionellosis or hospital-acquired pneumonia, the sample must be selectively cultured also for *Legionella* species, and a positive culture is diagnostic. Thus, respiratory cultures should be evaluated in relationship to the clinical illness of the patient.

table A-2. Anaerobic bacteria isolated from infected human, dog and cat bite wounds

	% of patient specimens positive			
			Human	
Isolate	Dog bite	Cat bite	CFI[a]	bite
Fusobacterium spp.	32	33	19	6
F. nucleatum	16	25	19	6
F. russii	2	14		
Other species	16	9		
Prevotella spp.	28	19		
P. melaninogenica	2	2	13	17
P. intermedia	8		31	22
P. buccae			19	17
P. heparinolytica	14	9		
P. zoogleoformans	4	2		
Other species	2	7	32	33
Porphyromonas spp.	28	30		
P. macacae	6	7		
P. cansulci	6	2		
P. gingivalis (P. gulae)	4	11		
P. canoris	4	9		
P. cangingivalis	4	4		
P. circumdentaria	2	5		
P. levii-like organisms	2			
Other species	4			
Bacteroides spp.	30	28		
B. tectus/ B. pyogenes	16	32		
B. forsythus	4			
B. ureolyticus	4		6	
Campylobacter gracilis	4			
Peptostreptococcus spp.	16	5	38	6
Veillonella spp.		2	25	6
Propionibacterium spp.	20	18		6
Filifactor villosus		5		
Eubacterium spp.	4	2	6	

[a] Clenched fist injury

Certain etiologic agents, such as viral agents, are not easily detected in respiratory secretions. For diagnosis of disease due to these agents, the protected bronchial brush catheter (PBC) is recommended (62). Patients who are unable to cooperate with medical personnel, or who have significant hypoxia, unstable cardiac conditions, or untreated asthma may be considered poor candidates for PBC or BAL cultures, which are typically done during bronchoscopy.

Processing the PBC Specimen

Note: Proper handling is essential to achieving meaningful results.

(1) Vigorously wipe the outer surface of the inner cannula (which is extending from the outer cannula) with 70% alcohol.
(2) Using sterile scissors, cut off the distal end of the inner cannula (right up to the tip of the brush). This cannula tip should be discarded.
(3) Then extend the brush through the cut end of the cannula and aseptically cut it off so that it falls into a tube containing transport medium. If only one sterile scissors is available, it should be wiped with alcohol before it is used to cut off the brush. The cap of the tube should be tightened securely.
(4) Hold the tube of transport medium containing the brush at room temperature and deliver to the laboratory immediately.

Laboratory Processing of the PBC or BAL Specimen

Media

Prepare two tubes of thioglycolate broth, 9.9 ml each, and agar plates for subculture, including three BAP, three KVLB, three PEA, or other anaerobic plates used. The aerobic plates, chocolate, BAP, and MacConkey, and *Legionella* media when indicated, are optionally placed in the chamber or on the bench.

Specimen Processing

(1) Initial processing should be performed in the anaerobic chamber. If the specimen is to be sonicated (VMGA II transport system), place the entire transport vial into the sonicator for 30 s before manipulating the sample in the chamber.
(2) If the transport tube contains glass beads for dispersal of the specimen, mix the vial or tube vigorously on a Vortex mixer in the chamber for a minimum of 20 s (30 s is recommended) before further manipulations.
(3) Use a 1.0 ml pipette and draw up enough of the specimen suspension to place 0.1 ml on each agar plate being inoculated, 0.1 ml into the first of the two 9.9 ml thiotubes (to make a 1:100 dilution of the original specimen), and a large drop on each necessary slide. Label the plates "1:10." Label the thioglycolate "1:100."
(4) Slides should always be prepared for Gram stain.
(5) Spread the inoculum on the plates using a pipette to run the suspension slowly onto the surface as the plate is rotated mechanically or manually using a sterile wire or glass spreader or with the aid of the spiral streaker instrument.

A

(6) Mix the 1:100 dilution thioglycolate tube thoroughly (invert or vortex) and use a fresh, sterile 1.0 ml pipette to repeat the procedure carried out on the original specimen. These plates, prepared from 0.1 ml of the 1:100 dilution should be labeled "1:10^3." The second 9.9 ml thioglycolate tube should be labeled "1:10^5."

(7) Streak the second set of plates as before.

(8) Thoroughly mix the second thioglycolate tube, and use a fresh, sterile pipette to place 0.1 ml on each of one final set of agar plates. Label these plates "1:10^7." This is the final dilution.

(9) Incubate the anaerobic plates in the chamber. Pull all other plates, the thioglycolate broths, and slides out of the chamber for appropriate handling. Alternatively, all nonanaerobic media and slides may be prepared on the bench from the thioglycolate tubes after the anaerobic plates have been inoculated in the chamber.

Interpretation of Results

(1) Hold anaerobic plates for 7–14 d before discarding as negative. Hold other plates as for routine respiratory specimens. Examine thioglycolate broth and subculture only if there is no growth on any agar plates or if Gram stain suggests an organism not recovered on the plates.

(2) The number of CFU in the original specimen is determined by counting colonies and multiplying by the dilution factor. (Example: If 24 colonies of staphylococci are present on the plates labeled "1:10^3," then the final CFU will be 2.4×10^4/ml original specimen.) Since the PBC brush is expected to hold only 0.01 ml, this amount of growth corresponds to 2.4×10^6 CFU/ml in the patient's secretions. For another style of PBC brush that holds only 0.001 ml, the 24 colonies would correspond to 2.4×10^7 CFU/ml. CFU greater than 1000/ml in the specimen (and thus $>10^5$/ml in the secretions) is considered to be significant. Count only the plates containing countable colonies (30–300). Do complete ID and susceptibilities on all organisms present in >1,000/ml in the original specimen.

(3) All colony morphotypes in less than significant numbers should be identified morphologically only.

COLLECTION OF ENDOMETRIAL SPECIMENS USING THE PIPELLE PROTECTED SUCTION CURETTE

Principle

The best diagnostic specimen for determining the microbial etiology of endometrial infections is a sterile tissue specimen obtained with minimal or no contamination with vaginal flora and avoiding atmospheric oxygen (to allow recovery of anaerobic bacteria) (96). If the specimen is contaminated from endogenous sources, the culture results may be uninterpretable. Because the infecting agents are usually found within tissue, a routinely collected swab specimen is almost never adequate for obtaining relevant microbiological data (219).

Specimen Collection

(1) Collect the specimen as described by the manufacturer.

(2) Remove the screwcap on the transport tube, aseptically cut off the distal tip of the Pipelle sheath, just behind the hole, and expel the collected material into the vial.

Note: To depress the plunger of the piston, you must feed it into the sheath by holding it close to the sheath opening and pushing it in 1 in at a time.

(3) Immediately replace the screwcap and close tight.

(4) Maintain at room temperature. Transport to the laboratory within 2 h of collection.

Specimen Processing

(1) Process the specimen in the anaerobic chamber if possible. If no chamber is available, work quickly on the bench top within a safety cabinet, as aerosols are possible.

(2) Use a wide-tipped sterile pipette to transfer the endometrial tissue into a sterile tissue grinder. Add thioglycolate broth if necessary to add volume. Alternatively, use a Stomacher (Tekmar Co.) to homogenize the material.

(3) Homogenize or grind the specimen to a smooth consistency. Use a pipette to place a drop or two on the surface of each agar plate and into thioglycolate enrichment broth. Suggested media include anaerobic blood agar, PEA, BBE, and KVLB agar plates for anaerobic isolation and 5% sheep blood, MacConkey, Thayer-Martin or Martin-Lewis, chocolate, and Columbia with colistin and nalidixic acid (CNA) agar plates for aerobic isolation. Mycoplasma medium (Shepard's A7 agar) should also be inoculated.

(4) Process additional specimen for additional studies as ordered (*Chlamydia*, mycobacteria, etc.)

(5) Spread a drop onto a slide as for a peripheral blood differential, with some very thin areas. Allow to air-dry, fix with methanol for 1 min, and Gram stain.

(6) Incubate the anaerobic plates for 7 d before discarding as negative. Hold other plates and process as for routine wound cultures from the genital tract (look for *Neisseria gonorrhoeae*). Examine thioglycolate broth and subculture only if there is no growth on primary plates.

(7) Identify all colony morphotypes and perform susceptibility studies as appropriate. (See "Vaginal Flora Procedure," below.)

VAGINAL FLORA

Routine cultures of vaginal secretions for identification of the predominant microflora are not useful clinically. For diagnosis of bacterial vaginosis, a Gram stain is more reliable than culture (16;188;220;249), even though anaerobic bacteria are involved in this syndrome. Quantitative bacteriological studies of vaginal secretion are important, however, for certain investigational studies (106;364). We recommend two methods, described below.

A

Vaginal Wash (311)

(1) In an anaerobic chamber, add a suspension of reduced saline containing 0.02% dithiothreitol (Sigma) and 0.0001% resazurin to sterile 7 ml red-top Vacutainer tubes (Becton-Dickinson).
(2) Wipe the Vacutainer tube stopper with 70% alcohol and withdraw 3.5 ml of saline into a 5 ml syringe through a large-bore needle.
(3) Obtain the specimen from the patient during speculum-assisted examination. Use only warm water to lubricate the speculum.
(4) Inject the saline into the vagina against the walls (needle removed from syringe). Use a cotton swab to sweep vaginal secretions into the saline pool in the posterior fornix, avoiding the cervix.
(5) Use the same syringe attached to a sterile pipette to re-aspirate the washings.
(6) Expel all air from the syringe, attach a 20 gauge needle, and inject entire contents into an empty, prereduced red-top tube.
(7) In the anaerobic chamber, prepare serial dilutions of the saline suspension for plating onto appropriate media.
(8) Remaining saline suspension may be used for short-chain fatty acid analysis.

Calibrated Loop (196)

(1) In an anaerobic chamber, prepare screwcap tubes containing 9.9 ml of sterile, prereduced brain heart infusion broth.
(2) Obtain the specimen from the patient during speculum-assisted examination. Use only warm water to lubricate the speculum.
(3) Use a 0.01 ml calibrated loop (platinum or plastic disposable) to collect a sample from the pooled secretions in the posterior fornix of the vagina. Insert the loop as vertically as possible.
(4) Maintain the brain heart infusion tube under a constant flow of oxygen-free gas (nitrogen is best), open the cap, and quickly inoculate the broth with the loopful of organisms. Immediately retighten the cap.
(5) In the anaerobic chamber vortex and then prepare serial dilutions of the broth suspension for plating onto appropriate media.

A

Processing Vaginal Specimens

However the vaginal secretions are obtained, they should be diluted serially and plated onto a variety of media. Table A-3 suggests media for culturing vaginal wash. Colony-forming units usually range up to 10^9/ml when the vaginal wash technique is used. Other methods may yield different quantitities and dilutions will need to be adjusted. Rota-Plate (Fisher) is a handy tool for evenly spreading a 0.1 ml inoculum. The spiral gradient instrument (Spiral Systems) is also very satisfactory. The spiral gradient system reduces the number of dilutions necessary to inoculate.

For detection of clostridia, 1.0 ml of the suspension may be added to 1.0 ml of ethanol, allowed to stand for 45 min, and then 0.1 ml can be inoculated to egg yolk agar as

table A-3. Suggested media for vaginal flora studies[1]

Medium	Incubation atmosphere	Dilutions plated [2]	Days of incubation	Purpose
Brucella PRAS blood agar	Anaerobic	10^{-2},10^{-4},10^{-6},10^{-7}	4–7	Total counts and predominant flora
BBE agar	Anaerobic	10^{-2},10^{-4}	2–7	*B. fragilis* group *Bilophila*
KVLB–2 agar	Anaerobic	10^{-2}, 10^{-4}, 10^{-6}, 10^{-7}	2–7	*Bacteroides,* Pigmented *Prevotella,* some *Porphyromonas* spp.
BL and BS agar	Anaerobic	10^{-2}, 10^{-4}, 10^{-6}	2–7	*Bifidobacterium*
Lactobacillus selective medium or MRS agar	Anaerobic	10^{-2},10^{-4},10^{-6}	2–3	*Lactobacillus*
Egg yolk agar (heated or ethanol–treated dilutions)	Anaerobic	10^{-2}, 10^{-4}	2–3	*Clostridium*
Rlk or SA agars	Anaerobic	10^{-2},10^{-4}, 10^{-6}	4–7	*Mobiluncus*
HV agar[3]	Air + 5–7% CO_2	10^{-2},10^{-4},10^{-6},10^{-7}	2–3	*G. vaginalis*
Chocolate agar	Air + 5–7% CO_2	10^{-2},10^{-4},10^{-6},10^{-7}	2–3	*Haemophilus, Neisseria,* and other capnophiles
Thayer–Martin agar	Air + 5–7% CO_2	10^{-2}	1–2	*N. gonorrhoeae*
Brucella or trypticase soy blood agar	Air + 5–7% CO_2	10^{-2},10^{-4},10^{-6}	2–3	Total counts and predominant capnophilic flora
MacConkey agar	Air	10^{-2},10^{-4}	1–2	Enterics, *Pseudomonas*
BEA[4] agar	Air	10^{-2},10^{-4}	1–2	*Enterococcus,* Group D *Streptococcus*
Mannitol salt agar	Air	10^{-2}, 10^{-4}	1–2	*Staphylococcus*
Sabouraud agar	Air	10^{-2}, 10^{-4}	2–7	Yeasts and fungi

[1] See end of Appendix A for interpretive information on media.

[2] 0.1 ml of each dilution is placed on each medium used.

[3] HBT agar has also been shown to be a satisfactory medium for the recovery of *G. vaginalis.* (HV and HTB are available from Remel and other sources.)

[4] Bacto Bile Esculin Azide -agar

described for other plates. Note that a 1:2 dilution of the suspension requires that the final CFU count be doubled. *Gardnerella vaginalis* is most easily detected on human blood bilayer Tween-80 agar plates (V-Agar, Hardy), on which it appears as small, translucent, β-hemolytic colonies. Since this organism may be picked from anaerobic plates, all aerotolerance tests on organisms resembling *Gardnerella* (catalase-negative, oxidase-negative, small gram-variable rods) should be carried out on human blood as well as chocolate agar.

All plates should be incubated anaerobically for at least 7 d before final examination to allow slow-growing bacteria (*Actinomyces, Propionibacterium propionicus*) to develop and to allow certain *Porphyromonas asaccharolytica* strains (which may be inhibited on KVLB) to show pigment. *Bacteroides ureolyticus* also may require prolonged incubation. Its colonies are small, translucent or transparent, and may pit the agar. BBE should be held for 7 d; after the first 48 h, however, only *Bilophila* species should be sought. These colonies appear small (around 1.0 mm diameter), translucent, usually with black centers, and the catalase reaction is strongly positive. Mobi agar (312) and Rlk or SA agars (302) have been used for isolation of *Mobiluncus* species. Colonies may take up to 5 d to appear; they are small and translucent. Descriptions of typical colonies on selective agars are found at the end of this appendix.

Bacterial Vaginosis

The syndrome of bacterial vaginosis (BV) has no specific etiology, but appears to be a major deviation from the normal mix and numbers of organisms present in the vaginal mucosa (16;151;306). Because anaerobes are the primary microbes in both the healthy vagina and in BV, a discussion of the presentation and laboratory diagnosis of BV is included in this manual. The condition is defined by a constellation of signs and symptoms, and in fact, some women with BV may not present with any symptoms at all. The syndrome is characterized by higher than normal numbers of certain organisms present in the normal vagina, primarily anaerobes, as well as by dramatically reduced numbers of vaginal H_2O_2-producing lactobacilli, the hallmark organism of normal vaginal secretions in healthy women (16;151).

Gram stain interpretation is the best laboratory test for diagnosis of BV. Presumptive diagnosis can be made clinically by the presence of a homogeneous grey malodorous vaginal discharge, the smell of which intensifies with alkalinization; pH of vaginal secretions >4.5; and visualization of clue cells on direct wet preparation or reduced amount of lactobacilli in Gram-stained smears of vaginal discharge (16;249). Clue cells are squamous epithelial cells shed from the vaginal lining and covered with large numbers of tiny bacteria, visualized as a stippled appearance to the cells. The epithelial cells also show shaggy and obscured edges due to heavy adherence of bacteria to the cell margins, which are sharp in normal vaginal secretions. Other aspects of wet preparation interpretation such as presence of lactobacilli or pus cells are difficult for all but very skilled observers. The odor is caused by the volatile fatty acids and amines (principally trimethylamine) produced by large numbers of anaerobic bacteria. Anaerobes associated with BV, including *Prevotella bivia, P. disiens, Mobiluncus mulieris, M. curtisii,* and *Peptostreptococcus* species, are present in as many as 3–4 logs higher numbers than in normal women. Other organisms associated with the syndrome include *Gardnerella vaginalis* and *Mycoplasma hominis.*

BV has recently gained attention because of its validated association with premature labor and delivery of low–birth weight infants (149). However, BV also predisposes women to a number of infectious processes, such as cuff cellulitis, postsurgical infection, and pelvic inflammatory disease (327). For these reasons, diagnosis of BV is important, and it is within the capability of any laboratory. Culture is not recommended; diagnosis of BV is best accomplished by Gram stain of vaginal discharge (16;249).

Laboratories offering diagnostic testing for BV should not ignore the role of yeast species and *Trichomonas vaginalis* as other agents of vaginal discharge (93). Infectious agents that also result in discharge, although usually of a more purulent nature, include *Chlamydia trachomatis* and *Neisseria gonorrhoeae* (16). Laboratory diagnosis of these two most common bacterial agents of sexually transmitted disease as well as human papillomavirus, probably the most prevalent agent of STD in the United States, is primarily based on molecular methods. Amplification of nucleic acid is the test of choice for chlamydia and gonorrhea (58).

Specimen Collection and Examination

Vaginal secretions can be collected by the patient or by a health care provider performing a vaginal examination with a speculum. The patient inserts a cotton or Dacron swab approximately 3 cm into the vaginal vault and rotates the swab to collect as much discharge as possible. The swab is placed into standard modified Stuart's transport medium or into a small volume (0.75–1.0 ml) of normal saline in a capped tube. If the practitioner is collecting the specimen, secretions pooling in the posterior fornix of the vagina should be sampled.

For the preferred method of laboratory diagnosis of BV, a smear of the material on the swab is prepared at the bedside. Rolling the swab across a large area on the slide allows deposition of a thin layer of material. The slide should be air dried and fixed with 95–100% methanol, and Gram stain can be performed at any time later. The Gram stain is scored for the presence of lactobacillus-like morphotypes, small gram-variable coccobacilli, and curved rods. The score is totaled and likelihood of BV is determined (188;249).

QUANTITATIVE STUDIES OF INTESTINAL FLORA: COLLECTION AND TRANSPORT

Fecal Specimens

The specimen should be passed naturally and the entire specimen collected into a plastic container. A small amount of blood on the outside of the specimen (from rectal bleeding) is allowable. If large amounts are present, the specimen cannot be considered for normal flora studies. Specimens contaminated with urine or menstrual flow are to be rejected.

The specimen container should be placed immediately with top loosened into an anaerobic chamber or an anaerobic jar if the specimen is to be transported. Total flora specimens should not be frozen as freezing kills many vegetative cells. Specimens should be maintained at room temperature no longer than 5 h if they cannot be processed immediately.

Other Specimens of Intestinal Contents

Specimens from the small bowel or other areas of the bowel are usually collected by syringe through a tube passed into the area to be sampled. The tube should be passed through the nose rather than the mouth. The specimen should be placed into a gassed-out tube or anaerobic transport tube.

Processing of Specimens

The directions that follow are used for fecal specimens. These may be modified as necessary for other specimens of intestinal contents.

(1) Weigh the specimen, container, and transporter and subtract the weight of the empty transporter and container from the total weight. Place the transporter (with specimen inside) into the anaerobic chamber.

(2) After the specimen is in the chamber, homogenize it as follows:

   (a) Place the specimen into a special plastic bag and homogenize in the Stomacher (Tekmar Co.; Model #3500 accommodates a 3.5 L sample). This method reduces potential aerosols and makes disposal of the unused specimen easier.

   (b) Alternatively, especially for larger specimens, a Waring stainless steel container with screw cover (Eberbach #8520) may be used for specimens ≥60 g. The Eberbach glass container (#8470) is used for stools 30–60 g, and the semimicro jar (Eberbach #8580) is used for solid specimens <30 g and for liquid specimens <40 g.

   (c) Blend nonliquid specimens six times for 30 s each time. Do not add diluent. Intermittent blending is necessary so that a sterile glass rod can be introduced into the blender between each blending operation to scrape the material from the sides of the container and place it at the bottom of the vessel. This procedure is not necessary with liquid specimens, for which a continuous 3 min blending is sufficient. Blending creates aerosols that can contaminate the interior of the chamber. Caution should be excercised when performing this procedure.

(3) Remove an aliquot of approximately 1 g of the homogenized specimen from the chamber and place it on a planchet (preweighed square of aluminum foil or plastic weigh-boat) for weighing and drying. Drying is carried out in a vacuum drying oven (15 in Hg vacuum, 170–180 °F) with calcium chloride over 48 h. Weigh the specimen again following drying. Determine the wet/dry weight ratio so that counts obtained may be corrected and expressed as number of organisms per gram of dry stool.

(4) Place another aliquot of approximately 1 g of the homogenized specimen in a sterile tube and weigh it. Add enough reduced 0.05% yeast extract solution to obtain a 10^{-1} dilution. From this dilution, 10^{-2} to 10^{-9} dilutions are made in reduced 0.05% yeast extract solution. See Table A-4 for appropriate dilutions.

   (a) Inoculate the first series onto appropriate anaerobic media in the chamber, then remove the dilution series from the chamber for inoculation of media for facultative and aerobic organisms. See Table A-4 and step 9, below, for suggested media.

   (b) Heat the second series at 80° for 10 min and then inoculate onto egg yolk agar for isolation of *Clostridium* (spore-forming) species.

   (c) Alternatively, prepare only one series of dilutions. Remove 1.0 ml quantities from each tube and place them into empty tubes that may be heated in a heating block within the chamber. If the ethanol method of spore testing is preferred, place a set of small tubes containing 1.0 ml ethanol into the chamber

table A-4. Suggested media for anaerobic bowel flora studies[1]

Medium	Dilutions to be Plated[2]: Feces	Dilutions to be Plated[2]: Small bowel[3]	Days of incubation	Purpose
PRAS Brucella blood agar	$10^{-7}, 10^{-8}, 10^{-9}$	$10^{-2}, 10^{-4}, 10^{-6}, 10^{-8}$	5–7	Total counts and predominant flora
BBE agar	$10^{-6}, 10^{-7}, 10^{-8}$	——	2–7	*B.fragilis* group and *Bilophila*
KVLB agar	$10^{-2}, 10^{-4}, 10^{-6}, 10^{-8}$	$10^{-2}, 10^{-4}, 10^{-6}, 10^{-8}$	3–5	*Bacteroides,* pigmented *Prevotella*
BL agar (non–selective)/BS agar	$10^{-4}, 10^{-6}, 10^{-7}, 10^{-8}$	$10^{-2}, 10^{-4}, 10^{-6}, 10^{-8}$	2–3	*Bifidobacterium*
Lactobacillus selective medium or MRS agar	$10^{-2}, 10^{-4}, 10^{-5}, 10^{-6}$	$10^{-2}, 10^{-4}, 10^{-6}$	2–3	*Lactobacillus*
Rifampin blood agar	$10^{-4}, 10^{-6}, 10^{-8}$	——	2–3	*F.mortiferum/varium* and certain *Eubacterium* and *Clostridium* species
CCFA or CCEY	$10^{-2}, 10^{-4}, 10^{-6}, 10^{-8}$	——	2–3	*C.difficile*[4]
Veillonella neomycin agar	$10^{-2}, 10^{-4}, 10^{-6}$	$10^{-2}, 10^{-4}, 10^{-6}$	2–3	*Veillonella* and other gram–negative cocci
JVN agar	$10^{-2}, 10^{-4}, 10^{-6}$	$10^{-2}, 10^{-4}, 10^{-6}$	2–3	*Fusobacterium* and *Leptotrichia*
Egg yolk agar (heated or ethanol-treated dilutions)	$10^{-2}, 10^{-4}, 10^{-6}$	——	2–3	*Clostridium* spp.
PMS agar	$10^{-4}, 10^{-6}, 10^{-8}$	$10^{-2}, 10^{-4}, 10^{-6}$	2–3	*Peptococcus* and *Megasphaera*
PS agar	$10^{-4}, 10^{-6}, 10^{-8}$	$10^{-2}, 10^{-4}, 10^{-6}$	2–3	*Peptostreptococcus*

[1] See end of Appendix A for interpretive information on media.

[2] 0.1 ml of each dilution is placed on each medium used.

[3] Selective media optional. Direct examination of specimen indicates whether they may or may not be helpful. If specimen is clear fluid, selective media are probably not necessary.

[4] If very low counts of *C. difficile* are to be detected (for epidemiologic studies) use pre-reduced peptone broth containing 39 µg/ml cefoxitin and

and add 1.0 ml from each of the primary dilution tubes. Mix the suspension with the ethanol and allow the mixture to stand for 30–45 min.

(d) Finally, plate the heated or ethanol-treated dilutions on egg yolk agar for isolation of spore-forming organisms. Samples from the untreated dilutions are plated onto anaerobic and aerobic media.

(5) Dilutions to be plated on the various media are indicated in Table A-4. Aside from the PRAS blood agar plates, which are prepared in the chamber, all other media are prepared on the bench (aerobic media can be purchased commercially) and brought into the chamber 24 h prior to use. PRAS blood agar and other PRAS plate media are available commercially (Anaerobe Systems). With a pipette, spread 0.1 ml of

each dilution to be plated over the medium indicated. This is done by the rotator pipette method.

(6) Place the plates in jars, seal the jars, remove them from the chamber, and incubate for the periods designated in Table A-4. If equipment permits, the plates may be incubated in the chamber.

(7) If one wishes to culture for *Methanobacterium,* use the following procedure:
   (a) Bring an unheated set of dilutions in Hungate-type tubes out of the chamber.
   (b) Steam PRAS tubes containing 4.9 ml of *Methanobacterium* medium (see Appendix C) to melt the agar.
   (c) Place the tubes in a 50 °C water bath and add 0.02 ml of 5% PRAS sodium sulfide (this may be stored for extended periods) to each tube just before inoculation.
   (d) Use a syringe that is first gassed-out with oxygen-free CO_2 for this purpose and for subsequent inoculation of tubes with 0.1 ml of each of the dilutions to be cultured.
   (e) After incubation for 1–2 weeks, measure methane gas production by analyzing the head-space gas (217).

(8) Microscopic examinations:
   (a) Prepare a Gram stain from the 10^{-2} dilution. Count bacterial morphotypes seen in 25 fields.
   (b) Make Petroff-Hausser chamber counts from the 10^{-3} dilution. Count the numbers of bacteria seen in 25 fields.
   (c) Prepare a darkfield examination as follows: Place a small drop from the 10^{-2} and 10^{-3} dilutions on separate cover slips and place the cover slips on microscope slides. Survey each wet preparation at 400× magnification and a make a detailed examination at 1,000× magnification. Note motility and describe each morphology. A minimum of 20 min should be spent in the examination.

(9) Use the dilutions for setting up aerobic cultures in a manner similar to that described for anaerobes. As a minimum, we recommend the following media: tryptic soy or Brucella blood agar (as a nonselective medium), deoxycholate or MacConkey agar (for gram-negative bacilli), PEA or colistin nalidixic acid agar (for gram-positive bacteria), cetrimide agar (for *Pseudomonas*), mannitol salt agar (for *Staphylococcus*), Pfizer selective *Enterococcus* agar (for group D streptococci), potato flake agar medium with chloramphenicol, 50 µg/ml, (for yeasts and molds; incubate at room temperature), and Mitis-Salivarius agar (for streptococci other than enterococci).

Alternative Processing Technique: Spiral System

The Spiral Plater Model D (Spiral Systems) may be used for plating fecal samples; it eliminates the need for extensive dilutions. This system dispenses 50 µl of sample onto a 100 mm diameter Petri dish in the pattern of a tight spiral; this results in a concentration gradient from the center to the periphery (high counts to low counts). We recommend an initial suspension of 10^5 CFU/ml to obtain a countable plate; experience with particular samples may dictate a need for more concentrated or more dilute suspensions. Based on an estimate of 10^{12} CFU/gm of stool, the following dilution scheme may be used:

(1) Make a 1:10 dilution (giving a concentration of 10^{11} CFU/gm) using 0.05% reduced yeast extract.

(2) Continue with three 100-fold dilutions:

Tube A	1:100	10^9	CFU/gm
Tube B	1:100	10^7	CFU/gm
Tube C	1:100	10^5	CFU/gm

(3) Plate the dilution from Tube C (10^5 CFU/gm). After incubation, count (with the aid of a colony viewer) the number of colonies of each morphotype.

Note: The colony viewer is equipped with a fixed, plastic counting grid that divides the 100 mm plate into precisely measured quadrants. The total count of each colony type is easily made as the volume of sample dispensed is known for the given areas counted.

Our experience demonstrates that the spiral plater is less time-consuming and requires less media than the serial dilution technique. This system can be modified to accommodate 150 mm diameter plates as well. Certain companies (e.g., Anaerobe Systems) offer a variety of media in plates with a level surface (required for this procedure) and will manufacture special plates upon request.

Identification of Isolates

After appropriate incubation periods, count the colonies and pick each morphotype to a blood agar plate or other appropriate medium to obtain a pure culture. Descriptions of typical colonies on selective agars used for fecal flora studies are found at the end of this appendix. Perform aerotolerance testing and Gram stain from the same colony. PRAS biochemicals and GLC are usually necessary to identify normal flora isolates. The total count is obtained by counting all colonies on the nonselective blood agar plate. The total count is multiplied by the wet/dry weight ratio and expressed as CFU/gram of stool, dry weight.

Alternatively, if species-level identification is not desired, characteristics such as Gram stain, disk identification reactions, growth on a specific selective medium, spore test, bile test, and endproduct analysis from a PYG or chopped meat broth may give sufficient information to place an organism in a particular group (e.g., *Clostridium* species, *B. fragilis* group, bile-sensitive gram-negative anaerobic bacilli, anaerobic gram-positive cocci, etc.). We have sometimes used this abbreviated identification scheme when looking for gross effects of antimicrobial agents on fecal flora. We also stock all the strains so that further identification may be done later if desired.

DESCRIPTIONS OF TYPICAL COLONIES ON SELECTIVE MEDIA FOR ANAEROBES

As with selective media for aerobes, selective media for anaerobes are not 100% inhibitory toward unwanted organisms; these may grow to a greater or lesser extent. Antibiotics in media, particularly cephalosporins, will deteriorate with time, especially if not stored in the refrigerator, thus allowing growth of unwanted organisms. If selective media are incubated for prolonged periods, unwanted organisms may start to grow.

In general, colonies less than or equal to 0.5 mm in diameter at 48 h should not be picked unless the typical colony morphology of the organism is that size (e.g., *Veillonella* and some *Peptostreptococcus* species).

Bacteroides Bile Esculin Agar (BBE)

Colonies of members of the *B. fragilis* group are >1 mm in diameter, circular, entire, raised, and appear grey-brown with brown to black halos. The one exception is *B. vulgatus,* which usually does not hydrolyze esculin and hence appears light grey. Occasional colonies of *Fusobacterium mortiferum* and *F. varium* that may grow on BBE agar are also >1 mm in diameter, but they are flat and irregular. Typical *Bilophila* colonies on BBE are small and convex with black centers and translucent margins. Small translucent colonies may represent *Sutterella.* This medium is highly selective, although occasionally yeast may grow; these appear as dry crinkly colonies and may be dark or grey in color. Also, rare strains of gentamicin-resistant *Enterobacteriaceae,* enterococci, staphylococci, or *Pseudomonas* may grow; their colony appearance is not predictable, but they are usually <1 mm in diameter.

Blood and Liver (BL) and *Bifidobacterium*-Selective (BS) Agars

Members of the genus *Bifidobacterium* appear as circular, tan to brown, medium-sized colonies. Even on the selective BS agar, *B. fragilis* group, *Clostridium* species, and *Lactobacillus* species have been isolated from stool specimens. BL agar is not selective but is an excellent medium for *Bifidobacterium.* All tan to brown colonies should be worked up as possible *Bifidobacterium* species. We have noted better recovery of *Bifidobacterium* from BL agar than from BS agar.

Cadmium Sulfate Fluoride Acriflavine Tellurite Agar (CFAT)

Colonies of *Actinomyces* species are 1.0–3.0 mm in diameter, cream to slightly greenish color, entire, convex or raised, and opaque. *A. odontolyticus* forms pink smooth colonies that later turn brown. Rough, pink colonies may be *A. graevenitzii.* This medium has been formulated for enhanced isolation of *A. naeslundii* and *A. viscosus* from mixed cultures.

Cefoxitin Cycloserine Egg Yolk (CCEY)

CCEY (LabM) medium is often used for selective isolation of *Clostridium difficile* after spore selection as it contains cholic acid to promote germination of spores after plating, *p*-hydroxyphenylacetic acid to promote formation of the *p*-cresol responsible for the typical horse stable colony odor, and blood to promote growth. Phenyl red is omitted as it autofluoresces under UV light (31). Colonies are irregular, greyish, and later form white centers and they fluoresce bright chartreuse under long-wave UV light. *Clostridium innocuum* may grow on this medium as may several other Clostridia and they must be differentiated by tests presented in Chapter 7.

Cycloserine Cefoxitin Fructose Agar (CCFA)

C. difficile appears as large yellow colonies with an irregular edge; *C. innocuum* also grows on CCFA, but colonies usually have an entire edge. A few other species of *Clostridium* may grow on this medium but the colonies appear more butyrous. The chartreuse fluorescence (under UV light) typical of *C. difficile* is best demonstrated on subculture to a blood agar plate, since the CCFA medium itself is fluorescent (methyl red). This medium deteriorates rapidly if not refrigerated and works best when fresh or reduced at least 4 h before use. Yeast, enterococci, and other enteric organisms will grow as small flat colonies. Commercial plates vary in their performance characteristics.

Egg Yolk Agar (EYA)

The numerous species of clostridia have a variety of colony morphologies. *C. perfringens* is lecithinase-positive (white precipitate around the colony). Other *Clostridium* species may produce lipase, which appears as a mother-of-pearl iridescence on and around the colony. Other clostridia may be both lipase- and lecithinase-negative and may not have distinctive colony morphology. It should be noted that for this medium to be selective for *Clostridium* species, the dilution must first be ethanol- or heat-treated to kill vegetative cells. Addition of 100 μg of neomycin will increase selectivity. Other typical lipase-positive species are *Prevotella intermedia/nigrescens, F. necrophorum,* and *A. neuii.* Certain *Porphyromonas* species of animal origin form mustard yellow colonies on EYA, as also does *Capnocytophaga ochraceae.*

Fusobacterium-Selective Medium (JVN) and *Fusobacterium* Neomycin Vancomycin (NV) Agar

Colonies of all species of *Fusobacterium* and *Leptotrichia buccalis* grow on JVN agar (josamycin, vancomycin, and norfloxacin) and NV agar in fastidious anaerobe agar base (32), and their appearance is similar to that on nonselective blood agar medium. Most other anaerobes are inhibited completely on JVN agar, or their colony size is greatly reduced; rare strains of members of the *B. fragilis* group are only slightly inhibited. Most facultative organisms are completely inhibited on JVN and NV agars.

Kanamycin Vancomycin Laked Blood (KVLB) Agar

Bacteroides and nonpigmented *Prevotella* grow on this medium and may be low convex or large and mucoid and vary in color from grey to yellow to tan to white. The pigmented *Prevotella* generally also grow on this medium. They may have a light tan color initially and become brown to black with prolonged incubation. *Fusobacterium* may occasionally grow on KVLB, especially in mixed culture. *Porphyromonas* is often inhibited by the high vancomycin concentration (7.5 μg/ml) in the medium. Therefore, for studies potentially including *Porphyromonas* species, vancomycin concentration should be reduced to 2 μg/ml. This reduction will increase also growth of accompanying flora, especially *Capnocytophaga* species from oral samples. *Clostridium innocuum* usually grows as a flat translucent colony on KVLB. Occasional strains of enterics and yeast may grow on this medium, but their colony morphology is not predictable. Since prolonged incubation may be desirable for pigment production, other organisms will probably grow.

Lactobacillus-Selective Medium (LBS)

This is a highly selective medium for a few species of lactobacilli. They may appear as white to grey to yellow to orange in color, and are usually low convex and less than 2–3 mm in diameter. One should not assume that this medium will grow all lactobacilli.

MRS (de Man, Rogosa, Sharpe) media

MRS is a widely used selective medium for isolation of lactobacilli. They form large, white opaque colonies usually with entire edges. Also many bifidobacteria grow on this medium as tiny, grey to white, entire to rough colonies (see Figure 6-30).

Rifampin (RIF) Blood Agar

C. ramosum typically appears as a flat, greyish green, 2–3 mm diameter colony. Isolates of *Eubacterium* may have a similar appearance. *Fusobacterium varium* has a fried egg colony and *F. mortiferum* has low convex, tan, translucent colonies.

Veillonella Neomycin Agar

Veillonella appear as small transparent colonies. *Bacteroides* and *Fusobacterium* also grow on this medium and their appearance is similar to that on nonselective blood agar. Most gram-positive and many facultative gram-negative rods are inhibited.

Trypticase Soy Bacitracin Vancomycin (TSBV) agar

TSBV has been designed for the selective isolation of *A. actinomycetemcomitans* from subgingival samples of periodontitis patients. This blood-free, transparent medium allows the visualization of the starlike inner structure of the whitish to grey colonies when viewed against light (preferably under a dissecting microscope) (see Figure 6-16). Other oral organisms, including fusobacteria, *Capnocytophaga, Selenomonas,* and some agar-pitting gram-negative organisms may grow on this medium.

appendix B

Gas-Liquid Chromatography for Metabolic Endproduct and Cellular Fatty Acid Methyl Ester Analysis

Anaerobic bacteria metabolize glucose and other nutrients to short-chain fatty and organic acids, as well as alcohols that are distinctive for genera and some species. These endproducts serve as a fingerprint that is useful for classification and, in conjunction with other tests, for identification. This is especially important for the gram-positive bacilli, which are otherwise difficult to identify to the genus level. Table 6-1 lists endproducts formed by genera of anaerobes isolated from clinical specimens and indigenous flora sources. Endproducts formed by species of anaerobes are listed in the identification tables in Chapter 6.

Although an organism's metabolic pathway is a genetically stable characteristic, various factors can influence the endproducts formed. An important variable is the ratio of the carbohydrate to peptone in the growth medium (345). Thus in peptone glucose broth, endproducts of glucose metabolism, such as acetate and butyrate, will accumulate first. Isoacids, formed by the metabolism of peptones, will appear more slowly. Media rich in peptones, such as chopped meat broth, will yield larger amounts of isoacids than media poor in peptones, such as thioglycolate broth. Carbon dioxide increases production of succinate in *Actinomyces* species. Quantities of endproducts increase with age of the culture until the nutrients are depleted and growth ceases, so the quantities of any product within the same species can vary accordingly. Culture media may contain trace amounts of some of these acids. Therefore, an uninoculated and incubated broth should be assayed and any endproducts detected subtracted from those found in the sample.

Short-Chain Fatty Acid Analysis

Chromatographs, Columns, and Operating Conditions

The separation of fatty acids by gas-liquid chromatography (GLC) is achieved by injecting a prepared sample through a septum in the proximal end of a packed column contained within a heated oven. The high temperature in the injector port volatilizes the sample, and a carrier gas moves the sample down the length of the column. The fatty acids move through the column at different speeds according to molecular weight and polarity. When a fatty acid reaches the detector at the distal end of the column, it causes an electrical response, which is converted by the recorder to a peak on the chart. The amount of the endproduct is proportional to the size of the peak. The retention time (the length of time from the injection of the sample to detection of a peak) determines the identity of the peak, as compared to a standard containing defined quantities of the fatty acids being sought. Increasing the carrier gas flow rate decreases the retention time for all peaks as they are forced through the column more rapidly; however, the resolution of the peaks is also decreased. Increasing the oven temperature has the same effect. Control standards are available from Supelco and other sources.

Chromatographs equipped with a thermal conductivity detector (TCD) use helium as a carrier gas and can be used to detect volatile and nonvolatile endproducts. Samples must be extracted into methanol or chloroform to eliminate water prior to injection into the column (318). Flame ionization detectors (FID) do not detect water; thus aqueous culture supernatants can be analyzed directly. However, they require nitrogen as a carrier gas, and air and hydrogen for the flame. Gas chromatographs are available in a wide range of sophistication and price. Some may include automatic injectors or an integrator and computer, which convert peak areas to milliequivalent (mEq) quantities.

Prepacked columns suitable for endproduct analysis are available from several sources. We use the SP-1000 and SP-1220, available from Supelco. Both columns may be used for the short-chain (1–7 carbon molecules) fatty acids and the methyl esters of the nonvolatile acids. It is essential to analyze the standard mixture each time the samples are run to determine the optimal conditions.

Preparation of Samples for Analysis

Incubate the test organism in a broth medium until good growth is achieved. Optimum quantities of endproducts are usually produced after 48 h with anaerobes that grow well. Acidify the broth culture to pH 2 by adding two drops of 50% H_2SO_4 to 5 ml of the broth. Centrifuge briefly to pellet the cell mass and remove the supernatant for analysis. Once acidified, the endproducts remain stable and the supernatants may be stored for several months.

Volatile Fatty Acids

The volatile fatty acids include formic, acetic, propionic, isobutyric, *n*-butyric, isovaleric, *n*-valeric, isocaproic, caproic, and heptanoic acids.

For detection in a chromatograph equipped with an FID and a SP-1220 column, inject 1 µl of the supernatant without further processing.

Nonvolatile Fatty Acids

The nonvolatile fatty acids useful for identification include lactic, succinic, and phenylacetic acids. For detection, their methyl derivatives must first be prepared as follows.

Pipette 1 ml of supernatant into a 13 × 100 mm glass tube.

(1) Add 1 ml of methanol and 0.1 ml of 50% H_2SO_4.
(2) Stopper the tube, shake or vortex vigorously, and heat to 60 °C for 30 min, or allow to stand at room temperature overnight.
(3) Working in a fume hood, add 0.5 ml chloroform, replace stopper, and mix well. If the sample was heated, allow to come to room temperature before adding chloroform.

(4) Allow to stand for a few minutes or centrifuge briefly to separate the chloroform and water layers.
(5) Draw the bottom chloroform layer into the syringe, being careful not to take up any water, wipe the outside of the needle, and inject into the column. The amount of sample to be injected will vary with the operating conditions. Test the standard mixture, methylated and extracted in the same way, to determine the optimum amount to inject.

Methyl Ester (FAME) Analysis

The long-chain fatty acids are 9–20 carbons long and reside in the cell membrane, unlike the short-chain fatty acids formed as a result of metabolism. The presence and quantity of long-chain fatty acids in bacteria produce fingerprints that are unique for genera and many species of anaerobes. Analysis requires a system such as the Microbial Identification System, which consists of a chromatograph containing a 25 × 0.2 mm methyl-phenyl-silicone fused silica capillary column, an automatic sampler with a sample tray, an integrator, a computer, and a printer (Agilent Technologies). The software is available from MIDI (Microbial ID Inc.). An extensive data base for anaerobes is available.

The organism to be analyzed is grown in a defined medium (PRAS PYG available from Anaerobe Systems) for a specified length of time (usually ≤48 h). The cells are harvested by centrifugation and then saponified with NaOH. The fatty acids are thus cleaved from the lipids and converted to their sodium salts. Methyl esters are prepared and extracted into an organic solvent phase, washed to remove residues, and then injected into the instrument.

The peaks obtained from the sample are compared to those of well-characterized strains in the database and a best-match identification with a statistical likelihood is given. The precision of identification varies and closely related strains may have similar profiles.

Procedure

The specific reagents and methods are fully described in a manual provided by the manufacturer.

(1) Inoculate 10 ml PYG or PYG with Tween (Anaerobe Systems) with several representative colonies from organisms grown on CDC or Brucella agar. Gram-positive organisms should be set up using PYG Tween.
(2) Incubate until maximum turbidity is reached (usually overnight).

Note: Do not leave at room temperature before or after centrifugation.

(3) Centrifuge for 30 min, discard the supernatant, and save the pellet. Recap the tube with a Teflon-lined cap only. If using a growth supplement, such as Tween or pyruvate, wash the pellet with 3 ml of 0.7% $MgSO_4$, recentrifuge, discard the supernatant, and save the pellet. Store tubes with pellets at −70 °C.

Label the tops of the caps with dot labels. Include a negative control (blank reagent tube) and two control strains, *B. fragilis* ATCC 25285 and *C. perfringens* ATCC 13124.

B

Before starting the procedure, turn a water bath on, set at 81 °C (allow 2 h to equilibrate). The water bath will drop to 80 °C after adding the tubes. Turn the heat block on and heat water to boiling (allow 30–45 min). The water level should be just slightly above the top of the rack that fits in the boiling water bath, allowing for evaporation. Prime reagent dispensers into a waste bottle before using.

Note: Wear gloves and work in the fume hood when using reagents. Only glass and Teflon can come in contact with the reagents.

(4) Add 1.0 (0.5) ml of reagent 1.
 (a) Tighten tube caps and vortex 5–10 s. Check the tubes for leaks. Heat in a covered boiling water bath for 5 min.
 (b) Remove after 5 min, allow to cool slightly, vortex again for 5–10 s, and heat again for 25 min in the covered boiling water bath.
 (c) Cool to room temperature by placing rack in a cool water bath and moving the rack back and forth.

(5) Add 1.0 (0.5) ml reagent 2a and 1.0 (0.5) ml of reagent 2b. (Mix these reagents well while dispensing, as they tend to layer.)
 (a) Vortex 5–10 s. Heat in 80 °C water bath for 10 min.
 (b) Cool in a cold water bath, moving rack back and forth until tubes are at room temperature.

(6) Add 1.25 (0.625) ml reagent 3. (Mix well before using.)
 (a) Mix end-over-end for 10 min (speed 6 to 7 on the Rota-Rack (Fisher)).

(7) Discard the aqueous bottom layer using glass pipettes.

(8) Add 3.0 (1.5) ml reagent 4 to the top phase.
 (a) Mix end-over-end for 5 min.

(9) Using glass pipettes, place top layer in GC vial taking care not to contaminate with the bottom layer.

(10) Load onto the automatic sampler.

Put all waste in MIS waste bottle, rinse dispensers from reagents 1 and 4 in deionized water, and store them with the tube portion in the water. The reagents (1 and 4) can be recapped. All other reagents can be stored with the dispensers intact on the bottle.

Dispensette III Bottle Top Dispensers can be obtained from Brand Tech Scientific (#4701121 for 2 ml, and #4701131 for 5 ml dispensers) or from VWR Distributors (#18901-136 for 2 ml, and #18901-138 for 5 ml dispensers).

appendix C

Biochemical Test Procedures

This appendix lists biochemical test procedures in alphabetical order.

Bile

The ability of an organism to grow in the presence of 20% bile (equal to 2% oxgall) can be tested with disks impregnated with 20% bile, in liquid medium containing 20% bile, or on BBE agar.

Performance of Test

Disk Test

Make a fresh subculture of the organism to be tested onto a supportive agar and place a bile disk in the heavy inoculum area. Incubate anaerobically for 24–48 h. Disks are available from Remel, Oxoid, and Rosco.

Broth Test

Inoculate two tubes of liquid medium (PRAS PYG or thioglycolate), one supplemented with 2% oxgall and 0.1% sodium deoxycholate and one unsupplemented, with two drops of actively growing culture of the organism. Incubate both tubes until the control tube (unsupplemented) shows good growth.

BBE Test

Culture on BBE agar 24–72 h (other ingredients in this agar may be inhibitory for some bile-resistant organisms; therefore restrict testing to suspected *B. fragilis* group and *Bilophila*).

Interpretation of Results

Sensitive (S)

- Disk test: Any zone of inhibition present around the bile disk.
- Broth test: Less growth in bile-containing tube than in control (not necessarily total inhibition).
- BBE: No growth on BBE agar. Some species may grow as pinpoint colonies compared to large colonies on non-selective media. This is interpreted as inhibition.

Resistant (R)

- Disk test: No zone of inhibition present around bile disk.
- Broth test: Equal or greater (stimulated) growth in bile-containing tube than in control tube.
- BBE: Growth on BBE agar; examine for presence of colonies and not simply the blackening of the agar, which may be due to esculin hydrolysis by preformed enzymes present in the inoculum.

CARBOHYDRATE FERMENTATION

See "PRAS Biochemical Inoculation" for inoculation procedure.

Performance of Test

(1) Invert the tubes to resuspend settled organisms and record the turbidity (1+ to 4+). At least 2+ growth should be observed to ensure reliable results.
(2) Measure the pH of each tube using a pH meter.

Interpretation of Results

(1) Positive: pH below 5.5 = "acid"; pH 5.6–5.8 = "weak acid." If tubes were inoculated with a syringe and without gassing with CO_2, the "weak acid" interpretation may include tubes with pH up to 6.3.

    Note: The pH of PY base should be above 6.2.

(2) Negative: pH 5.9 and above.

C

CATALASE TEST

For catalase testing of anaerobic bacteria, 15% H_2O_2 appears to be more sensitive than 3% H_2O_2.

Performance of Test

(1) Use 24–72 h growth on primary plates or on subculture plates, preferably from a medium that does not contain blood. Red blood cells contain catalase and may produce false-positive results. Carbohydrate-containing media may suppress catalase activity.
(2) Touch center of a pure colony with a loop or sterile wooden stick and transfer onto the surface of a clean, dry glass slide or Petri dish. If you must test growth from blood-containing medium, avoid touching the agar.
(3) Add one drop of 15% H_2O_2 onto the smear.
    (a) Do not introduce a metallic loop into the drop, because this often causes a false-positive reaction.

(4) Observe for immediate bubbling.
(5) An alternative is to apply a drop of 15% H_2O_2 on growth on a medium that does not contain blood and observe for bubble formation.

Interpretation of Results

(1) Positive: Immediate and sustained bubbling of the hydrogen peroxide (see Figure 4-8).
(2) Negative: No bubbling observed. Formation of rare bubbles after 20–30 s is considered a negative catalase test; some bacteria may possess enzymes other than catalase that can decompose H_2O_2.

CATALYST REGENERATION

After extended use, catalyst pellets may be boiled in mild acid to remove surface deposits.

Procedure

(1) Boil the catalysts in 1% H_2SO_4 for ≥2 h. The acid solution should be changed 1–2 times during boiling.
(2) Rinse the boiled catalysts with water until the water remains clear.
(3) Dry.

CONVERSION OF LACTATE TO PROPIONATE

Performance of Test

(1) Extract an actively growing PY lactate or chopped meat medium tube (test medium) and a PY base tube (control) for volatile fatty acid determination by gas chromatography.
(2) Compare the amounts of propionic acid produced.

Interpretation of Results

(1) Positive: More propionic acid detected in the test medium than in the PY base medium.
(2) Negative: Equal or smaller amount of propionic acid produced in test medium compared to PY.

CONVERSION OF THREONINE TO PROPIONATE

Performance of Test

(1) Extract an actively growing PY threonine tube (test medium) and a PY base tube (control) for volatile fatty acid determination by gas chromatography.
(2) Compare the amounts of propionic acid produced.

Interpretation of Results

(1) Positive: More propionic acid detected in the test medium than in the PY base medium.
(2) Negative: Equal or smaller amount of propionic acid produced in the test medium compared to PY.

DESULFOVIRIDIN TEST

Desulfomonas species and most members of *Desulfovibrio* species contain desulfoviridin pigment. The sirohydrochlorin chromophore of this pigment gives a characteristic red fluorescence when inspected under UV light immediately after addition of 2 N NaOH to a cell suspension (155).

Performance of Test

(1) Inoculate a tube of liquid medium supplemented with 1% pyruvate and 0.25% magnesium sulfate. Incubate until turbidity indicates good growth.
(2) Centrifuge tube to obtain a pellet.
(3) Pipette a heavy drop of the pellet onto a slide.
(4) Add a drop of 2 N NaOH on the pellet and immediately observe under long-wave UV (366 nm) light.

Interpretation of Results

(1) Positive: Red fluorescence.
(2) Negative: No fluorescence.

ESCULIN HYDROLYSIS

The esculin molecule is hydrolyzed to glucose and esculetin by a bacterial enzyme. Released esculetin then reacts with iron salt (ferric ammonium citrate) to form a dark brown or black complex, indicating a positive result. However, H_2S, which is produced by several organisms during metabolism, also reacts with iron to produce a black complex, which interferes with the interpretation of the esculin hydrolysis test. Therefore, all tubes showing darkening after the addition of the reagent need to be checked under UV light; intact esculin fluoresces white-blue under 366 nm light, whereas hydrolyzed esculin has lost its fluorescence.

Performance of Test

(1) Add five drops of 1% ferric ammonium citrate solution to a tube of actively growing organism in peptone yeast (PY) esculin broth.
(2) Observe for a color change and fluorescence under UV light.

Interpretation of Results

(1) Positive: Black or dark brown color development and no fluorescence under UV light (366 nm).
(2) Negative: No color development or positive fluorescence under UV light.

FLAGELLA STAIN

The flagella stain described here is that of Ryu, modified by Kodaka and coworkers (182).

Performance of Test

(1) Use 48–72 h culture on anaerobic blood agar; CDC blood agar is recommended.
(2) Filter distilled water through a 0.45 μm pore–size filter (for step 3).
(3) Drop a medium-size drop of the water on a slide and spread it to the size of a nickel using the pipette.

Note: Do not use slides cleaned with methanol or ethanol.

(4) Pick a colony with an inoculating needle, taking care not to pick up agar, and lightly touch the needle to the center of the drop.
(5) Let the preparations dry in air at ambient temperature.
(6) Make working stain solution: Mix 10 parts of solution 1 with 1 part of solution 2 (see Appendix C, "Reagents") and filter through glass wool twice.
(7) Flood slides with stain solution and stain for 4–5 min.
(8) Wash off majority of stain, then, with a strong stream of tap water, wash the front and the back of the slide (10–20 s).
(9) Air-dry the slide and examine microscopically under oil immersion lens beginning at the periphery. You may need to search many fields to find cells with intact flagella.

Interpretation of Results

(1) Positive: Flagella seen attached to the bacterial cells.
(2) Negative: No flagella seen.

FLUORESCENCE

The use of laked blood or rabbit blood in the medium has been found to enhance the detection of fluorescence of the black/brown pigmented *Porphyromonas* species and *Prevotella* species.

Performance of Test

(1) Use a Woods lamp to expose culture plate, patient sample, or infected site to UV light.

Note: Use cotton gloves to protect skin from UV light exposure!

Especially when examining colonies, keep moving the plate until you find the optimal angle. You may need to hold the colonies very close to the UV light source (within an inch or so), especially if the bulb is weak. Fluorescence may take a few seconds to develop.

(2) Note the presence and color of fluorescence. It often is necessary to reincubate plates for several days before fluorescence is visible.

Interpretation of Results (see Box 4-1 on page 64)

(1) Positive: Distinct brick-red color detected with Woods lamp (see Figure 4-7).

Note: Brick-red fluorescence is the only reliable color for presumptive identification of *Porphyromonas* species and pigmented *Prevotella* species. Other anaerobes may also have characteristic fluorescences.

(2) Negative: No color change detected.

GELATIN LIQUEFACTION

Performance of Test

(1) Refrigerate an actively growing (>2+ turbidity) PRAS gelatin tube along with an uninoculated tube (negative control) for at least 1 h.

(2) Remove tubes to room temperature and invert immediately. Observe the test and control tubes for liquefaction every 5 min (the solidified agar begins to melt and falls toward the inverted top of the tube). Record results.

Interpretation of Results

(1) Positive: Gelatin medium in the inoculated tube fails to solidify (drops to the top of the inverted tube immediately).

(2) Weak positive: The inoculated tube becomes liquid when it reaches room temperature (<30 min).

(3) Negative: Tube fails to liquefy when it reaches room temperature (>30 min).

GROWTH STIMULATION TEST (see Box 6-1 on page 130)

Growth supplements are required for growth or for stimulation of growth by certain anaerobes. If the organism fails to grow in a medium such as PYG or thioglycolate, supplements should be added.

- 0.5% arginine is required by *E. lenta.*
- Formate-fumarate (0.3 and 0.3%) is required by the *B. ureolyticus* group.
- 1% pyruvate stimulates the growth of *Bilophila* and *Veillonella* species.

C

- 1% pyruvate with 0.25% magnesium sulfate stimulates sulfate-reducing bacteria.
- N-acetylmuramic acid (0.001%) stimulates *B. forsythus*.
- Hemin stimulates pigmented gram-negative rods and the *B. fragilis* group.
- Sodium bicarbonate (1%) stimulates many gram-negative rods.
- 0.5% Tween-80 enhances the growth of gram-positive organisms.
- Serum (usually 1%) can be used to stimulate the growth of gram-positive and gram-negative anaerobes.

Performance of Test

Add supplement(s) to appropriate concentration and incubate for 48–72 h along with a tube without the supplement(s) as a control.

Interpretation of Results

Compare the growth of supplemented tubes and the control tube; if growth is substantially better (more turbid) in the presence of supplement, that is interpreted as "positive." Add the supplement providing the best enhancement to all the tubes inoculated for biochemical tests (i.e., PRAS biochemicals, motility).

INDOLE PRODUCTION (TRYPTOPHANASE)

To perform the indole test, growth medium that contains tryptophan is mandatory. Also, since the enzyme that degrades tryptophan is diffusible in agar, it is important to have only one culture of an organism per plate. Do not use a plate that has a nitrate disk on it; a positive nitrate test can interfere with the spot indole test by inducing false-negative results. It also has been reported that false-negative results may be obtained in indole nitrate broth medium when nitrite concentrations reach 0.25–0.75 mg/ml (nitrate test positive).

Performance of Test

Spot Test

(1) Inoculate an agar medium that contains sufficient tryptophan, such as Brucella blood agar or egg yolk agar.
(2) Place a piece of filter paper (No. 1 Whatman) in a clean Petri dish cover.
(3) Moisten the filter paper with paradimethylaminocinnamaldehyde reagent. The paper should be saturated, but not dripping wet.
(4) Remove several colonies from the agar with a wooden stick or loop and rub on the filter paper. When testing anaerobic bacteria, it is good to use a heavy inoculum, since sometimes the reaction can be weak. Several tests may be performed on a single filter paper, but once it dries, it must be replaced.
(5) Alternatively, a sterile paper disk can be placed on an area of heavy growth for 5 min, then removed to an empty Petri dish and a drop of reagent added to the disk.

Broth Test

(1) Inoculate organism into indole nitrate medium or chopped meat broth. Incubate until good growth occurs.
(2) Remove 2 ml of culture to a small tube for testing.
(3) Add 1 ml of xylene, stopper with a rubber stopper, shake vigorously, and let stand for 15 min with the cap on. It is good to open the cap to release gas pressure.
(4) Add 0.5 ml of Ehrlich's reagent slowly down the side of the tube.
(5) Alternatively, Kovacs reagent may be used without the xylene extraction step.

Interpretation of Results

Positive

- Spot test: Development of a blue or green color around the inoculum within 30 s. The dark-pigmented organisms can produce a greenish color, or sometimes the color development may be masked by the pigment and therefore the inoculum must be examined with care (see Figure 4-9).
- Broth test: Development of a pink or fuchsia ring in a layer directly under the Ehrlich's reagent within 15 min.

Negative

- Spot test: No color change or pinkish color. Late color development should be disregarded.
- Broth test: Development of a yellow ring. Late weak color development should be disregarded.

LECITHINASE

Lecithinase mediates the breakdown of lecithin to diglyceride and phosphorylcholine. On EYA, the formation of insoluble diglycerides causes a visible opacity around the colony.

Performance of Test

Subculture the organism onto EYA and incubate for 24–72 h.

Interpretation of Results

(1) Positive: White, opaque, diffuse zone around the colonies and extending into the medium surrounding the colonies (see Figure 4-11).
(2) Negative: No reaction on the agar. Compare negative plate with uninoculated plate, since lecithinase can diffuse throughout the whole agar and make interpretation difficult.

C

LIPASE

Bacterial lipases hydrolyze the breakdown of triglycerides into glycerol and free fatty acids. Fatty acids are mostly insoluble and cause opacity on EYA, producing an iridescent sheen on the colonies and the surface of EYA. Unlike lecithinase, lipase is not diffusible, and the reaction occurs only on the surface of the agar in the immediate vicinity of the colony.

Performance of Test

Subculture the organism to EYA and incubate for 48 h to 1 week.

Interpretation of Results

(1) Positive: An iridescent sheen on the surface of bacterial growth and on the agar surface around the colonies (observe under oblique light) (see Figure 4-12). The lipase reaction can take up to 1 week to occur for some isolates.

(2) Negative: No reaction on the agar.

MILK

Anaerobic bacteria may produce four different reactions in milk medium: acid, clotting, gas, and digestion.

Performance of Test

(1) Follow the same procedure to inoculate PRAS milk tubes as is described for inoculation of PRAS biochemical tests.

(2) Incubate tube at 35 °C for 4 d to 3 weeks.

(3) Observe for presence of acid, gas, clotting, and digestion; record.

Interpretation of Results

(1) Acid pH: Lactose fermentation is detected with a pH electrode at the end of the incubation period.

(2) Gas: Presence of bubbles, release of pressure upon loosening cap, or stormy fermentation (gas disrupting a clot).

(3) Clotting: Either lactose fermentation (coagulation of casein in low pH milieu) or hydrolysis of casein by a renninlike enzyme may occur.

(4) Digestion: Clearing of the milk in the tube; can take 4 d to 3 weeks. It may occur with curd formation or the curd may be digested. Be sure to differentiate digestion from the whey that is also produced. If the pH is strongly acid after incubation, then digestion has probably not taken place.

MOTILITY

Prepare a wet mount or use the hanging drop technique to examine an actively growing, warm culture in a liquid medium. Look for directed movement that can be differentiated from both Brownian movement and the flow of organisms in streams through the medium.

NITRATE REDUCTION TEST

Performance of Test

Disk Test

(1) Make a fresh subculture of the organism to be tested onto a supportive agar and place a nitrate disk (available commercially) in the heavy inoculum area. Incubate anaerobically for 24–48 h.

(2) Remove the disk from the surface of the plate and place it in a clean Petri dish or on a slide.

(3) Add 1 drop each of reagents A and B (see Appendix D). If no color develops in a few minutes, drop a small amount of zinc dust or zinc granules onto the surface of the disk and observe up to 5 min.

Broth Test

(1) Inoculate organism into indole nitrate medium. Incubate until good growth occurs.

(2) Remove 1 ml of culture to a small tube for testing.

(3) Add 0.2 ml each of reagents A and B. If no color develops in a few minutes, add a small amount of zinc dust or zinc granules to the tube and observe for 5 min.

Interpretation of Results

(1) Positive: Red or pink color development after adding the reagents **or** no color development after adding zinc (see Figure 4-10).

(2) Negative: No color development after adding the reagents **and** red color development after adding zinc.

PRAS BIOCHEMICAL INOCULATION

Performance of Test

(1) Use an actively growing broth culture (without carbohydrate) of at least 2+ turbidity. Invert the culture to provide an even suspension, and inoculate 2–10 drops into the appropriate PRAS biochemicals including a PY base. Inoculation should be done under anaerobic atmosphere (in a chamber or under continuous gas flow into the tube during manipulation) or through a rubber stopper using a needle and syringe. Alternatively, suspend cell paste into a broth medium (e.g., PY) to make a turbid suspension and use as the inoculum.

(2) If the organism grows poorly, be certain to add the necessary and appropriate supplements to each tube (see "Growth Stimulation Test," above). Tween and hemin

can be added directly to the inoculum broth just prior to inoculation, since only small amounts are needed.

(3) After the last tube has been inoculated, plate one drop of the inoculum onto a BA plate to check for purity and viability (incubate anaerobically), and one drop onto either BA or chocolate agar for CO_2 incubation.

(4) Incubate at 37 °C until good growth is seen (>2+). If 2+ growth is not reached after 1 week of incubation, a better supplement should be sought.

(5) Interpret the carbohydrate fermentation results by measuring acid production using a pH meter (see "Carbohydrate Fermentation," above), read other biochemical tests, and process appropriate tubes for GLC.

RAPID ENZYME TESTS (SEE CHAPTER 5)

The presence of chromogenic or fluorogenic glycosidases, alkaline phosphatase, aminopeptidases, decarboxylases, and dihydrolases can be detected in commercial identification kits, such as RapID ANA, Rapid ID 32A, Crystal, or API ZYM, or separately using Rosco Diagnostic Tablets, WEE-TABs, or 4-methylumbelliferyl (4-MU) substrates from Sigma (described below). Use appropriate positive and negative controls during test runs.

Performance of Test

Rosco Tablets

(1) Use actively growing (48–72 h) culture on an agar plate medium that does not contain carbohydrates.

(2) Make a heavy suspension of organism (>3.0 McFarland turbidity) in 0.25 ml of saline if only one test is to be performed, otherwise in 1.0–1.5 ml of saline, and dispense three to five drops to each substrate tube to be tested.

(3) Add the appropriate Rosco tablet, vortex the suspension, incubate at 37 °C for 4 h (alkaline phosphatase) or 4–24 h (glycosidases, aminopeptidases, decarboxylases, and dihydrolases).

Note: If decarboxylase (LDC, ODC) or dihydrolase (ADH), add three drops of paraffin to overlay before incubation.

WEE-TABs

(1) Use actively growing (48–72 h) culture on an agar plate medium that does not contain carbohydrates.

(2) Make a heavy suspension of organism (>3.0 McFarland turbidity) in distilled water and dispense the number of drops recommended by the manufacturer to each substrate. If only a single test is to be performed, add water drops to the tablet and heavily inoculate the strain directly into the tube.

(3) Incubate at 37 °C for 2–24 h. For each test follow the manufacturer's instructions.

4-MU Substrates

(1) Use actively growing (48–72 h) culture on an agar plate medium that does not contain carbohydrates.

(2) Make a heavy suspension of organism (>6.0 McFarland turbidity) in filter-sterilized TES buffer (pH 7.5) and dispense 20 µl onto a blank paper disk or inoculate a 10 µl loopful of colony material directly on the blank paper disk placed in an empty Petri dish.
(3) Add 20 µl of desired substrate in its 1:100 dilution in TES (made up from frozen stock solution that has been originally prepared by dissolving the substrate (Sigma) in dimethylsulfoxide 10 mg/ml and stored as a stock in the freezer).
(4) Place the lid on the Petri dish and incubate at 37 °C for 15–30 min.

Interpretation of Results: Rosco Tablets and WEE-TABs

Glycosidases

(1) Positive: Yellow color development. When testing β-N-acetylglucosaminidase, only a strong yellow color should be recorded as positive.
(2) Negative: No color or very pale yellow color.

Aminopeptidases

(1) Positive: After addition of aminopeptidase reagent, development of dark red color.
(2) Negative: No red color or change after reagent addition.

Dihydrolases/Decarboxylases

(1) Positive: Aniline red (ADH) or dark blue (LDC, ODC).
(2) Negative: No color or very pale color.

Interpretation of Results: 4-MU Substrates

View the Petri plate under long-wave UV light (see "Fluorescence" for equipment and examination specifications) and look for bright blue-white fluorescence.

(1) Positive: Blue-white fluorescence.
(2) Negative: No fluorescence.

REVERSE-CAMP TEST

Alpha-toxin-producing *Clostridium perfringens* and β-hemolytic *Streptococcus agalactiae* form a characteristic arrowhead-shaped, synergistic hemolytic pattern on blood agar.

Performance of Test

(1) Inoculate a blood agar plate with a single streak of suspected *C. perfringens*.
(2) Make a single streak of group B β-hemolytic *Streptococcus* at a 90° angle to within a few mm of the suspected *C. perfringens*.
(3) Incubate for 24–48 h.

Interpretation of Results

(1) Positive: Synergistic arrowhead-shaped hemolytic pattern present (see Figure 4-13).
(2) Negative: No synergistic action observed.

SPECIAL-POTENCY DISK IDENTIFICATION

Performance of Test

(1) Subculture the isolate onto a blood agar plate. To ensure an even, heavy lawn of growth, streak the first quadrant back and forth several times. Streak the other quadrants to yield isolated colonies.
(2) Place the antibiotic disks on the first and second quadrants well apart (~20 mm) from each other.
(3) If you have several isolates to test, first streak all the plates and then add the disks to them at the same time.
(4) Incubate the plate(s) anaerobically for 48–72 h at 35–37 °C.
(5) Examine for a zone of inhibition of growth surrounding the disk.

Interpretation of Results (see Box 4-2)

Vancomycin, Kanamycin, Colistin Disks

(1) Sensitive (S): Zone of inhibition is ≥10 mm.
(2) Resistant (R): Zone of inhibition is <10 mm.

SPS Disk

(1) Sensitive (S): Zone of inhibition is ≥12 mm. The value of 12 mm is the guideline according to the literature. In practice, *P. anaerobius* usually gives a very large zone (≥16 mm), whereas the other anaerobic cocci that appear to be sensitive to SPS produce a smaller zone. To presumptively identify *P. anaerobius,* examine the isolate for the presence of characteristic Gram and colonial morphology and overripe cantaloupe odor.

(2) Resistant (R): Zone of inhibition is <12 mm.

C

SPORE TEST

Spores can be observed in Gram stained preparations, under darkfield microscopy, or demonstrated with a spore test. Some commonly encountered *Clostridium* species, such as *C. perfringens* and *C. ramosum,* sporulate poorly and spores are rarely seen.

Performance of Test

(1) Use an actively growing broth culture of at least 2+ turbidity.
(2) Let the culture stand at room temperature for several days to 1 week.
(3) Plate one drop of the culture onto a BA plate for anaerobic incubation to check for purity and viability (pre-spore test), and one drop onto either BA or chocolate agar for CO_2 incubation (purity control).
(4) Add equal volumes of the culture to 95% ethanol and mix gently. Incubate at room temperature for 30 min.

(5) Subculture mixture on BA plate for anaerobic incubation (post-spore test).

(6) Incubate plates for 2 d.

(7) Examine for growth.

Interpretation of Results

(1) Positive: Growth present on both anaerobic plates (pre- and post-).

(2) Negative: Growth on the pre-spore test plate and no growth on the plate inoculated after the spore test.

TRYPSINLIKE ACTIVITY

Trypsinlike activity can be demonstrated using several methods including Rosco tablets, API ZYM system, BANA reagent (N-benzoyl-DL-arginine-2-naphthylamide) (192), and CAAM reagent (carbobenzoxy-L-arginine-7-amino-4-methylcoumarin amide-HCl) (298). The method utilizing Rosco tablets is described in this manual.

Performance of Test

(1) Use an actively growing (48–72 h) culture on an agar medium that does not contain carbohydrates. Make a heavy suspension of the organism (>2.0 McFarland turbidity) in 0.25 ml of saline.
(2) Add a Rosco trypsin tablet, vortex, and incubate at 37 °C for 4–18 h.
(3) After incubation, add a drop of aminopeptidase reagent (Rosco) and wait 5 min or examine under long-wave UV light.

Interpretation of Results

(1) Positive: Red color after adding aminopeptidase reagent or blue fluorescence under UV light.
(2) Negative: Yellow-orange color after adding aminopeptidase reagent or no fluorescence under UV light.

UREASE

Performance of Test

Rapid Test

Make a heavy suspension of the actively growing organism in 0.25 ml of saline and add a urea disk (WEE-TAB, Rosco). Alternatively, make a heavy suspension of the organism in 0.5 ml of urea broth (Appendix D). Incubate aerobically at 37 °C for 4–24 h.

Broth Test

(1) Inoculate a PY base medium and a PY medium supplemented with 30 mM urea. Incubate until good growth occurs.
(2) Determine the pH of both the PY and the PY with urea.

Interpretation of Results

Positive

- Rapid test: Bright red color in urea broth or urea disk tube.
- Broth test: pH of PY urea tube at least 0.3 higher than pH of PY base.

Negative

- Rapid test: No color change to yellow color in urea broth or urea disk tube.
- Broth test: ≤0.3 pH change in PY urea tube.

C

appendix D

Preparation of Media and Reagents

Quality Control Procedures

All media, reagents, and test systems should be monitored for acceptable performance. The Wadsworth quality control procedures have been established based on the recommendations of NCCLS Quality Assurance Document M22-A for commercially prepared microbiological culture media (245). Quality control (QC) procedures for antimicrobial susceptibility testing have been described in Chapter 8. Where tests such as nitrate reduction are common to aerobic, facultative, and anaerobic bacteria, results obtained for the aerobic or facultative bacteria will suffice.

All anaerobic systems should be tested daily for anaerobiosis using an indicator such as resazurin or methylene blue. When using resazurin, it must be protected from light to prevent inactivation. If the methylene blue strip is not white upon opening the foil packet, discard it and use a fresh one.

Table D-1 gives quality control procedures suggested for commonly used anaerobic media and reagents. Known positive and negative organisms are always used when testing commercially prepared or in-house prepared media and reagents. Each batch or lot number of plated or biochemical test medium should be tested. Antibiotic disks, other disks used for identification, additives for growth stimulation, and indole test reagents should be tested at weekly intervals. The following cultures should be maintained for use in quality control procedures:

B. fragilis, ATCC 25285
B. thetaiotaomicron, ATCC 29741
B. ureolyticus
P. melaninogenica, ATCC 25845
F. necrophorum
C. perfringens, ATCC 13124
E. lenta
B. ovatus, ATCC 8483
M. micros
P. anaerobius
E. coli, ATCC 25922
P. mirabilis
E. faecalis, ATCC 29212

table D-1. Quality control procedures for media and reagents

Media or Reagent	Test Organism	Results/Rationale
AGARS		
Blood agar	*B. fragilis*	growth
	P. levii	growth, pigment[1]
	C. perfringens	growth, double zone of β-hemolysis
	F. nucleatum	growth
	P. anaerobius	growth
	P. melaninogenica	growth, pigment[1]
BBE	*B. fragilis*	growth, black colonies
	P. melaninogenica	no growth, bile inhibition
	E. coli	no growth, gentamicin inhibition
KVLB	*P. melaninogenica*	growth, black pigment[1]
	C. perfringens	no growth, vancomycin inhibition
	E. coli	no growth or strong inhibition by kanamycin
PEA	*B. fragilis*	growth
	P. mirabilis	inhibition of swarming
EYA	*C. perfringens*	growth, lecithinase production
	F. necrophorum	growth, lipase production
CCFA, CCEY	*C. difficile*	growth, chartreuse fluorescence
	B. fragilis	inhibition of growth by cefoxitin
	E. coli	inhibition of growth by cycloserine
DISKS		
Colistin (10 µg)	*F. necrophorum*	susceptible
	B. fragilis	resistant
Kanamycin (1 mg)	*C. perfringens*	susceptible
	B. fragilis	resistant
Vancomycin (5 µg)	*C. perfringens*	susceptible
	B. fragilis	resistant
Nitrate	*B. ureolyticus*	nitrate reduced
	B. fragilis	nitrate not reduced
SPS	*P. anaerobius*	susceptible, zone ≥12 mm
	P. asaccharolyticus	resistant, zone <12 mm
GROWTH STIMULATION TESTS		
Base medium + F/F	*B. ureolyticus*	growth enhanced compared to that in base medium
Base medium + pyruvate	*B. wadsworthia*	growth enhanced compared to that in base medium
Base medium + arginine	*E. lenta*	growth enhanced compared to that in base medium
Base medium + Tween-80	*M. micros*	growth enhanced compared to that in base medium
MISCELLANEOUS		
Urea broth	*B. ureolyticus*	urease produced
	B. fragilis	urease not produced
Indole test reagents	*F. necrophorum*	indole produced
	B. fragilis	indole not produced
Thioglycolate	*B. fragilis*	growth in 24 hours
PY base	*B. fragilis*	growth in 24 hours

[1] Pigment production may require more than 48 hours incubation.

Some of these strains are available as freeze-dried disks (Anaerobe Systems, BBL, and American Type Culture Collection). Well-documented laboratory strains may be used where ATCC strains are not listed or not available.

MEDIA AND REAGENTS

Most of the media suggested for use are available in dehydrated form and are prepared according to the directions of the manufacturer. Supplements are listed for each medium when enrichment is advisable or required. Most agar bases may be dispensed before autoclaving into screwcapped tubes or 100 ml bottles and stored until needed. They can then be remelted, supplemented, and poured into plates. Many prepared agar media are available from local media suppliers. For instance, PRAS blood agar, kanamycin vancomycin laked blood agar, Bacteroides bile esculin, egg yolk, and phenylethyl alcohol agar plates are available from Anaerobe Systems. The shelf life of media stored under anaerobic conditions is generally longer than when they are stored in air.

Media used for biochemical tests and fermentation reactions are PRAS, prepared as described in the VPI Anaerobe Laboratory Manual (154), can be supplemented with vitamin K_1 and hemin; they can be purchased from Remel and Anaerobe Systems. The use of thioglycolate-based medium as recommended by the CDC (87) is an acceptable alternative to the use of PRAS media; however, all reactions listed in this manual are based upon results obtained in PRAS media.

Arginine (10%)

Ingredients

Arginine HCl	10 g
Distilled water	100 ml

Preparation

(1) Dissolve the arginine HCl in water.
(2) Sterilize by filtration.
(3) Store in the refrigerator.

Use

Add 0.5 ml of arginine solution to 10 ml of freshly steamed thioglycolate medium for growth stimulation of *E. lenta*.

Bacteroides Bile Esculin Agar (BBE)

Ingredients

Trypticase soy agar	40 g
Oxgall	20 g

D

Esculin	1 g
Ferric ammonium citrate	0.5 g
Hemin solution (5 mg/ml)	2 ml
Gentamicin solution (40 mg/ml)	2.5 ml
Distilled water	1,000 ml

Preparation

(1) Combine all the ingredients.
(2) Adjust the pH to 7.0.
(3) Boil or steam to dissolve.
(4) Autoclave at 121 °C for 15 min.
(5) Cool to 50 °C and pour plates.

Shelf Life

- Plates: 2 weeks at room temperature; 3 months at refrigerator temperature.
- Bottles: 1 year at refrigerator temperature.

BL (Blood and Liver) Agar

Ingredients

BL agar (Nissui Pharmaceutical Co.)	58 g
Distilled water	1,000 ml
Defibrinated horse blood	50 ml

Preparation

(1) Combine agar base and water.
(2) Heat to dissolve.
(3) Autoclave at 115 °C for 20 min.
(4) Cool to 50 °C, add BS solution (Nissui Pharmaceutical Co.) and defibrinated horse blood, and pour plates.

Shelf Life

Two weeks at refrigerator temperature.

BS (*Bifidobacterium*-Selective) Agar

Ingredients

BL agar (Nissui Pharmaceutical Co.)	58 g
BS solution (Nissui Pharmaceutical Co.)	50 ml

Defibrinated horse blood	50 ml
Distilled water	1,000 ml

Preparation

(1) Combine agar base and water.
(2) Heat to dissolve.
(3) Autoclave at 115 °C for 20 min.
(4) Cool to 50 °C, add BS solution and defibrinated horse blood, and pour plates.

Shelf life

Two weeks at refrigerator temperature.

Brucella Blood Agar

Ingredients

Brucella agar (BBL or Difco)	43 g
Hemin solution (5 mg/ml)	1 ml
Vitamin K_1 solution (1 mg/ml)	1 ml
Distilled water	1,000 ml
Sterile defibrinated sheep blood	50 ml

Preparation

(1) Combine all ingredients except sheep blood.
(2) Boil to dissolve.
(3) Autoclave at 121 °C for 15 min.
(4) Cool to 50 °C, add sheep blood, and pour plates.

Shelf life

- Plates: 2 weeks in refrigerator.
- Bottles: 1 year at refrigerator temperature.

Brucella Blood Agar for Susceptibility Testing

Ingredients

Brucella agar (BBL or Difco)	43 g
Vitamin K_1 solution (1 mg/ml)	1 ml
Distilled water	1000 ml
Sterile laked sheep blood	50 ml

Hemin (5 μg/ml) may be added to improve growth of some organisms. Similarly, other supplements may be added to enhance growth of fastidious anaerobes, but care should be taken to ensure that quality control organisms are within range.

Preparation

(1) Combine all ingredients except laked blood.
(2) Boil to dissolve.
(3) Autoclave at 121 °C for 15 min.
(4) Media may be refrigerated in bottles or flasks, and used as needed for susceptibility tests. Blood should be added when the test plates are poured (see Chapter 8).

Shelf Life

For media without added blood, shelf life is up to 1 month refrigerated.

Brucella Broth

Brucella broth is used in broth dilution susceptibility tests.

Ingredients

Brucella broth (BBL or Difco)	28 g
Vitamin K_1 solution (1 mg/ml)	1 ml
Hemin solution (5 mg/ml)	1 ml
Sodium bicarbonate (20 mg/ml)	5 ml
Distilled water	1,000 ml

Preparation

(1) Dissolve ingredients.
(2) Dispense in convenient volumes.
(3) Autoclave at 121 °C for 15 min.

Shelf Life

One year at refrigerator temperature.

Cadmium Fluoride Acriflavine Tellurite Agar (CFAT)

Ingredients

Trypticase soy broth	30 g
Glucose	5 g
Agar	15 g

Cadmium sulfate	13 mg
Sodium fluoride	80 mg
Neutral acriflavine	1.20 mg
Basic fuchsin	0.25 mg
Distilled water	1,000 ml
Sterile potassium tellurite	2.5 mg
Sterile defibrinated sheep blood	50 ml

Preparation

(1) Combine all ingredients except potassium tellurite and sheep blood.
(2) Boil to dissolve.
(3) Autoclave at 121 °C for 15 min.
(4) Cool to 50 °C, add potassium tellurite and sheep blood, then pour plates.

Shelf Life

Shelf life is unknown.

Cefoxitin Cycloserine Egg Yolk Agar (CCEY)

Ingredients

Brazier's CCEY agar—LAB 160 (Lab M, Bury England)	48 g
Distilled water	1,000 ml
Cefoxitin cycloserine selective supplement X093 (8 μg/ml cefoxitin and 250 μg/ml cycloserine)	2 bottles
Egg yolk emulsion	40 ml
Lysed horse or sheep blood	10 ml

Preparation

(1) Dissolve the powder in distilled water and mix.
(2) Let dissolve for 10 min.
(3) Cool to 47 °C.
(4) Add 2 bottles of supplement X093, egg yolk emulsion, and blood; then pour the plates.

Shelf Life

Seven days at 2–8 °C in the dark.

Chocolate Bacitracin Agar

Ingredients

Blood agar base	40 g
Distilled water	1,000 ml
Sterile defibrinated sheep blood	50 ml
Bacitracin	300 mg

Preparation

(1) Combine blood agar base and water.
(2) Boil to dissolve.
(3) Autoclave at 121 °C for 15 min.
(4) Cool to 50 °C. Add blood.
(5) Raise temperature to 70–80 °C, mixing constantly but gently until blood turns a chocolate color.
(6) Cool to 50 °C, add bacitracin, and pour plates.

Shelf Life

- Plates: 1 week at refrigerator temperature.
- Bottles: 1 year at refrigerator temperature.

Cycloserine Cefoxitin Fructose Agar (CCFA)

Ingredients

Proteose peptone No. 2 (Difco)	40 g
Disodium phosphate	5 g
Monopotassium phosphate	1 g
Fructose	6 g
Agar	20 g
Neutral red (1% in ethanol)	3 g
Distilled water	1,000 ml
Cycloserine	500 mg
Cefoxitin	16 mg

Preparation

(1) Combine all ingredients except cycloserine and cefoxitin.
(2) Adjust pH to 7.6.

(3) Boil to dissolve.
(4) Autoclave at 121 °C for 15 min.
(5) Cool to 50°, add cycloserine and cefoxitin, then pour plates.

Shelf Life

- Plates: 8 weeks at refrigerator temperature.
- Bottles: 1 year at refrigerator temperature.

Egg Yolk Agar (EYA)

Ingredients

Proteose peptone No. 2 (Difco)	40 g
Disodium phosphate	5 g
Monopotassium phosphate	1 g
Sodium chloride	2 g
Magnesium sulfate	0.1 g
Glucose	2 g
Vitamin K_1 (10 mg/ml)	1 ml
Hemin solution (5 mg/ml)	1 ml
Agar	20 g
Distilled water	1,000 ml
Egg yolk emulsion (Difco or laboratory preparation, see below)	50 ml
or Egg yolk emulsion (Oxoid)	74 ml

Preparation

(1) Combine all ingredients except egg yolk, vitamin K_1, and hemin.
(2) Adjust pH to 7.6.
(3) Boil to dissolve.
(4) Autoclave at 121 °C for 15 min.
(5) Cool to 50 °C; add egg yolk emulsion, vitamin K_1, and hemin; and pour plates.

Shelf Life

- Plates: 1 week at refrigerator temperature.
- Bottles: 1 year at refrigerator temperature.

D

Egg Yolk Emulsion

Ingredients

Fresh eggs

Sterile normal saline

Preparation

(1) Cleanse fresh eggs thoroughly with an alcohol pad.
(2) Punch an airhole in one end and a hole approximately 10 mm in diameter in the other end.
(3) Carefully allow the egg white to drip out, assisting as necessary with sterile syringe and needle.
(4) When white has been removed, aspirate yolk into a sterile syringe.
(5) Place into sterile flask with glass beads and an equal volume of normal saline.
(6) Mix thoroughly and test sterility by plating a drop on Brucella blood agar and inoculating 1 ml into supplemented thioglycolate medium. Prepared egg yolk emulsion is available from Difco and Oxoid.

Ehrlich's Reagent

Ingredients

p-dimethylaminobenzaldehyde	1 g
Ethyl alcohol (95%)	95 ml
Hydrochloric acid (concentrated)	20 ml

Preparation

(1) Dissolve *p*-dimethylaminobenzaldehyde in alcohol.
(2) Slowly add hydrochloric acid.
(3) Store in a dark bottle refrigerated for up to 1 year.

Ferric Ammonium Citrate (1%)

Ingredients

Ferric ammonium citrate	1 g
Distilled water	100 ml

Preparation

(1) Dissolve ferric ammonium citrate in water.
(2) Store in a dark bottle refrigerated for up to 1 year.

Fildes Enrichment

Available from BBL and Difco.

Flagella Stain (Ryu)

Ingredients: Solution 1

5% Carbolic acid (Mallinckrodt)	10 ml
Tannic acid (J.T. Baker Chem. Co)	2 g
Saturated aluminum potassium sulfate 12-hydrate (~13.2 g/100 ml) (J.T. Baker Chem. Co)	10 ml

Preparation: Solution 1

(1) Mix 5% carbolic acid (phenol) solution, tannic acid, and saturated aluminum potassium sulfate 12-hydrate solution. Use a brown glass bottle.
(2) Store at room temperature.

Note: When using carbolic acid crystals, first melt crystals in 50 °C water bath, then pipette 5 ml of liquefied carbolic acid into 95 ml of distilled water. Use a glass bottle.

Solution 2

Saturated solution of crystal violet in ethanol (12 g per 100 ml). Store at room temperature.

Preparation: Final Stain

(1) Mix 10 parts of solution 1 with 1 part of solution 2 in a brown glass bottle.
(2) Filter the final stain twice through glass wool before using.

Formate Fumarate Additive

Ingredients

Sodium formate	3 g
Fumaric acid	3 g
Distilled water	50 ml
Sodium hydroxide	20 pellets

Preparation

(1) Combine ingredients, stirring until pellets are dissolved and fumaric acid is in solution.
(2) Bring pH to 7.0 with 4 N sodium hydroxide.
(3) Sterilize by filtration.

Use

Add 0.5 ml of the F/F additive to 10 ml of culture medium to stimulate growth of *B. ureolyticus* and other fastidious gram-negatives.

Fusobacterium-Selective (JVN) Agar

Ingredients

Ingredient	Amount
Fastidious anaerobe agar (FAA) (Lab M)	46 g
Josamycin (Fluka)	3 mg
Vancomycin	4 mg
Norfloxacin	1 mg
Distilled water	1,000 ml
Sterile defibrinated horse blood	50 ml

Preparation

(1) Combine agar base and distilled water.
(2) Bring to boil to dissolve.
(3) Autoclave at 121 °C for 15 min.
(4) Cool to 50 °C, add blood, vancomycin, norfloxacin, and josamycin; then pour plates.

Shelf Life

- Plates: 2 weeks at refrigerator temperature.
- Bottles: 1 year at refrigerator temperature.

D

Fusobacterium Neomycin Vancomycin (NV) Agar

Ingredients

Ingredient	Amount
Fastidious anaerobe agar (FAA) (Lab M, England)	46 g
Vancomycin (7.5 mg/ml)	1 ml
Neomycin (100 mg/ml)	1 ml
Distilled water	1,000 ml
Sterile defibrinated sheep blood	50 ml

Preparation

(1) Combine all ingredients except sheep blood and vancomycin.
(2) Adjust pH to 7.0–7.2.
(3) Autoclave at 121 °C for 15 min.
(4) Cool to 50 °C, add blood and vancomycin, then pour plates.

Shelf Life

- Plates: 10 days at refrigerator temperature.
- Bottles: 1 year at refrigerator temperature.

Hemin Solution (5 mg/ml)

Ingredients

Hemin (bovine)	0.5 g
Commercial ammonia water or sodium hydroxide (1 N)	10 ml
Distilled water	90 ml

Preparation

(1) Dissolve hemin in ammonia water or sodium hydroxide (1 N). Bring volume to 100 ml with water.
(2) Autoclave at 121 °C for 15 min.

Use

Add to medium as a supplement in a final concentration of 5 μg/ml.

Shelf Life

One month at refrigerator temperature.

Kanamycin Solution (100 mg/ml)

Ingredients

Kanamycin	1 g
Sterile phosphate buffer, pH 8	10 ml

Preparation

(1) Dissolve kanamycin in phosphate buffer.
(2) Store in refrigerator for up to 1 year.

Kanamycin Vancomycin Laked Blood Agar (KVLB)

Ingredients

Brucella agar (BBL or Difco)	43 g
Hemin solution (5 mg/ml)	1 ml
Vitamin K_1 solution (10 mg/ml)	1 ml
Kanamycin solution (100 mg/ml)	0.75 ml
Distilled water	1,000 ml

Vancomycin solution (7.5 mg/ml)	1 ml
Laked sheep blood (blood frozen overnight, then thawed)	50 ml

Preparation

(1) Combine all ingredients except vancomycin and laked sheep blood.
(2) Boil to dissolve.
(3) Autoclave at 121 °C for 15 min.
(4) Cool to 50 °C, add vancomycin and laked sheep blood, then pour plates.

Shelf Life

- Plates: 2 weeks at refrigerator temperature.
- Bottles: 1 year at refrigerator temperature.

Lactobacillus-Selective Medium (LBS)

Ingredients

LBS agar (Rogosa-BBL)	8.4 g
Tomato juice (grocery store)	40 ml
Distilled water	60 ml
Acetic acid (glacial)	0.13 ml

Preparation

(1) Mix ingredients.
(2) Steam for 20–30 min until agar dissolves or heat with frequent agitation and boil for 1 min. Do not autoclave.
(3) Cool to 45 °C and pour plates.

Shelf Life

Plates can be stored for 24 h at room temperature.

Laked Blood

Freeze blood in glass or plastic containers filled ≤3/4 full (to avoid cracking) for a minimum of overnight. Thaw when ready to use.

Magnesium Sulfate (10%)

Ingredients

Magnesium sulfate	5 g
Sterile distilled water	50 ml

Preparation

(1) Dissolve the magnesium sulfate in water.
(2) Sterilize by filtration or autoclave at 121 °C for 15 min.
(3) Store in refrigerator for up to 6 months.

Use

Add 0.25 ml of magnesium sulfate along with 1% pyruvate to 10 ml of medium to stimulate the growth of sulfate-reducing organisms.

McFarland Standards

Ingredients

Barium chloride, 1.175% aqueous ($BaCl_2 \cdot 2\ H_2O$; i.e., 0.048 M $BaCl_2$)
Sulfuric acid, 1% aqueous (0.36 M)

Preparation

(1) Add amounts of the two solutions indicated in Table D-2 to tubes or ampules.
(2) Seal.
Note: Tubes or ampules should have the same diameter as the tube to be used in subsequent density determinations.

0.5 McFarland Standard

Ingredients

Barium chloride, 1.175% aqueous	0.5 ml
Sulfuric acid, 1% aqueous	99.5 ml

Preparation

(1) Mix the two solutions.
(2) Vortex and immediately dispense 4–6 ml into an appropriate tube or ampule and seal.

table D-2. Preparation of McFarland nephelometer standards

Tube No.	1.175% barium chloride (ml)	1% sulfuric acid (ml)	Corresponding approximate density of bacteria/ml
1	0.1	9.9	3×10^8
2	0.2	9.8	6×10^8
3	0.3	9.7	9×10^8
4	0.4	9.6	1.2×10^9
5	0.5	9.5	1.5×10^9
6	0.6	9.4	1.8×10^9
7	0.7	9.3	2.1×10^9
8	0.8	9.2	2.4×10^9
9	0.9	9.1	2.7×10^9
10	1.0	9.0	3×10^9

Note: The tube or ampule should have the same diameter as the tube to which it will be compared. Unless the standard is contained in a heat-sealed tube or ampule it should be replaced every 6 months. A 0.5 McFarland standard is assumed to be equivalent to approximately 1.5 $\times$ 10^8 organisms/ml.

(3) Store in the dark at room temperature.
(4) Vigorously agitate the standard before each use.

Methanobacterium Medium (PRAS Roll Tubes)

Note: A gassing cannula apparatus must be used for roll-tube preparation.

Ingredients

Rumen fluid (Animal Technologies, Inc.)	15 ml
Sodium chloride	0.2 g
Ammonium chloride	0.1 g
Monopotassium phosphate	0.1 g
Magnesium chloride, 6-hydrate	0.1 g
Calcium chloride	0.1 g
Ammonium molybdate, 4-hydrate	0.002 g
Cobalt chloride, 6-hydrate	0.002 g
Sodium bicarbonate	0.4 g
Resazurin solution	0.8 ml
L-cysteine hydrochloride	0.05 g
Distilled water	200 ml

Preparation

(1) Combine all ingredients except cysteine.
(2) Steam for 15 min (resazurin will stay pink).
(3) Cool while gassing with CO_2.
(4) Add cysteine, dissolve, and adjust pH to 6.8. At this time, indicator will be colorless.
(5) Switch from CO_2 to 70% H_2 and 30% CO_2 (explosive mixture).
(6) Place 0.125 g agar in each roll tube.
(7) Dispense in 4.9 ml amounts, gassing with 70% H_2 and 30% CO_2.
(8) Autoclave tubes in a media press at 121 °C for 15 min.
(9) Just before use, add 0.02 ml of sterile 5% Na_2S solution.

Shelf Life

Three months at room temperature.

MRS (de Man, Rogosa, Sharpe) Medium

Ingredients

Peptone (casein/meat 50/50)	10 g
Yeast extract	5 g
Glucose	20 g
Dipotassium phosphate	2 g
Tween-80	1 ml
Sodium acetate	5 g
Ammonium citrate	2 g
Magnesium sulfate	0.1 g
Manganese sulfate	0.05 g
Agar	12 g
Distilled water	1,000 ml

Preparation

(1) Combine all ingredients and mix.
(2) Adjust pH to 6.2–6.6.
(3) Autoclave at 121 °C for 15 min.
(4) Cool and pour plates.

Shelf Life

Six to eight weeks at 2–8 °C.

Several commercial formulations that include the above ingredients are available.

Neomycin Solution (100 mg/ml)

Ingredients

Neomycin	1 g
Sterile phosphate buffer, pH 8	10 ml

Preparation

(1) Dissolve neomycin in phosphate buffer.
(2) Store in refrigerator for up to 1 year.

Nessler Reagent

Ingredients

Mercuric iodide	100 g
Potassium iodide	70 g
Sodium hydroxide (8 N)	500 ml
Distilled water	~500 ml

Preparation

(1) Dissolve mercuric iodide and potassium iodide in 100–200 ml distilled water.
(2) Add this solution slowly, with stirring, to the 8 N sodium hydroxide at 20–25 °C.
(3) Adjust volume to 1 L with distilled water.
(4) Store in a rubber-stoppered bottle in the dark for up to 1 year.

Nitrate Disks

Ingredients

Potassium nitrate	30 g
Sodium molybdate dihydrate	0.1 g
Distilled water	100 ml
Sterile $\frac{1}{4}$ in filter paper disks (BBL)	

Preparation

(1) Dissolve the nitrate and molybdate in the water.
(2) Sterilize by filtration.
(3) Dispense 20 μl quantities of the solution onto the disks arranged in a single layer in clean Petri dishes.
(4) Allow the disks to dry at room temperature for 72 h.
(5) Store at room temperature for up to 6 months.

Nitrate Reagents

Ingredients: Solution A

Sulfanilic acid	0.5 g
Glacial acetic acid	30 ml
Distilled water	120 ml

Ingredients: Solution B

1,6-Cleve's acid	0.2 g
Glacial acetic acid	30 g
Distilled water	120 ml

Preparation

(1) Dissolve ingredients of each solution in distilled water in separate containers.
(2) Store refrigerated for up to 3 months.

Oxgall (40%)

Ingredients

Oxgall	40 g
Distilled water	100 ml

Preparation

(1) Dissolve oxgall in water.
(2) Autoclave at 121 °C for 15 min.
(3) Store refrigerated up to 1 year.

Use

Add 0.5 ml of oxgall solution to 10 ml of thioglycolate medium for the bile inhibition test.

Paradimethylaminocinnamaldehyde

Ingredients

Paradimethylaminocinnamaldehyde (Aldrich)	1 g
Hydrochloric acid (10%)	100 ml

Preparation

(1) Dissolve paradimethylaminocinnamaldehyde in the hydrochloric acid solution.
(2) Store in a dark bottle in the refrigerator for up to 2 months.

Use

Use for the spot-indole test.

Phenylethyl Alcohol Blood Agar (PEA)

Ingredients

Phenylethyl alcohol agar	42.5 g
Vitamin K_1 solution (10 mg/ml)	1 ml

Distilled water	1,000 ml
Sterile defibrinated sheep blood	50 ml

Preparation

(1) Combine ingredients except sheep blood.
(2) Boil to dissolve.
(3) Autoclave at 121 °C for 15 min.
(4) Cool to 50 °C, add sheep blood, then pour plates.

Shelf Life

- Plates: 2 weeks at refrigerator temperature.
- Bottles: 1 year at refrigerator temperature.

Pyruvate (20%)

Ingredients

Pyruvic acid, sodium salt	20 g
Distilled water	100 ml

Preparation

(1) Dissolve the pyruvate in water.
(2) Sterilize by filtration or autoclave at 121 °C for 15 min.
(3) Store in refrigerator.

Resazurin

Ingredients

Resazurin (Difco, Sigma)	1 tablet
Distilled water	44 ml

Preparation

(1) Dissolve tablet in water.
(2) Store at room temperature.

Use

This indicator is used to prepare PRAS biochemicals.

Rifampin (1 mg/ml)

Ingredients

Rifampin	100 mg

Absolute ethanol	20 ml
Distilled water	80 ml

Preparation

(1) Dissolve the rifampin in the alcohol.
(2) Add the distilled water.
(3) Store in refrigerator for up to 2 months.

Rifampin Blood Agar

Ingredients

Brucella agar (BBL or Difco)	43 g
Hemin solution (5 mg/ml)	1 ml
Vitamin K_1 solution (10 mg/ml)	1 ml
Distilled water	1,000 ml
Sterile defibrinated sheep blood	50 ml
Rifampin solution (1mg/ml, see above)	50 ml

Preparation

(1) Combine all ingredients except sheep blood and rifampin.
(2) Boil to dissolve.
(3) Autoclave at 121 °C for 15 min.
(4) Cool to 50 °C, add rifampin and sheep blood, then pour plates.

Shelf Life

- Plates: 1 week at refrigerator temperature.
- Bottles: 1 year at refrigerator temperature.

Ringer's Solution

Ingredients

Sodium chloride	9 g
Calcium chloride, dihydrate	0.25 g
Potassium chloride	0.4 g
Distilled water	1,000 ml

Preparation

(1) Combine all ingredients.
(2) Mix to dissolve.

Use

Use to prepare Ringer's solution with metaphosphate, or Ringer's dilution solution.

One-Quarter Strength Ringer's Solution with Metaphosphate

Ingredients

Ringer's solution	50 ml
Resazurin solution (see above)	0.8 ml
Distilled water	150 ml
Sodium metaphosphate	0.2 g
L-cysteine hydrochloride	0.1 g

Preparation

(1) Combine all ingredients.
(2) Mix to dissolve; dispense.
(3) Autoclave at 121 °C for 15 min.

Ringer's Dilution Solution

Ingredients

Ringer's solution	200 ml
Resazurin solution	0.8 ml
L-cysteine hydrochloride	0.1 g

Preparation

(1) Combine all ingredients.
(2) Mix to dissolve, then dispense 9 ml into disposable Hungate anaerobic culture tubes (Bellco).
(3) Autoclave at 121 °C for 15 min.

Skim Milk (20%)

Ingredients

Skim milk powder (BBL)	200 g
Distilled water	1,000 ml

Preparation

(1) Combine ingredients.
(2) Dispense 0.25–0.5 ml into half-dram vial.
(3) Autoclave at 110 °C for 10 min with quick exhaust.
(4) Store refrigerated for up to 6 months.

Sodium Bicarbonate (20 mg/ml)

Ingredients

Sodium bicarbonate	2 g
Distilled water	100 ml

Preparation

(1) Dissolve sodium bicarbonate in water.
(2) Sterilize by filtration.
(3) Store refrigerated for up to 6 months.

Use

Add 0.5 ml of sodium bicarbonate solution to 10 ml of medium.

Sodium Polyanethol Sulfonate (SPS) Disks

Ingredients

Sodium polyanethol sulfonate (Sigma)	5 g
Distilled water	100 ml
Sterile $\frac{1}{4}$ in filter paper disks (BBL)	

Preparation

(1) Dissolve sodium polyanethol sulfonate in water.
(2) Sterilize by filtration.
(3) Dispense 20 μl onto each filter paper disk.
(4) Allow to dry at room temperature for 72 h.
(5) Store at room temperature for up to 6 months. (Disks are available from Anaerobe Systems, Remel, and Difco.)

TE Buffer (50 mM)

Ingredients

N-tris (hydroxymethyl) methyl-2-aminoethan (Sigma)	11.46 g
Distilled (Milli-Q) water	1,000 ml

Preparation

(1) Adjust pH to 7.5, adding 1 M NaOH (about 2–3 ml).
(2) Filter sterilize through 0.2 μm filter (such as Minisart 27 mm, 0.2 μm).

Use

Use to dilute 4-MU reagents and to prepare the inoculum (see "Rapid Enzyme Tests").

Thioglycolate, Supplemented

Ingredients

Thioglycolate medium without indicator (BBL)	30 g
Hemin solution (5 mg/ml)	1 ml
Vitamin K_1 (1 mg/ml)	0.1 ml
Distilled water	1,000 ml

Preparation

(1) Combine ingredients.
(2) Boil to dissolve.
(3) Dispense into tubes containing a marble chip (Fisher), filling the tubes two-thirds to three-fourths full.
(4) Autoclave at 118–121 °C for 15 min.

Use

Just prior to use, boil or steam for 5 min; cool. Supplement with normal rabbit or horse serum (10%) or Fildes enrichment (5%) if needed.

Shelf Life

Six months at room temperature.

Tryptic Soy Serum Bacitracin Vancomycin (TSBV) Agar

Ingredients

Trypticase soy agar (BBL)	40 g
Yeast extract	1 g
Distilled water	1,000 ml
Sterile horse serum	100 ml
Bacitracin (7.5 mg/ml)	1 ml
Vancomycin (5 mg/ml)	1 ml
Vitamin K_1 (1 mg/ml)	1 ml
Hemin (5 mg/ml)	1 ml

Preparation

(1) Combine agar, yeast extract, and water.
(2) Adjust pH to 7.2.
(3) Boil to dissolve.

(4) Autoclave at 121 °C for 15 min.
(5) Cool to 56 °C, add sterile horse serum, bacitracin, and vancomycin, and pour plates.

Shelf Life

Unknown.

Tween-80

Ingredients

Tween-80 (polysorbate 80)	10 ml
Distilled water	90 ml

Preparation

(1) Mix ingredients as much as possible.
(2) Autoclave at 121 °C for 15 min and mix thoroughly while warm.

Use

Add 0.5 ml of Tween solution to 10 ml of medium to enhance growth of most gram-positive bacteria.

Urea Broth

Ingredients

Urea broth powder (Difco)	38.7 g
Distilled water	1,000 ml

Preparation

(1) Combine ingredients.
(2) Sterilize by filtration.
(3) Dispense into sterile tubes.
(4) Alternative method: Dilute commercial 10× urea broth (Difco or BBL) 1:10 in distilled water.
(5) Store refrigerated for up to 6 months.

Vancomycin Stock Solution (7.5 mg/ml)

Ingredients

Vancomycin	75 mg
Hydrochloric acid (N/20)	5 ml
Distilled water	5 ml

D

Preparation

(1) Dissolve vancomycin in hydrochloric acid.

Note: To make 4 mg/ml or 5 mg/ml stock solutions (for JVN and TSBV agars), use 40 mg or 50 mg of vancomycin, respectively.

(2) Add distilled water.
(3) Store in refrigerator for up to 1 month or in freezer (−20 °C) for up to 1 year.

Veillonella Neomycin Agar (VNA)

Ingredients

Veillonella agar (Difco)	36 g
Neomycin solution (100 mg/ml)	1 ml
Distilled water	1,000 ml
Vancomycin solution (7.5 mg/ml)	1 ml

Preparation

(1) Combine all ingredients except vancomycin.
(2) Boil to dissolve.
(3) Autoclave at 121 °C for 15 min.
(4) Cool to 50 °C, add vancomycin, and pour plates.

Shelf Life

- Plates: 2 weeks at refrigerator temperature.
- Bottles: 1 year at refrigerator temperature.

Vitamin K_1 Solution (1 mg/ml)

Ingredients

Vitamin K_1 (Sigma)	0.1 g
Absolute ethanol	100 ml

Preparation

(1) Weigh out the vitamin K_1.
(2) Add this to a tube or bottle containing the ethanol.
(3) Store in the refrigerator in a tightly closed container, protected from light.
(4) Dilutions of this solution may be made in sterile distilled water.

Use

To supplement any media, add to a final concentration of 1 μl/ml.

VPI Salts Solution

Ingredients

Calcium chloride (anhydrous)	0.2 g
Magnesium sulfate (anhydrous)	0.2 g
Dipotassium phosphate	1 g
Monopotassium phosphate	1 g
Sodium bicarbonate	10 g
Sodium chloride	2 g
Distilled water	1,000 ml

Preparation

(1) Mix calcium chloride and magnesium sulfate in 300 ml water until dissolved.
(2) Add 500 ml water and while stirring add remaining salts.
(3) Continue stirring until all salts are dissolved.
(4) Add 200 ml water; mix.
(5) Store in refrigerator.

appendix E

Stocking and Shipping Cultures

Stocking Cultures

Stock cultures should be prepared as soon as an organism is isolated in pure culture. Isolates can be put into stock from liquid or solid media. In either case, a young, actively growing culture should be used.

Supplemented thioglycolate medium or chopped meat broth incubated for 24–48 h, depending upon the growth rate of the isolate, can be used to prepare stock cultures. Add 0.5 ml of the broth culture to an equal volume of sterile 20% skim milk prepared in an unbreakable screwcapped vial such as 2 ml Microtube with cap (72.694; Sarstedt, Inc. NC). Freeze and maintain the stock culture at –70 °C. Plate out a portion of the broth culture to check the purity of the isolate.

If stock cultures are prepared from solid media, the growth taken from the plate or slant must be suspended carefully and mixed thoroughly in the skim milk. We have found that sterile wooden sticks work very well for this purpose.

Lyophilization (or freeze-drying) is an excellent method for preservation of stock cultures. Bacterial suspensions in a liquid medium (cooked meat broth) can be directly lyophilized; alternatively, a healthy culture grown on solid medium can be suspended in 20% skim milk and subsequently lyophilized. Both techniques allow for adequate recovery of the organism.

For toxin-producing organisms, stocking under anaerobic conditions at –70 °C may be important.

Shipping Cultures

To ensure adequate recovery of the isolate, a fresh culture must be used to inoculate the transport system. Cryotubes (1.8 ml) with 1.5 ml Brucella agar slant stab inoculated with the bacterial strain closed by heat-sealing in an anaerobic pouch maintain the viability of most anaerobic bacteria up to 10 days. This system has been extensively used when sending bacterial strains between Finland and the United States by fast delivery. A stoppered agar slant (chopped meat glucose) injected with a liquid culture suspension is simple and eliminates the need for paraffin-sealed tubes. If the technique is available, lyophilized preparations are preferred for their ease of maintenance and transportation. For shorter distance transportation, cultures may also be injected into specimen transport vials (Port-a-Cul, B-D Microbiology Systems; Anaerobe Systems tube). Alternatively, grow the organism on an agar medium and use a swab to collect the growth and inoculate PRAS semisolid transport medium (Anaerobe Systems).

Specimen mailer systems for handling diagnostic materials, complete with protective mailer, absorbent material, waterproof tape, labels, and shipping carton, are commercially available (Polyfoam Packers Corp., O. Berk International, Inc.). Verify that the system conforms to federal and postal regulations.

Regulations concerning packaging and shipment of etiologic agents are detailed in the Code of Federal Regulations, Section 72.25 of Part 72, Title 42, amended.

E

appendix F

Sources of Supplies

Media and Reagents

AB Biodisk North American, Inc.
200 Centennial Ave.
Piscataway, NJ 08854-3910
Tel: (800) 874-8814/(732) 457-0408
www.abbiodisk.com
Fax: (732)457-8980
etest@abbiodisk.se

Abbott Laboratories
100 Abbott Park Rd.
Abbott Pk, IL 60064
Tel: (800) 323-9100/(847) 937-6100
Fax: (847) 937-3130
www.abbott.com

Alexon-Trend
14000 Unity Street NW
Ramsey, MN 55303
Tel: (800) 366-0096/(800) 255-6730
Fax: (763) 323-7858
www.alexon-trend.com

Dade Behring MicroScan
1584 Enterprise Blvd.
W. Sacramento, CA 95691
Tel: (800) 677-7226
Fax: (916) 373-2081
www.dadebehring.com

Anaerobe Systems
15906 Concord Circle
Morgan Hill, CA 95037
Tel: (800) 443-3108/(408) 782-7557
Fax: (408) 782-3031
www.anaerobesystems.com

Animal Technologies, Inc.
2016 E. Erwin
Tyler, TX 75702
Tel: (903) 592-1363
Fax: (903) 593-0960
www.animaltechnologies.com

Bartels Immunodiagnostics
(See Intracel)

Becton, Dickinson Microbiology
Systems
7 Loveton Circle
Sparks, MD 21152
Tel: (800) 675-0908/(410) 316-4000
www.bd.com or www.bd.com/clinical/

bioMerieux Vitek, Inc.
595 Anglum Road
Hazelwood, MO 63042
Tel: (800) 634-7656/(314) 731-8500
Fax: (314) 731-8700
www.biomerieux-vitek.com
www,biomerieux-USA.com

Biosite Diagnostics, Inc.
11030 Roselle St.
San Diego, CA 92121
Tel: (858) 455-4808
Fax: (858) 455-4815
www.biosite.com

Biowhittaker, Inc.
Cambrex Corp.
8830 Biggs Ford Rd.
Walkersville, MD 21793
Tel: (800) 638-8174/(301)898-7025
Fax: (301) 845-8338
www.cambrex.com

Eastman Kodak Co.
343 State St.
Rochester, NY 14650
Tel: (716) 724-4000
www.kodak.com

Eiken Chemical Co., LTD
33-8 Hongo 1-chome, Bunkyo-ku
Tokyo 113-8408, Japan
www.eiken.co.jp

Hardy Diagnostics
1430 West McCoy Ln.
Santa Maria, CA 93455
Tel: (800) 266-2222
Fax: (805) 346-2760
www.hardy diagnostics.com

Intracel
1330 Piccard Dr.
Rockville, MD 20850
Tel: (800) 542-2281/(301)840-2161
Fax: (301) 840-2161
www.intracel.com

Key Scientific Products
1402D Chisolm Trail
Round Rock, TX 78681
Tel: (512) 218-1913
Fax: (512) 218-8580
www.keyscientific.com

Lab M Ltd.
Topley House
POB 19, Bury
Lancashire, BL9 6AU
United Kingdom
Fax: +44-061-7629322
www.IDGPLC.com

Mallinckrodt Science Products
675 McDonnell Blvd.
St. Louis, MO 63134
Tel: (314) 654-2000
www.mallinckrodt.com

Merck, E.
Frankfurterstrasse 250, D-6100
Darmstadt 1
West Germany
Tel: 49-6151-72-0
www.merck.de

Meridian Diagnostics, Inc.
3471 River Hills Dr.
Cincinnati, OH 45244
Tel: (513) 271-3700
Fax: (513) 272-5432
www.meridianbioscience.com

MSI/MicroMedia Systems
2330 Denison Ave.
Cleveland, OH 44109
Tel: (800) 423-6496

Nissui Pharmaceutical Co., LTD
2-11-1 Sugamo, Toshima-ku
Tokyo 170-0002, Japan
Tel: 81-3918-8161
www.nissui-pharm.co.jp

Oxoid USA, Inc.
800 Proctor Ave.
Ogdensburg, NY 13669
Tel: (800) 567-8378/(613)226-1318
Fax: (613) 226-3728
www.oxoid.com

Pittman and Moore, Inc.
Contact: Shering-Plough
Tel: (800) 521-5767 and
Mallinckrodt Veterinary
Tel: (847) 949-3300

PML Microbiologicals
27120 SW 95th Ave.
Wilsonview, OR 97070
Tel: (800) 628-7014
Fax: (800) 765-4415
www.PMLmicro.com

Porton (Maran & Co.)
190 Clarence Gate Gardens
London NW96AD
England
Fax: +44 (171) 723 8059
www.maran.co.UK

Trek (Sensititre) Diagnostic Systems, Inc.
25760 First St.
Westlake, OH 44145
Tel: (800) 871-8909
Fax: (440) 808-0400
www.trekds.com

Remel
12076 Santa Fe Dr.
Lenexa, KS 66215
Tel: (800) 255-6730
Fax: (800) 621-8251
www.remel.com

Rohm GmbH & Co. KG
64275 Darmstadt
Germany
Fax: +49-6151-1802

Rosco Diagnostica
Taastrupgaardsvej 30
Taastrup, Denmark, DK-2630
Fax: +45 43527374
info@as-rosco.DK

TechLab, Inc.
VPI Corporate Reseach Center
1861 Pratt Drive, Suite 1030
Blacksburg, VA 25060
Tel: (540) 953-1664
Fax: (540) 953-1665
www.techlabinc.com

Sigma-Aldrich Corporation
3050 Spruce St.
St. Louis, MO 63178
Tel: (800) 325-3010/(314) 771-5765
Fax: (314) 652-9930
www.sigmaaldrich.com

Equipment and Supplies

Agilent Technologies
395 Page Mill Rd
Palo Alto, CA 94304
Tel: (800) 227-9770/(650) 752-5000
www.agilent.com

Bellco Glass, Inc.
340 Edrudo Rd.
Vineland, NJ 08360
Tel: (800) 257-7043/(856) 691-1075
Fax: (856) 691-3247
www.bellcoglass.com

Brand Tech Scientific
11 Bokum Rd.
Essex, CT 06426
Tel: (888) 522-2726/(860) 767-2562
Fax: (860) 767-2563
www.brandtech.com

CMI-Promex, Inc.
7 Benjamin Green Rd.
P.O. Box 418
Pedricktown, NJ 08067
Tel: (856) 351-1000
Fax: (856) 351-1659
www.CMI-promex.com

Coy Laboratory Products, Inc.
14500 Coy Dr.
Grass Lake, MI 49240
Tel: (734) 475-2200
Fax: (734) 475-1846
www.coylab.com

Don Whitley
14 Oatley Road
Shipley, West Yorkshire
England. BD177SE
Tel: 44 (1247) 595128
Fax: +44 (01274) 531197
www.dwscientific.co.UK

Engelhard Industries
Gas Equipment Division
101 Wood Ave. So.
Iselin, NJ 08830
Tel: (732) 205-5000
www.engelhard.com

Fluka-RdH
Sigma-Aldrich, Inc.
P.O. Box 2060
Milwaukee, WI 53201
Tel: (800) 558-9160
Fax: (800) 962-9591/(800) 325-5052
www.sigmaaldrich.com

Forma Scientific, Inc.
401 Millcreek Rd.
P.O. Box 649
Millcreek Rd.
Marietta, OH 45750
Tel: (800) 848-3080/(740) 373-4763
Fax: (740) 373-6770
www.forma.com or www.thermo.com

MART Microbiology B.V.
P.O. Box 165
7130 AD Lictenvoorde
The Netherlands
Tel: 00-31-544-396600
Fax: 00-31-544-396611
www.anoxomat.com
e-mail: info@anoxomat.com

Matheson Tri-Gas Products
www.mathesongas.com
www.matheson-trigas.com

Microbial ID, Inc.
115 Barksdale Professional Ctr.
Newark, DE 19711
Tel: (800) 276-8068/(302) 737-4297
Fax: (302) 737-7781
www.microbialid.com

Micromass UK Limited
Floats Road
Wythenshawe
Manchester M23 LZ
Tel: +44 (0) 161 945 4170,
(978) 524-8200 (USA)
Fax: +44 (0) 161 998 8915,
(978) 524-8210
www.micromass.co.UK

O. Berk International Inc.
3 Milltown Ct.
Union, 07083
Tel: (800) 631-7392/(908) 851-9500
Fax: (908) 851-9367
www.oberk.com

Packard Instrument Co.
Perkin Elmer
800 Research Pky
Meriden, CT 06450
Tel: (800) 323-1891
www.perkin elmer.com
www.packardbioscience.com

Polyfoam Packers Corp.
2320 S. Foster
Wheeling, IL 60090
Tel: (800) 323-7442/(847) 398-0110
www.polyfoam.com

Ruskinn Technology Limited
Business & Innovation Centre
Angel Way, Listerhills
Bradford, West Yokshire BD7 1BX
United Kingdom
Fax: 44-1943879721
www.ruskinn.com

Sarstedt, Inc.
P.O. Box 468
Newton, NC 28658
Tel: (800) 257-5101/(828) 465-4000
Fax: (828) 465-4003
www.sarstedt.com

Scientific Device Laboratory, Inc.
P.O. Box 88
Glenview, IL 60025
Tel: (847) 803-9495
www.scientificdevice.com

Sheldon Manufacturing, Inc.
P.O. Box 627
300 N. 26th Ave.
Cornelius, OR 97113
Tel: (800) 322-4897/(503) 640-3000
Fax: (503) 640-1366
www.shellab.com

Spectra-Physics
1335 Terra Bella Ave.
P.O. Box 7012
Mountain View, CA 94039
Tel: (650) 961-2550
www.spectraphysics.com

Spiral Biotech Inc.
An Advanced Instruments Company
7830 Old Georgetown Road
Bethesda, MD 20814
Tel: (800) 554-1620/(301) 657-1620
www.spiralbiotech.com
www.aicompanies.com/sbi.products.htm

Tekmar-Dohrman
4736 Socialville Foster Rd.
Mason, OH 45040
Tel: (800) 543-4461/(513) 229-7000
Fax: (513) 247-7050
www.tekmar.com

Unimar, Inc.
CooperSurgical
475 Danbury Rd.
Wilton, CT 06897
Tel: (800) 243-6608
Fax: (800) 262-0105
www.coopersurgical.com

UVP, Inc.
2066 West 11th Street
Upland, CA 91786
Tel: (800) 452-6788/(909) 946-3197
Fax: (909) 946- 3597
uvp@uvp.com

Yamanouchi Pharmaceuticals
#5, 2-Chome, Nihon bashi-Hancho,
Chuo-ko
Tokyo 103-8411, Japan
www.yamanouchi.com

Reference Cultures

American Type Culture Collection
(ATCC)
10801 University Blvd.
Manassas, VA 20110-2209
Tel: (703) 365-2700
www.atcc.com

National Collection of Type Cultures
(NCTC)
Central Public Health Laboratory
Colindale Avenue
London NW9 5HT, England
www.phls.co.UK

Culture Collection University of
Göteborg (CCUG)
Department of Clinical Bacteriology,
University of Göteborg, S-41346
Göteborg, Sweden
www.ccug.gu.SE

Japan Collection of Microorganisms (JCM)
The Institute of Physical and Chemical
Research (RIKEN)
2-1 Hirosawa
Wako, Saitama 351-0198, Japan
Tel: 81-48-467-9560
Fax: 81-48-462-4617

Deutche Sammlung von Microorganismen
und Zellkulturen (DSMZ)
(German Collection of Microorganisms
and Cell Cultures)
GmBH, Mascheroder Weg
16, D-38124, Braunschweig, Germany
Fax: +49(0) 531-2616-418
www.dsmz.de

references

1. **Ahmed, S., L. Gruer, C. McGuigan, et al.** 2000. Update: *Clostridium novyi* and unexplained illness among injecting-drug users—Scotland, Ireland, and England, April–June 2000. Morbid. Mortal. Weekly Rep. **49**:543–545.
2. **Al-Barrak, A., J. Embil, B. Dyck, K. Olekson, D. Nicoll, M. Alfa, and A. Kabani.** 1999. An outbreak of toxin A negative, toxin B positive *Clostridium difficile*–associated diarrhea in a Canadian tertiary-care hospital. Canada Communicable Dis. Rep. **25**:1–3.
3. **Alderson, D., A. J. Strong, H. R. Ingham, and J. B. Selkon.** 1981. Fifteen-year review of the mortality of brain abscess. Neurosurgery **8**:1–6.
4. **Alexander, C. J., D. M. Citron, J. S. Brazier, and E. J. Goldstein.** 1995. Identification and antimicrobial resistance patterns of clinical isolates of *Clostridium clostridioforme, Clostridium innocuum,* and *Clostridium ramosum* compared with those of clinical isolates of *Clostridium perfringens.* J. Clin. Microbiol. **33**:3209–3215.
5. **Alexander, C. J., D. M. Citron, S. Hunt Gerardo, M. C. Claros, D. Talan, and E. J. C. Goldstein.** 1997. Characterization of saccharolytic *Bacteroides* and *Prevotella* isolates from infected dog and cat bite wounds in humans. J. Clin. Microbiol. **35**:406–411.
6. **Alfa, M., A. Kabani, D. Lylerly, S. Moncrief, L. M. Neville, A. Al-Barrak, G. K. H. Harding, B. Dyck, K. Olekson, and J. Embil.** 2000. Characterization of a toxin A-negative, toxin B-positive strain of *Clostridium difficile* responsible for a nosocomial outbreak of a *Clostridium difficile*–associated diarrhea. J. Clin. Microbiol. **38**:2706–2714.
7. **Alfageme, I., F. Munoz, N. Pena, and S. Umbria.** 1993. Empyema of the thorax in adults. Etiology, microbiologic findings, and management. Chest **103**:839–843.
8. **Ali, R. W., G. Bancescu, Ø. Nielsen, and N. Skaug.** 1995. Viability of four putative periodontal pathogens and enteric rods in the anaerobic transport medium VMGA III. Oral Microbiol. Immunol. **10**:365–371.
9. **Allen, S. D., C. L. Emery, and J. A. Siders.** 1997. *Clostridium,* p. 662–666. *In* P. R. Murray (ed.), Manual of clinical microbiology, 7th ed. American Society for Microbiology, Washington, D.C.
10. **Altemeier, W. A.** 1940. The anaerobic streptococci in tubo-ovarian abscess. Am. J. Obstet. Gynecol. **39**:1038–1042.
11. **Altemeier, W. A., and W. D. Fullen.** 1971. Prevention and treatment of gas gangrene. JAMA **217**:806–813.

12. **Avgustin, G., R. J. Wallace, and H. J. Flint.** 1997. Phenotypic diversity among ruminal isolates of *Prevotella ruminicola:* proposal of *Prevotella brevis* sp. nov., *Prevotella bryantii* sp. nov., and *Prevotella albenis* sp. nov. and redefinition of *Prevotella ruminicola*. Int. J. Syst. Bacteriol. **47**:284–288.

13. **Bailey, G. D., and D. N. Love.** 1993. *Fusobacterium pseudonecrophorum* is a synonym for *Fusobacterium varium.* Int. J. Syst. Bacteriol. **43**:819–821.

14. **Bandoh, K., K. Ueno, K. Watanabe, and N. Kato.** 1993. Susceptibility patterns and resistance to imipenem in the *Bacteroides fragilis* group species in Japan: a 4-year study. Clin. Infect. Dis. **16 Suppl. 4**:S382–S386.

15. **Barnham, M., and N. Weightman.** 1998. *Clostridium septicum* infection and hemolytic uremic syndrome. Emerg. Infect. Dis. **4**:321–324.

16. **Baron, E. J., G. H. Cassell, L. B. Duffy, D. A. Eschenbach, J. R. Greenwood, S. M. Harvey, N. E. Madinger, E. M. Peterson, and K. B. Waites.** 1993. Laboratory diagnosis of female genital tract infections, p. 1-1. American Society for Microbiology, Washington, D.C.

17. **Baron, E. J., and S. M. Finegold.** 1990. Bailey & Scott's Diagnostic Microbiology, 8th ed. C.V. Mosby Co., St. Louis.

18. **Baron, E. J., C. A. Strong, M. McTeague, M.-L. Vaisanen, and S. M. Finegold.** 1995. Survival of anaerobes in original specimens transported by overnight mail services. Clin. Infect. Dis. **20**:S174–S177.

19. **Baron, E. J., P. Summanen, J. Downes, M. C. Roberts, H. Wexler, and S. M. Finegold.** 1989. *Bilophila wadsworthia,* a unique gram-negative anaerobic rod recovered from appendicitis specimens and human feces. J. Gen. Microbiol. **135**: 3405–3411.

20. **Bartlett, J. G.** 1993. Anaerobic bacterial infections of the lung and pleural space. Clin. Infect. Dis. **16**:S248–S255.

21. **Bartlett, J. G., and S. L. Gorbach.** 1976. Anaerobic infections of the head and neck. Otolaryngol. Clin. North Am. **9**:655–678.

22. **Bartlett, J. G., S. L. Gorbach, and S. M. Finegold.** 1974. The bacteriology of aspiration pneumonia. Am. J. Med. **56**:202–207.

23. **Bartlett, J. G., and B. F. Polk.** 1984. Bacterial flora of the vagina: Quantitative study. Rev. Infect. Dis. **6**:S67–S72.

24. **Becker, G. D., G. J. Parell, D. F. Busch, S. M. Finegold, and M. J. Acquarelli.** 1978. Anaerobic and aerobic bacteriology in head and neck cancer surgery. Arch. Otolaryngol. **104**:591–594.

25. **Beebe, J. L. and E. W. Koneman.** 1995. Recovery of uncommon bacteria from blood: association with neoplastic disease. Clin. Microbiol. Rev. **8**:336–356.

26. **Bennion, R. S., J. E. Thompson, E. J. Baron, and S. M. Finegold.** 1990. Gangrenous and perforated appendicitis with peritonitis—treatment and bacteriology. Clin. Ther. **12**:31–44.

27. **Bjerkestrand, G., A. Digranes, and A. Schreiner.** 1975. Bacteriological findings in transtracheal aspirates from patients with chronic bronchitis and bronchiectasis: a preliminary report. Scand. J. Respir. Dis. **56**:201–207.

28. **Boone, J. H., and R. J. Carman.** 1997. *Clostridium perfringens:* food poisoning and antibiotic-associated diarrhea. Clin. Microbiol. Newsl. **19**:65–67.

29. **Brander, M. A., and H. R. Jousimies-Somer.** 1992. Evaluation of the RapID ANA II and API ZYM systems for the identification of *Actinomyces* species from clinical specimens. J. Clin. Microbiol. **30**:3112–3116.

30. **Brazier, J. S.** 1986. Yellow fluorescence of fusobacteria. Lett. Appl. Microbiol. **2**:125–126.

31. **Brazier, J. S.** 1993. Role of laboratory investigations of *Clostridium difficile* diarrhea. Clin. Infect. Dis. **16**:S228–S233.

32. **Brazier, J. S., D. M. Citron, and E. J. C. Goldstein.** 1991. A selective medium for *Fusobacterium* spp. J. Appl. Bacteriol. **71**:343–346.

33. **Brazier, J. S., E. J. C. Goldstein, D. M. Citron, and M. I. Ostovari.** 1990. Fastidious anaerobe agar compared with Wilkins-Chalgren agar, brain heart infusion agar, and brucella agar for susceptibility testing of *Fusobacterium* species. Antimicrob. Agents Chemother. **34**:2280–2282.

34. **Brazier, J. S., and T. V. Riley.** 1988. UV red fluorescence of *Veillonella* spp. J. Clin. Microbiol. **26**:383–384.

35. **Brazier, J. S., and S. A. Smith.** 1989. Evaluation of the Anoxomat: a new technique for anaerobic and microaerophilic clinical bacteriology. J. Clin. Pathol. **42**:640–644.

36. **Brazier, J. S., S. L. Stubbs, and B. I. Duerden.** 1999. Metronidazole resistance among clinical isolates belonging to the *Bacteroides fragilis* group: time to be concerned? [letter]. J. Antimicrob. Chemother. **44**:580–581.

37. **Brook, I.** 1980. Anaerobic and aerobic bacteriology of decubitus ulcers in children. Am. Surg. **46**:624–626.

38. **Brook, I.** 1981. Bacteriologic features of chronic sinusitis in children. JAMA **246**:967–969.

39. **Brook, I.** 1987. Comparison of two transport systems for recovery of aerobic and anaerobic bacteria from abscesses. J. Clin. Microbiol. **25**:2020–2022.

40. **Brook, I.** 1988. Microbiology of non-puerperal breast abscesses. J. Infect. Dis. **157**:377–379.

41. **Brook, I.** 1989. Microbiology of infected pilonidal sinuses. J. Clin. Pathol. **42**:1140–1142.

42. **Brook, I.** 1992. Aerobic and anaerobic bacteriology of intracranial abscesses. Pediatr. Neurol. **8**:210–214.

43. **Brook, I., and P. Burke.** 1992. The management of acute, serous and chronic otitis media: the role of anaerobic bacteria. J. Hosp. Infect. **22**:75–87.

44. **Brook, I., and E. H. Frazier.** 1993. Aerobic and anaerobic microbiology of empyema: a retrospective review in two military hospitals. Chest **103**:1502–1507.

45. **Brook, I., and E. H. Frazier.** 1993. Anaerobic osteomyelitis and arthritis in a military hospital: a 10-year experience. Am. J. Med. **94**:21–28.

46. **Brook, I., and E. H. Frazier.** 1993. Role of anaerobic bacteria in liver abscesses in children. Pediatr. Infect. Dis. **12**:743–747.

47. **Brook, I., and E. H. Frazier.** 1997. The aerobic and anaerobic bacteriology of perirectal abscesses. J. Clin. Microbiol. **35**:2974–2976.

48. **Brook, I., E. H. Frazier, and P. A. Foote.** 2000. Microbiology of chronic maxillary sinusitis: comparison between specimens obtained by sinus endoscopy and by surgical drainage. J. Med. Microbiol. **46**:430–432.

49. **Brook, I., E. H. Frazier, and M. E. Gher.** 1991. Aerobic and anaerobic microbiology of periapical abscess. Oral Microbiol. Immunol. **6**:123–125.

50. **Brook, I., S. Grimm, and R. B. Kielich.** 1981. Bacteriology of acute periapical abscess in children. J. Endod. **7**:378–380.

51. **Brook, I., D. H. Thompson, and E. H. Frazier.** 1994. Microbiology and management of chronic maxillary sinusitis. Arch. Otolaryngol. Head Neck Surg. **120:**1317–1320.

52. **Brook, I., and P. Yocum.** 1989. Quantitative bacteria culturea and beta-lactamase activity in chronic suppurative otitis media. Ann. Otol. Rhinol. Laryngol. **98**:293–297.

53. **Broughton, W. A., J. B. Bass, and M. B. Kirkpatrick.** 1987. The technique of protected brush catheter bronchoscopy. J. Crit. Ill. **2**:63–70.

54. **Bryan, C. S., C. E. Dew, and K. L. Reynolds.** 1983. Bacteremia associated with decubitus ulcers. Arch. Inter. Med. **143**:2093–2095.

55. **Buchanan, A. G.** 1982. Clinical laboratory evaluation of a reverse CAMP test for presumptive identification of *Clostridium perfringens*. J. Clin. Microbiol. **16**:761–762.

56. **Burlage, R. S., and P. D. Ellner.** 1985. Comparison of the PRAS II, AN-Ident, and RapID-ANA systems for identification of anaerobic bacteria. J. Clin. Microbiol. **22**:32–35.

57. **Campion, M., A. T. Evangelista, and J. Mortensen.** 1999. Evaluation of five enzyme immunoassay methods for the detection of *Clostridium difficile* toxins, p. 105. Abstr. 99th Gen. Meet. Am. Soc. Microbiol. American Society for Microbiology, Washington, D.C.

58. **Carroll, K. C., W. E. Aldeen, M. Morrison, R. Anderson, D. Lee, and S. Mottice.** 1998. Evaluation of the Abbott LCx ligase chain reaction assay for detection of *Chlamydia trachomatis* and *Neisseria gonorrhoeae* in urine and genital swab specimens from a sexually transmitted disease clinic population. J. Clin. Microbiol. **36**:1630–1633.

59. **Cavallaro, J. J., L. S. Wiggs, and J. M. Miller.** 1997. Evaluation of the BBL crystal anaerobe identification system. J. Clin. Microbiol. **35**:3186–3191.

60. **Centers for Disease Control.** 1995. Wound botulism—California. Morbid. Mortal. Weekly Rep. **44**:889–892.

61. **Centers for Disease Control.** 1998. Tetanus among injecting-drug users—California, 1997. Morbid. Mortal. Weekly Rep. **47**:149–151.

62. **Chastre, J., J. Y. Fagon, P. Soler, M. Bornet, Y. Domart, J. L. Trouillet, C. Gibert, and A. J. Hance.** 1988. Diagnosis of nosocomial bacterial pneumonia in intubated patients undergoing ventilation: comparison of the usefulness of bronchoalveolar lavage and the protected specimen brush. Am. J. Med. **85**:499–506. (Erratum Am. J. Med. **86(2)**:258, 1989.)

63. **Cheeseman, S. L., S. J. Hiom, A. J. Weightman, and W. G. Wade.** 1996. Phylogeny of oral asaccharolytic *Eubacterium* species determined by 16S ribosomal DNA sequence comparison and proposal of *Eubacterium infirmum* sp. nov., and *Eubacterium tardum* sp. nov. Int. J. Syst. Bacteriol. **46**:1120–1124.

64. **Chow, A. W., J. E. Galpin, and L. B. Guze.** 1977. Clindamycin for treatment of sepsis caused by decubitus ulcers. J. Infect. Dis. **135(Suppl.)**:S65–S68.

65. **Chow, A. W., K. L. Malkasian, J. R. Marshall, and L. B. Guze.** 1975. The bacteriology of acute pelvic inflammatory disease. Am. J. Obstet. Gynecol. **122**:876–879.

66. **Chow, A. W., J. R. Marshall, and L. B. Guze.** 1977. A double-blind comparison of clindamycin with penicillin plus chloramphenicol in treatment of septic abortion. J. Infect. Dis. **135**:S35–S39.

67. **Chow, A. W., S. M. Roser, and F. A. Brady.** 1978. Orofacial odontogenic infections. Ann. Intern. Med. **88**:392–402.

68. **Chun, C. H., J. D. Johnson, M. Hofstetter, and M. J. Raff.** 1986. Brain abscess. A study of 45 consecutive cases. Medicine (Baltimore) **65**:415–431.

69. **Citron, D. M., E. J. Baron, S. M. Finegold, and E. J. C. Goldstein.** 1990. Short prereduced anaerobically sterilized (PRAS) biochemical scheme for identification of clinical isolates of bile-resistant *Bacteroides* species. J. Clin. Microbiol. **28**:2220–2223.

70. **Citron, D. M., S. Hunt Gerardo, M. C. Claros, F. Abrahamian, D. Talan, and E. J. C. Goldstein.** 1996. Frequency of isolation of *Porphyromonas* species from infected dog and cat bite wounds in humans and their characterization by biochemical tests and arbitrarily primed–polymerase chain reaction fingerprinting. Clin. Infect. Dis. **23**:S78–S82.

71. **Citron, D. M., M. I. Ostavari, A. Karlsson, and E. J. C. Goldstein.** 1991. Evaluation of the epsilometer (E-test) for susceptibility testing of anaerobic bacteria. J. Clin. Microbiol. **29**:2197–2203.

72. **Citron, D. M., Y. A. Warren, M. K. Hudspeth, and E. J. Goldstein.** 2000. Survival of aerobic and anaerobic bacteria in purulent clinical specimens maintained in the Copan Venturi Transystem and Becton Dickinson Port-a-cul transport systems. J. Clin. Microbiol. **38**:892–894.

73. **Civen, R., H. Jousimies-Somer, M. Marina, L. Borenstein, H. Shah, and S. M. Finegold.** 1995. A retrospective review of cases of anaerobic empyema and update of bacteriology. Clin. Infect. Dis. **20**:S224–S229.

74. **Cockerill, F. R. I., J. G. Hughes, E. A. Vetter, et al.** 1997. Analysis of 281,797 consecutive blood cultures collected over an 8-year period: trends in microorganisms isolated and the value of anaerobic culture of blood. Clin. Infect. Dis. **24**:403–418.

75. **Coleman, N., G. Speirs, J. Khan, V. Broadbent, D. G. Wight, and R. E. Warren.** 1993. Neutropenic enterocolitis associated with *Clostridium tertium.* J. Clin. Pathol. **46**:180–183.

76. **Collins, D. M., and P. A. Lawson.** 2000. The genus *Abiotrophia* (Kawanura et. al.) is not monophyletic: proposal of *Granulicatella* gen. nov., *Granulicatella adiacens* comb. nov., *Granulicatella elegans* comb. nov. and *Granulicatella balaenopteraea* comb. nov. Int. J. Syst. Bacteriol. **50**:365–369.

77. **Collins, M. D., P. A. Lawson, A. Willems, et al.** 1994. The phylogeny of the genus *Clostridium:* proposal of five new genera and eleven new species combinations. Int. J. Syst. Bacteriol. **44**:812–826.

78. **Collins, M. D., D. N. Love, J. Karjalainen, A. Kanervo, B. Forsblom, A. Willems, S. Stubbs, E. Sarkiala, G. D. Bailey, D. I. Wigney, and H. Jousimies-Somer.** 1994. Phylogenetic analysis of members of the genus *Porphyromonas* and description of *Porphyromonas cangingivalis* sp. nov., and *Porphyromonas cansulci* sp. nov. Int. J. Syst. Bacteriol. **44**:674–679.

79. **Collins, M. D., and S. Wallbanks.** 1992. Comparative sequence analyses of the 16S rRNA genes of *Lactobacillus minutus, Lactobacillus rimae* and *Streptococcus parvulus:* proposal for the creation of a new genus *Atopobium*. FEMS Microbiol. Lett. **95**:235–240.

79a. **Collins, M. D., L. Hoyles, S. Kalfas, G. Sundquist, T. Monsen, N. Nikolaitchouk, and E. Falsen.** 2000. Characterization of *Actinomyces* isolates from infected root canals of teeth. Description of *Actinomyces radicidentis,* sp. nov. J. Clin. Microbiol. **38**:3399–3403.

80. **Cox, M. E.** 1997. Explosive potential of gas mixtures commonly used in anaerobic chambers. Clin. Infect. Dis. **25**:S140–S140.

81. **Crociani, F., B. Biavati, A. Alessandrini, C. Chiarini, and V. Scardovi.** 1996. *Bifidobacterium inopinatum* sp. nov. and *Bifidobacterium denticolens* sp. nov., two new species isolated from human dental caries. Int. J. Syst. Bacteriol. **46**:564–571.

82. **Croco, J. L., M. E. Erwin, J. M. Jennings, L. R. Putnam, and R. N. Jones.** 1995. Evaluation of the Etest for determinations of antimicrobial spectrum and potency against anaerobes associated with bacterial vaginosis and peritonitis. Clin. Infect. Dis. **20 Suppl 2**:S339–S341.

83. **Curk, M.-C., J.-C. Hubert, and F. Bringel.** 1996. *Lactobacillus paraplantarum* sp. nov., a new species related to *Lactobacillus plantarum.* Int. J. Syst. Bacteriol. **46**:595–598.

84. **Davis, B., and D. M. Systrom.** 1998. Lung abscess: pathogenesis, diagnosis and treatment. Curr. Clin. Top. Infect. Dis. **18**:252–273.

85. **Dix, K., S. M. Watabe, S. McArdle, D. I. Lee, C. Randolph, B. Moncla, and D. E. Schwartz.** 1990. Species-specific oligonucleotide probes for the identification of periodontal bacteria. J. Clin. Microbiol. **28**:319–323.

86. **Dixon, J. M.** 1992. Outpatient treatment of non-lactational breast abscess. Br. J. Surg. **79**:56–57.

87. **Dowell, V. R., Jr., and T. M. Hawkins.** 1974. Laboratory methods in anaerobic bacteriology. U.S. Government Printing Office, Washington, D.C.

88. **Downes, J., D. A. Spratt, M. A. Munson, E. Könönen, E. Tarkka, H. Jousimies-Somer, and W. Wade.** 2001. Characterization of *Eubacterium*-like species from oral infections. J. Med. Microbiol. In press.

89. **Downes, J., A. King, J. Hardie, and I. Phillips.** 1999. Evaluation of the Rapid ID 32A system for the identification of anaerobic Gram-negative bacilli, excluding *Bacteroides fragilis* group. Clin. Microbiol. Infect. **5**:319–326.

90. **Downes, J., J. I. Mangels, J. Holden, M. J. Ferraro, and E. J. Baron.** 1990. Evaluation of two single-plate incubation systems and the anaerobic chamber for the cultivation of anaerobic bacteria. J. Clin. Microbiol. **28**:246–248.

91. **Downes, J., B. Olsvik, S. J. Hiom, D. A. Spratt, S. L. Cheeseman, I. Olsen, A. J. Weightman, and W. G. Wade.** 2000. *Bulleidia extructa* gen. nov., sp. nov., isolated from the oral cavity. Int. J. Syst. Evolut. Microbiol. **50**:979–983.

92. **Downes, J., M. A. Munson, D. A. Spratt, H. Jousimies-Somer, and W. G. Wade.** 2000. Novel *Eubacterium*-like taxa from oral infections. Americam Society for Microbiology General Meeting. R7, p. 628.

93. **Draper, D., R. Parker, E. Patterson, W. Jones, M. Beutz, J. French, K. Borchardt, and J. McGregor.** 1993. Detection of *Trichomonas vaginalis* in pregnant women with the InPouch TV culture system. J. Clin. Microbiol. **31**:1016–1018.

94. **Draper, D. L., and A. L. Barry.** 1977. Rapid identification of *Bacteroides fragilis* with bile and antibiotic disks. J. Clin. Microbiol. **5**:439–443.

95. **Drasar, D. R., and M. J. Hill.** 1975. Human intestinal flora. Academic Press, London.

96. **Driscoll, C.** 1990. Minimally invasive endometrial sampling. Patient Care **Feb. 15**:206–208.

97. **Duguid, H.** 1982. *Actinomyces* and intrauterine devices. JAMA **248**:1579–1580.

98. **Duncan, A. J., R. J. Carman, G. J. Olsen, and K. H. Wilson.** 1993. Assignment of the agent of Tyzzer's disease to *Clostridium piliforme* comb. nov. on the basis of 16S rRNA sequence analysis. Int. J. Syst. Bacteriol. **43**:314–318.

99. **Durmaz, B., H. R. Jousimies-Somer, and S. M. Finegold.** 1995. Enzymatic profiles of *Prevotella, Porphyromonas,* and *Bacteroides* species obtained with the API ZYM system and Rosco diagnostic tablets. Clin. Infect. Dis. **20**:S192–S194.

100. **Edmiston, C. E., Jr., A. P. Walker, C. J. Krepel, and C. Gohr.** 1990. The nonpuerperal breast infection: aerobic and anaerobic microbial recovery from acute and chronic disease. J. Infect. Dis. **162**:695–699.

101. **Edwards, R., and D. Greenwood.** 1996. Mechanisms responsible for reduced susceptibility to imipenem in *Bacteroides fragilis.* J. Antimicrob. Chemother. **38**:941–951.

102. **Ehrenkranz, N. J., B. Alfonso, and D. Nerenberg.** 1990. Irrigation-aspiration for culturing draining decubitus ulcers: correlation of bacteriological findings with a clinical inflammatory scoring index. J. Clin. Microbiol. **28**:2389–2393.

102a. **Engelkirk, P. G., J. Duben-Engelkirk, and V. R. Dowell, Jr.** 1992. Principles and practice of anaerobic bacteriology. Star Publishing Co., Belmont, CA.

103. **England, D. M., and J. E. Rosenblatt.** 1977. Anaerobes in human biliary tracts. J. Clin. Microbiol. **6**:494–498.

104. **Eschenbach, D. A.** 1993. Bacterial vaginosis and anaerobes in obstetric-gynecologic infection. Clin. Infect. Dis. **16**:S282–S287.

105. **Eschenbach, D. A., T. M. Buchanan, H. M. Pollock, P. S. Forsyth, E. R. Alexander, J. S. Lin, S. P. Wang, B. B. Wentworth, W. M. MacCormack, and K. K. Holmes.** 1975. Polymicrobial etiology of acute pelvic inflammatory disease. N. Engl. J. Med. **293**:166–171.

106. **Eschenbach, D. A., P. R. Davick, B. L. Williams, S. J. Klebanoff, K. Young-Smith, C. M. Critchlow, and K. K. Holmes.** 1989. Prevalence of hydrogen peroxide-producing *Lactobacillus* species in normal women and women with bacterial vaginosis. J. Clin. Microbiol. **27**:251–256.

107. **Etoh, Y., F. E. Dewhirst, B. J. Paster, A. Yamamoto, and N. Goto.** 1993. *Campylobacter showae* sp. nov., isolated from the human oral cavity. Int. J. Syst. Bacteriol. **43**:631–639.

108. **Euzéby, J. P.** 1998. Taxonomic note: necessary correction of specific and subspecific epithets according to Rules 12c and 13b of the International Code of Nomenclature of Bacteria (1990 Revision). Int. J. Syst. Bacteriol. **48**:1073–1075.

109. **Ezaki, T., N. Li, Y. Hashimoto, H. Miura, and H. Yamamoto.** 1994. 16S ribosomal DNA sequences of anaerobic cocci and proposal of *Ruminococcus hansenii* comb. nov. and *Ruminococcus productus* comb. nov. Int. J. Syst. Bacteriol. **44**:130–136.

110. **Farrow, J. A. E., P. A. Lawson, H. Hippe, U. Gaulitz, and M. D. Collins**. 1995. Phylogenetic evidence that the gram-negative nonsporulating bacterium *Tissierella praeacuta* is a member of the *Clostridium* subphylum of the gram-positive bacteria and description of *Tissierella creatinini* sp. nov. Int. J. Syst. Bacteriol. **45**:436–440.

111. **Fedorko, D. P., and E. Willems.** 1997. Use of cycloserine-cefoxitin-fructose agar and L-proline-aminopeptidase (PRO disc) in the rapid identification of *Clostridium difficile.* J. Clin. Microbiol. **35**:1258–1259.

112. **Finegold, S. M.** 1977. Anaerobic bacteria in human disease. Academic Press, New York.

113. **Finegold, S. M.** 1989. Anaerobic pulmonary infection. Hosp. Pract. [Off] **24**:103–133.

114. **Finegold, S. M., and M. A. C. Edelstein.** 1988. Coping with anaerobes in the 80's, p. 1–10. *In* J. M. Hardie and S. P. Borriello (eds.), Anaerobes today. John Wiley & Sons, Chichester, England.

115. **Finegold, S. M., and W. L. George.** 1989. Anaerobic infections in humans. Academic Press, San Diego, Ca.

116. **Finegold, S. M., and H. Jousimies-Somer.** 1997. Recently described clinically important anaerobic bacteria: medical aspects. Clin. Infect. Dis. **25**:S88–S93.

117. **Finegold, S. M., H. R. Jousimies-Somer, and H. Wexler.** 1993. Current perspectives on anaerobic infections: diagnostic approaches. Clin. Infect. Dis. **7**:257–275.

118. **Finegold, S. M., V. L. Sutter, and G. E. Mathisen.** 1983. Normal indigenous intestinal flora, p. 3–31. *In* D. J. Hentges (ed.), Human intestinal microflora in health and disease. Academic Press, New York.

119. **Forsblom, B., D. N. Love, and H. R. Jousimies-Somer.** 1997. Characterization of anaerobic, gram-negative, non-pigmented saccharolytic rods from subgingival sites of dogs. Clin. Infect. Dis. **25**:S100–S106.

119a. **Fournier, D., C. Mouton, P. LaPierre, T. Kato, K. Okuda, and C. Menard.** 2001. *Porphyromonas gulae* sp. nov., an anaerobic, Gram-negative coccobacillus from the gingival sulcus of various animal hosts. Int. J. Syst. Evol. Bacteriol. **51:**1179–1189.

120. **Fukumoto, O. K., and Y. Takazoe.** 1988. Enumeration of cultivable black-pigmented *Bacteroides* species in human subgingival dental plaque and fecal samples. Oral Microbiol. Immunol. **3**:28–31.

121. **Funke, G., N. Alvarez, C. Pascual, E. Falsen, E. Akervall, L. Sabbe, L. Schouls, and N. Weiss.** 1997. *Actinomyces europaeus* sp. nov., isolated from human clinical specimens. Int. J. Syst. Bacteriol. **47**:687–692.

122. **Funke, G., S. Stubbs, A. von Graevenitz, and M. D. Collins.** 1994. Assignment of human derived CDC group 1 coryneform bacteria and CDC group 1-like coryneform bacteria to the genus *Actinomyces* as *Actinomyces neuii* subsp. *neuii* sp. nov., subsp. nov., and *Actinomyces neuii* subsp. *anitratus* subsp. nov. Int. J. Syst. Bacteriol. **44**:167–171.

123. **George, W. L., V. L. Sutter, D. Citron, and S. M. Finegold.** 1979. Selective and differential medium for isolation of *Clostridium difficile.* J. Clin. Microbiol. **9**:214–219.

124. **Gerding, D. N.** 1995. Foot infections in diabetic patients: the role of anaerobes. Clin. Infect. Dis. **20**:S283–S288.

125. **Gibson, M. M., E. M. Ellis, K. A. Graeme-Cook, and C. F. Higgins.** 1987. OmpR and EnvZ are pleiotropic regulatory proteins: positive regulation of the tripeptide permease (tppB) of *Salmonella typhimurium.* Mol. Gen. Genet. **207**:120–129.

126. **Goldstein, E. J., D. M. Citron, C. V. Merriam, K. Tyrrell, and Y. Warren.** 1999. Activity of gatifloxacin compared to those of five other quinolones versus aerobic and anaerobic isolates from skin and soft tissue samples of human and animal bite wound infections. Antimicrob. Agents Chemother. **43**:1475–1479.

127. **Goldstein, E. J., D. M. Citron, M. C. Vreni, K. Tyrrell, and Y. Warren.** 1999. Activities of gemifloxacin (SB 265805, LB20304) compared to those of other oral antimicrobial agents against unusual anaerobes [In Process Citation]. Antimicrob. Agents Chemother. **43**:2726–2730.

128. **Goldstein, E. J., D. M. Citron, Y. Warren, K. Tyrrell, and C. V. Merriam.** 1999. In vitro activity of gemifloxacin (SB 265805) against anaerobes. Antimicrob. Agents Chemother. **43**:2231–2235.

129. **Goldstein, E. J. C., D. M. Citron, and C. A. Nesbit.** 1996. Diabetic foot infections. Diabetes Care **19**:638–641.

130. **Goldstein, E. J. C., D. M. Citron, B. Wield, U. Blachman, V. L. Sutter, T. A. Miller, and S. M. Finegold.** 1978. Bacteriology of human and animal bite wounds. J. Clin. Microbiol. **8**:667–672.

131. **Goodman, A. D.** 1977. Isolation of anaerobic bacteria from the root canal systems of necrotic teeth by use of a transport solution. Oral Surg. **43**:766–770.

132. **Gorbach, S. L.** 1975. Management of anaerobic infections: Intra-abdominal sepsis. Ann. Intern. Med. **83**:377–379.

133. **Gorbach, S. L., J. W. Mayhew, J. G. Bartlett, H. Thadepalli, and A. B. Onderdonk.** 1976. Rapid diagnosis of anaerobic infections by direct gas-liquid chromatography of clinical specimens. J. Clin. Invest. **57**:478–484.

134. **Gorbach, S. L., H. Thadepalli, and J. Norsen.** 1974. Anaerobic microorganisms in intraabdominal infections, p. 399–407. *In* A. Balows, R. M. DeHaan, V. R. Dowell, Jr.,

and L. B. Guze (eds.), Anaerobic bacteria: role in disease. Charles C. Thomas, Springfield, Ill.

135. **Gruner, E., A. von Graevenitz, and M. Altwegg.** 1992. The API ZYM system: tabulated review from 1977 to date. J. Microbiol. Methods **16**:101–118.

136. **Hall, B. B., R. H. Fitzgerald, and J. E. Rosenblatt.** 1983. Anaerobic osteomyelitis. J. Bone. Joint. Surg. **65**:30.

137. **Hall, B. B., J. E. Rosenblatt, and R. H. Fitzgerald, Jr.** 1984. Anaerobic septic arthritis and osteomyelitis. Orthop. Clin. North Am. **15**:505–516.

138. **Hall, W. L., A. I. Sobel, C. P. Jones, and R. T. Parker.** 1967. Anaerobic postoperative pelvic infections. Obstet. Gynecol. **30**:1–7.

138a. **Hall, W., G. L. O'Neill, J. T. Magee, and B. I. Duerden.** 1999. Development of amplified 16S ribosomal DNA restriction analysis for identification of *Actinomyces* species and comparison with pyrolysis-mass spectrometry and conventional biochemical tests. J. Clin. Microbiol. **37**:2255–2261.

139. **Hamory, B. H., M. A. Sande, A. Sydnor, Jr., D. L. Seale, and J. M. Gwaltney, Jr.** 1979. Etiology and antimicrobial therapy of acute maxillary sinusitis. J. Infect. Dis. **139**:197–202.

140. **Hanff, P. A., J. A. Rosol-Donoghue, C. A. Spiegel, K. H. Wilson, and L. H. Moore.** 1995. *Leptotrichia sanguinegens* sp. nov., a new agent of postpartum and neonatal bacteremia. Clin. Infect. Dis. **20**:S237–S239.

141. **Hansen, M. V., and L. P. Elliott.** 1980. New presumptive identification test for *Clostridium perfringens*: reverse CAMP test. J. Clin. Microbiol. **12**:617–619.

142. **Hartroth, B., I. Seyfahrt, and G. Conrads.** 1999. Sampling of periodontal pathogens by paper points: evaluation of basic parameters. Oral Microbiol. Immunol. **14**:326–330.

143. **Haug, J. B., S. Harthug, T. Kalager, A. Digranes, and C. O. Solberg.** 1994. Bloodstream infections at a Norwegian university hospital, 1974–1979 and 1988–1989: changing etiology, clinical features, and outcome. Clin. Infect. Dis. **19**:246–256.

144. **Hecht, D. W., and H. M. Wexler.** 1996. In vitro susceptibility of anaerobes to quinolones in the United States. Clin. Infect. Dis. **23**:S2–S8.

145. **Heimdahl, A., G. Hall, M. Hedberg, H. Sandberg, P-O. Söder, et al.** 1990. Detection and quantitation by lysis-filtration of bacteremia after different oral procedures. J. Clin. Microbiol. **28**:2209.

146. **Heimdahl, A., L. von Konow, and C. E. Nord.** 1988. Beta-lactamase producing anaerobic bacteria in orofacial infections. Rev. Esp. Quimioterap. **1**:57–60.

147. **Hentges, D. J.** 1993. The anaerobic microflora of the human body. Clin. Infect. Dis. **16**:S175–S180.

148. **Hill, G. B., K. K. St. Claire, and L. T. Gutman.** 1995. Anaerobes predominate among the vaginal microflora of prepubertal girls. Clin. Infect. Dis. **20**:S269–S270.

149. **Hillier, S. L.** 1995. Association between bacterial vaginosis and preterm delivery of a low-birth-weight infant. N. Engl. J. Med. **333**:1737–1742.

150. **Hillier, S. L., M. A. Krohn, E. Cassen, T. R. Easterling, L. K. Rabe, and D. A. Eschenbach.** 1995. The role of bacterial vaginosis and vaginal bacteria in amniotic

fluid infection in women with preterm labor with intact membranes. Clin. Infect. Dis. **20**:S276–S278.

151. **Hillier, S. L., M. A. Krohn, L. K. Rabe, S. J. Klebanoff, and D. A. Eschenbach.** 1993. The normal vaginal flora, H_2O_2-producing lactobacilli, and bacterial vaginosis in pregnant women. Clin. Infect. Dis. **16**:S273–S281.

152. **Hillier, S. L., and R. Lau.** 1997. Vaginal microflora in postmenopausal women who have not received estrogen replacement therapy. Clin. Infect. Dis. **25**:S123–S126.

153. **Hirasawa, M., and K. Takada.** 1994. *Porphyromonas gingivicanis* sp. nov. and *Porphyromonas crevioricanis* sp. nov., isolated from beagles. Int. J. Syst. Bacteriol. **44**:637–640.

154. **Holdeman, L. V., E. P. Cato, and W. E. C. E. Moore.** 1977. Anaerobe laboratory manual. Virginia Polytechnic Institute and State University, Blacksburg, Va.

155. **Holt, J. G., and N. R. E. Krieg.** 1984. Bergey's manual of systematic bacteriology, 9th ed., subvol. 1. The Williams and Wilkins Co., Baltimore.

156. **Hudspeth, M. K., S. Hunt Gerardo, D. M. Citron, and E. J. C. Goldstein.** 1997. Growth characteristics and a novel method for identification (WEE-TAB System) of *Porphyromonas* isolated from infected dog and cat bite wounds in humans. J. Clin. Microbiol. **35**:2450–2453.

157. **Hungate, R. E.** 1969. A roll tube method for cultivation of strict anaerobes, p. 117–132. *In* J. R. Norris and D. W. Robbons (eds.), Methods in microbiology, vol. 3B. Academic Press, New York.

158. **Iino, Y., E. Hoshino, S. Tomioka, T. Takasaka, Y. Kaneko, and R. Yuasa.** 1983. Organic acids and anaerobic microorganisms in the contents of the cholesteatoma sac. Ann. Otol. Rhinol. Laryngol. **92**:91–96.

159. **Iwu, C., T. W. MacFarlane, D. MacKenzie, and D. Stenhouse.** 1990. The microbiology of periapical granulomas. Oral Surg. Oral Med. Oral Pathol. **69**:502–505.

160. **Jalava, J., and E. Eerola.** 1999. Phylogenetic analysis of *Fusobacterium alocis* and *Fusobacterium sulci* based on 16S rRNA gene sequences: proposal of *Filifactor alocis* (Cato, Moore and Moore) comb. nov. and *Eubacterium sulci* (Cato, Moore and Moore) comb. nov. Int. J. Syst. Bacteriol. **49**:1375–1379.

161. **Johnson, J. L., L. V. H. Moore, B. Kaneko, and W. E. C. Moore.** 1991. *Actinomyces georgiae* sp.nov., *Actinomyces gerencseriae* sp. nov., designation of two genospecies of *Actinomyces naeslundii,* and *viscosus* serotype II in *A. naeslundii* genospecies 2. Int. J. Syst. Bacteriol. **40**:273–286.

162. **Johnson, L. L., L. V. McFarland, P. Dearing, V. Raisys, and F. D. Schoenknecht.** 1989. Identification of *Clostridium difficile* in stool specimens by culture-enhanced gas-liquid chromatography. J. Clin. Microbiol. **27**:2218–2221.

163. **Johnson, S., F. Lebahn, L. R. Peterson, and D. N. Gerding.** 1995. Use of anaerobic collection and transport swab device to recover anaerobic bacteria from infected foot ulcers in diabetics. Clin. Infect. Dis. **20**:S289–S290.

164. **Jokipii, A. M. M., and L. Jokipii.** 1981. Metronidazole, tinidazole, ornidazole and anaerobic infections of the middle ear, maxillary sinus and central nervous system. Scand. J. Infect. Dis. **26**:129.

165. **Jones, D. B., and N. M. Robinson.** 1977. Anaerobic ocular infections. Trans. Am. Acad. Ophthalmol. Otolaryngol. **83**:309–331.

166. **Jousimies-Somer, H., S. Savolainen, A. Mäkitie, and J. Ylikoski.** 1993. Bacteriologic findings in peritonsillar abscess in young adults. Clin. Infect. Dis. **16**:S292–S298.

167. **Jousimies-Somer, H. R.** 1995. Update on the taxonomy and the clinical and laboratory characteristics of pigmented anaerobic gram-negative rods. Clin. Infect. Dis. **20**:S187–S191.

168. **Jousimies-Somer, H. R., A. Bryk, S. Asikainen, A. Kanervo, A. Takala, and E. Kononen.** 1999. Oral colonization of infants with *Veillonella* species. Anaerobe **5**:251–253.

169. **Jousimies-Somer, H. R., P. Laine, E. Könönen, E. Tarkka, and C. Lindqvist.** Microbiologic findings of failing dental implants. 1998. Abstr. 2nd World congr. Anaerob. bacteria and infect, 7.026, p. 112.

170. **Jousimies-Somer, H. R., S. Savolainen, and J. S. Ylikoski.** 1988. Bacteriological findings of acute maxillary sinusitis in young adults. J. Clin. Microbiol. **26 (no. 10)**:1919–1925.

171. **Jousimies-Somer, H. R., P. Summanen, and S. M. Finegold.** 1995. *Bacteroides levii*-like organisms isolated from clinical specimens. Clin. Infect. Dis. **20**:S208–S209.

172. **Jousimies-Somer, H. R., P. H. Summanen, and S. M. Finegold.** 1999. *Bacteroides, Porphyromonas, Prevotella, Fusobacterium,* and other anaerobic gram-negative rods and cocci, p. 690–711. *In* P. R. Murray (ed.), E. J. Baron, M. A. Pfaller, F. C. Tenover, and R.H. Yolken. Manual of clinical microbiology. 7th ed. American Society for Microbiology, Washington, D.C.

173. **Jungkind, D., J. Millan, S. Allen, J. Dyke, and E. Hill.** 1986. Clinical comparison of a new automated infrared blood culture system with the Bactec 460 system. J. Clin. Microbiol. **23**:262–266.

174. **Kader, H. A., D. A. Piccoli, A. F. Jawad, K. L. McGowan, and E. S. Maller.** 1998. Single toxin detection is inadequate to diagnose *Clostridium difficile* diarrhea in pediatric patients. Gastroenterology **115**:1329–1334.

175. **Kageyama, A., Y. Benno, and T. Nakase.** 1999. Phylogenetic and phenotypic evidence for the transfer of *Eubacterium aerofaciens* to the genus *Collinsella* as *Collinsella aerofaciens* gen. nov., comb. nov. Int. J. Syst. Bacteriol. **49**:557–565.

176. **Kato, H., N. Kato, K. Watanabe, N. Iwai, H. Nakamura, T. Yamamoto, K. Suzuki, S.-M. Kim, Y. Chong, and E. B. Wasito.** 1998. Identification of toxin A-negative, toxin B-positive *Clostridium difficile* by PCR. J. Clin. Microbiol. **36**:2178–2182.

177. **Kato, N., C.-Y. Ou, H. Kato, S. L. Bartley, V. K. Brown, V. R. Dowell, Jr., and K. Ueno.** 1991. Identification of toxigenic *Clostridium difficile* by the polymerase chain reaction. J. Clin. Microbiol. **29**:33–37.

178. **Kawamura, Y., X. G. Hou, F. Sultana, S. Liu, H. Yamamoto, and T. Ezaki.** 1995. Transfer of *Streptococcus adjacens* and *Streptococcus defectivus* to *Abiotrophia* gen. nov. as *Abiotrophia adiacens* comb. nov. and *Abiotrophia defectiva* comb. nov., respectively. Int. J. Syst. Bacteriol. **45**:798–803.

179. **King, A., J. Downes, C.-E. Nord, and I. Phillips.** 1999. Antimicrobial susceptibility of non-Bacteroides group anaerobic Gram-negative bacilli in Europe. Clin. Microbiol. Infect. **5**:404–416.

180. **Kitch, T. T., and P. C. Appelbaum.** 1989. Accuracy and reproducibility of the 4-hour ATB 32A method for anaerobe identification. J. Clin. Microbiol. **27**:2509–2513.

181. **Koch, C. L., P. Derby, and V. R. Abratt.** 1998. In-vitro antibiotic susceptibility and molecular analysis of anaerobic bacteria isolated in Cape Town, South Africa. J. Antimicrob. Chemother. **42**:245–248.

182. **Kodaka, H., A. Y. Armfield, G. L. Lombard, and V. R. Dowell, Jr.** 1982. Practical procedure for demonstrating bacterial flagella. J. Clin. Microbiol. **16**:948–952.

183. **Kornman, K. S., and W. J. Loesche.** 1978. A new medium for isolation of *Actinomyces viscosus* and *Actinomyces naeslundii* from dental plaque. J. Clin. Microbiol. **7**:514–518.

184. **Könönen, E.** 2000. Development of oral bacterial flora in young children. Ann. Med. **32**:107–112.

185. **Könönen, E., E. Eerola, E. V. G. Frandsen, J. Jalava, and J. Mättö.** 1998. Phylogenetic characterization and proposal of a new pigmented species to the genus *Prevotella: Prevotella pallens* sp. nov. Int. J. Syst. Bacteriol. **48**:47–51.

186. **Könönen, E., A. Kanervo, A. Takala, S. Asikainen, and H. Jousimies-Somer.** 1999. Establishment of oral anaerobes during first year of life. J. Dent. Res. **78**:1634–1639.

187. **Könönen, E., J. Mättö, M.-L. Väisänen-Tunkelrott, E. V. G. Frandsen, I. Helander, S. M. Finegold, and H. R. Jousimies-Somer.** 1998. Biochemical and genetic characterization of *Prevotella intermedia/Prevotella nignescens*-like organisms. Int. J. Syst. Bacteriol. **48**:39–46.

188. **Krohn, M. A., S. L. Hillier, and D. A. Eschenbach.** 1989. Comparison of methods for diagnosing bacterial vaginosis among pregnant women. J. Clin. Microbiol. **27**:1266–1271.

189. **Külekçi, G., D. Í. Yaylali, H. Koçak, Ç. Kasapoglu, and O. Z. Gümrü.** 1996. Bacteriology of dentoalveolar abscesses in patients who have received empiric antibiotic therapy. Clin. Infect. Dis. **23**:S51–S53.

190. **Kyne, L., M. Warny, A. Qamar, and C. P. Kelly.** 2000. Asymptomatic carriage of *Clostridium difficile* and serum levels of IgG antibody against toxin A. N. Engl. J. Med. **342**:390–397.

191. **Labbe, A. C., A. M. Bourgault, J. Vincelette, P. L. Turgeon, and F. Lamothe.** 1999. Trends in antimicrobial resistance among clinical isolates of the *Bacteroides fragilis* group from 1992 to 1997 in Montreal, Canada [In Process Citation]. Antimicrob. Agents Chemother. **43**:2517–2519.

192. **Laughon, B. E., S. A. Syed, and W. J. Loesche.** 1982. Rapid identification of *Bacteroides gingivalis.* J. Clin. Microbiol. **15**:345–346.

192a. **Lawson, A. J., S. L. W. On, J. M. J. Logan, and J. Stanley.** 2001. *Campylobacter hominis* sp. nov., from the human gastrointestinal tract. Int. J. Syst. Evol. Bacteriol. **51:**651–660.

193. **Lawson, P. A., E. Falsen, E. Akervall, P. Vandamme, and M. D. Collins.** 1997. Characterization of some *Actinomyces*-like isolates from human clinical specimens: reclassification of *Actinomyces suis* (Soltys and Spratling) as *Actinobaculum suis* comb. nov. and description of *Actinobaculum schalii* sp. nov. Int. J. Syst. Bacteriol. **47**:899–903.

193a. **Lawson, P. A., N. Nikolaitchouk, E. Falsen, K. Westling, and M. D. Collins,** 2001. *Actinomyces funkei* sp. nov., isolated from the human clinical specimens. Int J. Syst. Evol. Bacteriol. **51:**853–855.

194. **Leach, R. D., S. J. Eykyn, I. Phillips, and B. Corrin.** 1979. Anaerobic subareolar breast abscess. Lancet **1**:35–7.

195. **Lehman, W. L., Jr., W. W. Jones, M. D. Allo, and R. M. Johnston.** 1985. Human bite infections of the hand: adjunct treatment with hyperbaric oxygen. Infect. Surg. **4**:460–465.

196. **Levison, M. E., I. Trestman, R. Quach, C. Sladowski, and C. N. Floro.** 1979. Quantitative bacteriology of the vaginal flora in vaginitis. Am. J. Obstet. Gynecol. **133**:139–44.

197. **Lewis, R. P., V. L. Sutter, and S. M. Finegold.** 1978. Bone infections involving anaerobic bacteria. Medicine (Baltimore) **57**:279–305.

198. **Li, N., Y. Hashimoto, S. Adnan, H. Miura, H. Yamamoto, and T. Ezaki.** 1992. Three new species of the genus *Peptostreptococcus* isolated from humans: *Peptostreptococcus vaginalis* sp. nov., *Peptostreptococcus lacrimalis* sp. nov., *Peptostreptococcus lactolyticus* sp. nov. Int. J. Syst. Bacteriol. **42**:602–605.

199. **Limaye, A. P., D. K. Turgeon, B. T. Cookson, and T. R. Fritsche.** 2000. Pseudomembraneous colitis caused by toxin A^-B^+ strain of *Clostridium difficile.* J. Clin. Microbiol. **38**:1696–1697.

200. **Livingston, S. J., S. D. Kominos, and R. B. Yee.** 1978. New medium for selection and presumptive identification of the *Bacteroides fragilis* group. J. Clin. Microbiol. **7**:448–453.

201. **Llera, J. L., R. C. Levy, and J. L. Staneck.** 1984. Cutaneous abscess: natural history and management in an outpatient facility. J. Emerg. Med. **1**:489–493.

202. **Loesche, W. J., and D. E. Lopatin.** 1998. Interactions between periodontal disease, medical diseases and immunity in the older individual. Periodontology 2000 **16**:80–105.

203. **Loesche, W. J., S. A. Syed, E. Schmidt, and E. C. Morrison.** 1985. Bacterial profiles of subgingival plaques in periodontitis. J. Periodontol. **56**:447–456.

204. **Lombard, G. L., D. N. Whaley, and V. R. Dowell, Jr.** 1982. Comparison of media in the Anaerobe-Tek and Presumpto plate systems and evaluation of the Anaerobe-Tek system for identification of commonly encountered anaerobes. J. Clin. Microbiol. **16**:1066–1072.

205. **Lorber, B., and R. M. Swenson.** 1974. Bacteriology of aspiration pneumonia. A prospective study of community and hospital acquired cases. Ann. Intern. Med. **81**:329–331.

206. **Love, D. N.** 1995. *Porphyromonas macacae* comb. nov., a consequence of *Bacteroides macacae* being a senior synonym of *Porphyromonas salivosa.* Int. J. Syst. Bacteriol. **45**:90–92.

207. **Love, D. N., J. Karjalainen, A. Kanervo, B. Forsblom, E. Sarkiala, G. D. Bailey, D. Wigney, and H. Jousimies-Somer.** 1994. *Porphyromonas canoris* sp. nov., an asaccharolytic, black pigmented species from periodontal pockets of dogs. Int. J. Syst. Bacteriol. **44**:204–208.

208. **Lylerly, D. M., L. M. Neville, D. T. Evans, J. Fill, S. Allen, W. Greene, R. Sautter, P. Hnatuck, D. J. Torpey, and R. Schwalbe.** 1998. Multicenter evaluation of the *Clostridium difficile* TOX A/B TEST. J. Clin. Microbiol. **36**:184–190.

209. **MacLennan, J. D.** 1962. The histotoxic clostridial infections of man. Bacteriol. Rev. **26**:177–276.

210. **Maiden, M. F. J., A. Tanner, and P. J. Macuch.** 1996. Rapid characterization of periodontal bacterial isolates by using fluorogenic substrate tests. J. Clin. Microbiol. **44**:376–384.

211. **Malnick, H.** 1997. *Anaerobiospirillum thomasii* sp. nov., an anaerobic spiral bacterium isolated from the feces of cats and dogs and from diarrheal feces of humans, and emendation of the genus *Anaerobiospirillum*. Int. J. Syst. Bacteriol. **47**:381–384.

212. **Malnick, H., A. Jones, and J. C. Vickers.** 1989. Anaerobiospirillum: Cause of a "new" zoonosis? [letter]. Lancet **1**:1145–1146.

213. **Malnick, H., M. E. M. Thomas, H. Lotay, and M. Robbins.** 1983. Anaerobiospirillum species isolated from humans with diarrhoea. J. Clin. Pathol. **36**:1097–1101.

214. **Mangels, J. I., M. Cox, and L. H. Lindberg.** 1984. Methanol fixation: an alternative to heat fixation of smears before staining. Diagn. Microbiol. Infect. Dis. **2**:129–137.

215. **Mangels, J. I., and B. P. Douglas.** 1989. Comparison of four commercial brucella agar media for growth of anaerobic organisms. J. Clin. Microbiol. **27**:228–2271.

216. **Mangels, J. I., I. Evaldson, and M. Cox.** 1993. Rapid presumptive identification of *Bacteroides fragilis* group organisms with use of 4-methylumbelliferone-derivative substrates. Clin. Infect. Dis. **16**:S319–S321.

217. **Mårdh, P.-A., L. Larsson, and G. Odham.** 1981. Head-space gas chromatography as a tool in the identification of anaerobic bacteria and diagnosis of anaerobic infections. Scand. J. Infect. Dis. **Suppl. 26**:14–18.

218. **Marquette, C. H., H. Georges, F. Wallet, P. Ramon, R. N. D. Saulnier, D. Mathieu, A. Rime, and A. B. Tonnel.** 1993. Diagnostic efficiency of endotracheal aspirates with quantitative bacterial cultures in intubated patients with suspected pneumonia. Comparison with the protected specimen brush. Am. Rev. Respir. Dis. **148**:138.

219. **Martens, M. D., et al.** 2000. Transcervical uterine cultures with a new endometrial suction curette: a comparison of three sampling methods in postpartum endometritis. Obstet. Gynecol. **74**:273–276.

220. **Mazulli, T., A. E. Simor, and D. E. Low.** 1990. Reproducibility of interpretation of gram-stained vaginal smears for the diagnosis of bacterial vaginosis. J. Clin. Microbiol. **28**:1506–1508.

221. **Mättö, J., S. Asikainen, M.-L. Väisänen, M. Saarela, P. Summanen, S. M. Finegold, and H. R. Jousimies-Somer.** 1997. *Porphyromonas gingivalis, Prevotella intermedia* and *Prevotella nigrescens* in extraoral and some odontogenic infections. Clin. Infect. Dis. **25**:S194–S198.

222. **Meislin, H. W., S. A. Lerner, M. H. Graves, M. D. McGehee, F. E. Kocka, J. A. Morello, and P. Rosen.** 1977. Cutaneous abscesses. Anaerobic and aerobic bacteriology and outpatient management. Ann. Intern. Med. **87**:145–149.

223. **Miranda, C., J. C. Alados, J. M. Molina, C. Dominguez, J. A. Miranda, et al.** 1993. Posthysterectomy wound infection. A review. Diagn. Microbiol. Infect. Dis. **17**:41–44.

224. **Mitchelmore, I. J., A. J. Prior, P. Q. Montgomery, and S. Tabaqchali.** 1995. Microbiological features and pathogenesis of peritonsillar abscesses. Eur. J. Clin. Microbiol. Infect. Dis. **14**:870–877.

225. **Mombelli, A., and N. P. Lang.** 1998. The diagnosis and treatment of peri-implantitis. Periodontology 2000 **17**:63–76.

226. **Moncla, B. J., P. Braham, K. Dix, S. Watabe, and D. Schwartz.** 1990. Use of oligonucleotide DNA probes for the identification of *Bacteroides gingivalis*. J. Clin. Microbiol. **28**:324–327.

227. **Moncla, B. J., P. Braham, L. K. Rabe, and S. L. Hillier.** 1991. Rapid presumptive identification of black-pigmented gram-negative anaerobic bacteria by using 4-methylumbelliferone substrates. J. Clin. Microbiol. **29**:1955–1958.

228. **Montgomerie, J. Z., E. Chan, D. S. Gilmore, H. N. Canawati, and F. L. Sapico.** 1991. Low mortality among patients with spinal cord injury and bacteremia. Rev. Infect. Dis. **13**:871.

229. **Moore, L. V. H., J. L. Johnson, and W. E. C. Moore.** 1994. Description of *Prevotella tannerae* sp. nov. and *Prevotella enoeca* sp. nov. from human gingival crevice and emendation of the description of *Prevotella zoogleoformans*. Int. J. Syst. Bacteriol. **44**:599–602.

230. **Moore, L. V. H., and W. E. Moore.** 1991. Anaerobe laboratory manual update. Virginia Polytechnic Institute and State University, Blacksburg, Va.

231. **Moore, W. E. C.** 1987. Microbiology of periodontal disease. J. Periodont. Res. **22**:335–341.

232. **Moore, W. E. C., E. P. Cato, and L. V. Holdeman.** 1969. Review. Anaerobic bacteria of the gastrointestinal flora and their occurrence in clinical infections. J. Infect. Dis. **119**:641–649.

233. **Moore, W. E. C., L. V. Holdeman, E. P. Cato, I. J. Good, E. P. Smith, R. R. Ranney, and K. G. Palcanis.** 1984. Variation in periodontal floras. Infect. Immun. **46**:720–726.

234. **Morello, J. A., C. Leitch, S. Nitz, J. W. Dyke, and M. Andruszewski.** 1994. Detection of bacteremia by Difco ESP blood culture system. J. Clin. Microbiol. **32**:811–818.

235. **Mory, F., A. Lozniewski, S. Bland, A. Sedallian, G. Grollier, F. Girard-Pipau, M. F. Paris, and L. Dubreuil.** 1998. Survey of anaerobic susceptibility patterns: a French multicentre study. Int. J. Antimicrob. Agents **10**:229–236.

236. **Möller, A. J. R.** 1966. Microbiological examination of root canals and periapical tissues of human teeth. Odontol. Tidskr. Scand. Dent. J. **74**:365–368.

237. **Mulligan, M. E., D. M. Citron, B. T. McNamara, and S. M. Finegold.** 1982. Impact of cefoperazone therapy on fecal flora. Antimicrob. Agents Chemother. **22**:226–230.

238. **Murdoch, D. A.** 1998. Gram-positive anaerobic cocci. Clin. Microbiol. Rev. **11**:81–120.

239. **Murdoch, D. A., M. D. Collins, A. Willems, J. M. Hardie, K. A. Young, and J. T. Magee.** 1997. Description of three new species of the genus *Peptostreptococcus* from human clinical specimens: *Peptostreptococcus harei* sp. nov., *Peptostreptococcus ivorii* sp. nov., and *Peptostreptococcus octavius* sp. nov. Int. J. Syst. Bacteriol. **47**:781–787.

240. **Murdoch, D. A., and H. N. Shah.** 1999. Reclassification of *Peptostreptococcus magnus* (Prevot 1933) Holdeman and Moore 1972 as *Finegoldia magna* comb. nov. and

Peptostreptococcus micros (Prevot 1933) Smith 1957 as *Micromonas micros* comb. nov. Anaerobe **5**:555–559.

241. **Murray, P. R.** 1978. Growth of clinical isolates of anaerobic bacteria on agar media: effects of media composition, storage conditions, and reduction under anaerobic conditions. J. Clin. Microbiol. **8**:708–714.

242. **Murray, P. R., E. J. Baron, M. A. Pfaller, F. C. Tenover, and R. H. Yolken.** 1999. Manual of clinical microbiology. American Society for Microbiology, Washington, D.C.

243. **Nakazawa, F., S. E. Poco, T. Ikeda, M. Sato, S. Kalfas, G. Sundqvist, and E. Hoshino.** 1999. *Cryptobacterium curtum* gen. nov., sp. nov., a new genus of Gram-positive rod isolated from human oral cavities. Int. J. Syst. Bacteriol. **49**:1193–2000.

244. **Nakazawa, F., M. Sato, S. E. Poco, T. Hashimura, T. Ikeda, S. Kalfas, G. Sundqvist, and E. Hoshino.** 2000. Description of *Mogibacterium pumilum* gen. nov., sp. nov. and *Mogibacterium vescum* gen. nov., sp. nov., and reclassification of *Eubacterium timidum* (Holdeman et al. 1980) as *Mogibacterium timidum* gen. nov., comb. nov. Int. J. Syst. Evolut. Bacteriol. **50**:679–688.

245. **National Committee for Clinical Laboratory Standards.** 1990. Quality assurance for commercially prepared microbiological culture media, Vol. 10, No. 14. Approved Standard, NCCLS Publication M22-A. NCCLS, Villanova, Pa.

246. **National Committee for Clinical Laboratory Standards.** 1997. Methods for antimicrobial susceptibility testing of anaerobic bacteria, 4th ed. Approved Standard, NCCLS Document M11-A4. NCCLS, Wayne, Pa.

246a. **National Committee for Clinical Laboratory Standards.** 2000. Methods for antimicrobial susceptibility testing of anaerobic bacteria, 5th ed. Approved Standard, NCCLS Document M11-A5. NCCLS, Villanova, Pa.

247. **Newman, M. G., and T. N. Sims.** 1979. The predominant cultivable microbiota of the periodontal abscess. J. Periodontol. **50**:350–354.

248. **Nichols, R. L., and J. W. Smith.** 1975. Intragastric microbial colonization in common disease states of the stomach and duodenum. Ann. Surg. **182**:557–561.

248a. **Nikolaitchouk, N., L. Hoyles, E. Falsen, J. M. Grainger, and H. D. Collins.** 2000. Characterization of *Actinomyces* isolates from samples from the human urogenital tract: description of *Actinomyces urogentialis* sp. nov. Int J. Syst. Evol. Microbiol. **50**: 1649–1654.

249. **Nugent, R. P., M. A. Krohn, and S. L. Hillier.** 1991. Reliability of diagnosing bacterial vaginosis is improved by standardized method of Gram stain interpretation. J. Clin. Microbiol. **29**:297–301.

250. **Nyfors, S., E. Kononen, A. Takala, and H. Jousimies-Somer.** 1999. Beta-lactamase production by oral anaerobic gram-negative species in infants in relation to previous antimicrobial therapy. Antimicrob. Agents Chemother. **43**:1591–1594.

251. **O'Donnell, J. A., and L. E. Asbel.** 1999. *Bacteroides fragilis* bacteremia and infected aortic aneurysm presenting as fever of unknown origin: diagnostic delay without routine anaerobic blood cultures. Clin. Infect. Dis. **29**:1309–1311.

252. **Olsen, I., J. L. Johnson, L. V. Moore, et al.** 1992. *Lactobacillus uli* sp. nov. and *Lactobacillus rimae* sp. nov. from the human gingival crevice and emended descriptions of *Lactobacillus minutus* and *Streptococcus parvulus*. Int. J. Syst. Bacteriol. **41**:261–266.

253. **Onodera, Y., and K. Sato.** 1999. Molecular cloning of the gyrA and gyrB genes of *Bacteroides fragilis* encoding DNA gyrase [In Process Citation]. Antimicrob. Agents Chemother. **43**:2423–2429.

254. **Parker, R. T., and C. P. Jones.** 1966. Anaerobic pelvic infections and developments in hyperbaric oxygen therapy. Am. J. Obstet. Gynecol. **96**:645–659.

255. **Pascual Ramos, C., G. Foster, and M. D. Collins.** 1997. Phylogenetic analysis of *Actinomyces* based on 16S rRNA gene sequence: description of *Arcanobacterium phocae* sp. nov., *Arcanobacterium bernardiae* comb. nov., and *Arcanobacterium pyogenes* comb. nov. Int. J. Syst. Bacteriol. **47**:46–53.

256. **Paster, B. J., F. E. Dewhirst, I. Olsen, and G. I. J. Frazer.** 1994. Phylogeny of *Bacteroides, Prevotella,* and *Porphyromonas* spp. and related bacteria. J. Bacteriol. **176**:725–732.

257. **Peach, S., and L. Hayek.** 1974. The isolation of anaerobic bacteria from wound swabs. J. Clin. Pathol. **27**:578–582.

258. **Perry, J. L.** 1997. Assessment of swab transport systems for aerobic and anaerobic organism recovery. J. Clin. Microbiol. **35**:1269–1271.

259. **Perry, L. D., J. H. Brinser, and H. Kolodner.** 1982. Anaerobic corneal ulcers. Ophthalmology **89**:636–642.

260. **Peterson, L. R.** 1997. Effect of media on transport and recovery of anaerobic bacteria. Clin. Infect. Dis. **25**:S134–S136.

261. **Peterson, L. R., P. J. Kelly, and H. A. Nordbrock.** 1996. Role of culture and toxin detection in laboratory testing for diagnosis of *Clostridium difficile*-associated diarrhea. Eur. J. Clin. Microbiol. Infect. Dis. **15**:330–336.

262. **Poulet, P. P., D. Duffaut, and J. P. Lodter.** 1999. Evaluation of the Etest for determining the in-vitro susceptibilities of *Prevotella intermedia* isolates to metronidazole [letter]. J. Antimicrob. Chemother. **43**:610–611.

263. **Poxton, I. R., R. Brown, A. F. Sawyerr, and A. Ferguson.** 1997. The mucosal anaerobic gram-negative bacteria of the human colon. Clin. Infect. Dis. **25**:S111–S113.

264. **Rabe, L. K., D. Sheiness, and S. L. Hillier.** 1995. Comparison of the use of oligonucleotide probes, 4-methulumbelliferyl derivatives, and conventional methods for identifying *Prevotella bivia.* Clin. Infect. Dis. **20**:S195–S197.

265. **Ramos, C. P., E. Falsen, N. Alvarez, E. Akervall, B. Sjoden, and M. D. Collins.** 1997. *Actinomyces graevenitzii* sp. nov., isolated from human clinical specimens. Int. J. Syst. Bacteriol. **47**:885–888.

266. **Rantakokko-Jalava, K., S. Nikkari, J. Jalava, E. Eerola, M. Skurnik, O. Meurman, O. Ruuskanen, A. Alanen, E. Kotilainen, P. Toivanen, and P. Kotilainen.** 2000. Direct amplification of rRNA genes in diagnosis of bacterial infections. J. Clin. Microbiol. **38**:32–39.

267. **Rao, R., and C. S. Bhaskaran.** 1984. Bacteriology of chronic suppurative otitis media with special reference to anaerobes. Indian J. Pathol. Microbiol. **27**:341–346.

268. **Rasmussen, B. A., and K. Bush.** 1997. Carbapenem-hydrolyzing beta-lactamases [see comments]. Antimicrob. Agents Chemother. **41**:223–232.

269. **Rasmussen, B. A., K. Bush, and F. P. Tally.** 1997. Antimicrobial resistance in anaerobes. Clin. Infect. Dis. **24 Suppl. 1**:S110–S120.

270. **Rautio, M., E. Eerola, J. Jalava, and H. R. Jousimies-Somer.** 1997. Phylogenetic description of a bile-resistant pigmenter, probably conforming to a new genus and species. Rev. Med. Microbiol. **8**:S103–S103.

271. **Rautio, M., H. Saxén, M. Lönnroth, M.-L. Väisänen, R. Nikku, and H. R. Jousimies-Somer.** 1997. Characteristics of an unusual anaerobic pigmented gram-negative rod isolated from normal and inflamed appendices. Clin. Infect. Dis. **25**:S107–S110.

272. **Räisänen, S., Paatsama, J., Tarkka, E., Meurman, J., Lindqvist, C., and Jousimies-Somer, H.** 1998. Bacteriology of severe odontogenic infections. Proceedings of the 2nd World Congress on Anaerobic Bacteria and Infections. ISAB, 78.

273. **Reysset, G., A. Haggoud, and M. Sebald.** 1992. Genetics of resistance of *Bacteroides* species to 5-nitroimidazole. Clin. Infect. Dis. **16, Suppl. 4**:S401–S403.

274. **Roberts, M. C.** 1995. Distribution of tetracycline and macrolide-lincosamide-streptogramin B resistance genes in anaerobic bacteria. Clin. Infect. Dis. **20 Suppl. 2**:S367–S369.

275. **Rodloff, A. C., S. L. Hillier, and B. J. Moncla.** 1999. *Peptostreptococcus, Propionibacterium, Lactobacillus, Actinomyces,* and other non-sporeforming anaerobic gram-positive bacteria, p. 672–689. *In* P. R. Murray (ed.), Manual of clinical microbiology. American Society for Microbiology, Washington D.C.

276. **Rodriques-Jovita, M., M. D. Collins, B. Sjoden, and E. Falsen.** 1999. Characterization of a novel *Atopobium* isolate from the human vagina: description of *Atopobium vaginae* sp. nov. Int. J. Syst. Bacteriol. **49**:1573–1376.

277. **Rosebury, T.** 1962. Microorganisms indigenous to man. McGraw-Hill, New York.

278. **Rosenblatt, J. E.** 1997. Can we afford to do anaerobic cultures and identification? A positive point of view. Clin. Infect. Dis. **25**:S127–S131.

279. **Rosenblatt, J. E., and I. Brook.** 1993. Clinical relevance of susceptibility testing of anaerobic bacteria. Clin. Infect. Dis. **16 Suppl. 4**:S446–S448.

280. **Rosenblatt, J. E., and D. R. Gustafson.** 1995. Evaluation of the Etest for susceptibility testing of anaerobic bacteria. Diagn. Microbiol. Infect. Dis. **22**:279–284.

281. **Rosseel, P., and S. Lauwers.** 1997. Evaluation of six different commercial blood culture media for the isolation of anaerobic bacteria. Clin. Infect. Dis. **25**:S141–S142.

282. **Rotheram, E. B., Jr., and S. F. Schick.** 1969. Nonclostridial anaerobic bacteria in septic abortion. Am. J. Med. **46**:80–89.

283. **Rupnik, M.** 2000. *Clostridium difficile* toxinotypes. Abstr. VIII-F3, p. 164. Anaerobe 2000. Int. Congr. Confed. Anaerobe Soc.

284. **Sabbaj, J., V. L. Sutter, and S. M. Finegold.** 1972. Anaerobic pyogenic liver abscess. Ann. Intern. Med. **77**:629–638.

285. **Salonen, J. H., E. Eerola, and O. Meurman.** 1998. Clinical significance and outcome of anaerobic bacteremia. Clin. Infect. Dis. **26**:1413–1417.

286. **Salyers, A. A., and N. B. Shoemaker.** 1996. Resistance gene transfer in anaerobes: new insights, new problems. Clin. Infect. Dis. **23 Suppl 1**:S36–S43.

287. **Sapico, F. L., H. N. Canawati, J. L. Witte, J. Z. Montgomerie, F. W. Wagner, Jr., and A. N. Bessman.** 1980. Quantitative aerobic and anaerobic bacteriology of infected diabetic feet. J. Clin. Microbiol. **12**:413–420.

288. **Sarkonen, N., E. Könönen, P. Summanen, A. Kanervo, A. Takala, and H. Jousimies-Somer.** 2000. Oral colonization with *Actinomyces* species in infants by two years of age. J. Dent. Res. **79**:864–867.

289. **Sarkonen, N., E. Könönen, P. Summanen, M. Könönen, and H. Jousimies-Somer.** 2001. Phenotypic identification of *Actinomyces* and related species isolated from human sources. J. Clin. Microbiol. **39**: 3955–61.

290. **Schreckenberger, P. C., D. M. Celig, and W. M. Janda.** 1988. Clinical evaluation of the Vitek ANI card for identification of anaerobic bacteria. J. Clin. Microbiol. **26**:225–230.

291. **Schreiner, A.** 1979. Anaerobic pulmonary infections. Scand. J. Infect. Dis. Suppl. 77–79.

292. **Seydoux, C. H., and P. Francioli.** 1992. Bacterial brain abscess: factors influencing mortality and sequelae. Clin. Infect. Dis. **15**:394–401.

293. **Shah, H., M. D. Collins, I. Olsen, B. J. Paster, and F. E. Dewhirst.** 1995. Reclassification of *Bacteroides levii* (Holdeman, Cato, and Moore) in the genus *Porphyromonas* and *Porphyromonas levii* comb. nov. Int. J. Syst. Bacteriol. **45**:586–588.

294. **Shah, H. N., and S. E. Gharbia.** 1992. Biochemical and chemical studies on strains designated *Prevotella intermedia* and proposal of a new pigmented species, *Prevotella nigrescens* sp. nov. Int. J. Syst. Bacteriol. **42**:542–546.

295. **Sharp, C. S., A. N. Bessman, F. W. Wagner, Jr., D. Garland, and E. Reece.** 1979. Microbiology of superficial and deep tissues in infected diabetic gangrene. Surg. Gynecol. Obstet. **149**:217–219.

296. **Sheppard, A., C. Cammarata, and D. H. Martin. 1990.** Comparison of different medium bases for the semiquantitative isolation of anaerobes from vaginal secretions. J. Clin. Microbiol. **28**:455–457.

297. **Shimada, K., T. Inamatsu, and M. Yamashiro.** 1977. Anaerobic bacteria in biliary disease in elderly patients. J. Infect. Dis. **135**:850–854.

298. **Slots, J.** 1987. Detection of colonies of *Bacteroides gingivalis* by a rapid fluorescence assay for trypsin-like activity. Oral Microbiol. Immunol. **2**:139–141.

299. **Slots, J., and H. S. Reynolds.** 1982. Longwave UV light fluorescence for identification of black-pigmented *Bacteroides* spp. J. Clin. Microbiol. **16**:1148–1151.

300. **Slots, J., and M. A. Taubman.** 1992. Contemporary oral microbiology and immunology. Mosby-Year Book, St. Louis.

301. **Smith, C. J.** 1985. Characterization of *Bacteroides ovatus* plasmid pBI136 and structure of its clindamycin resistance region. J. Bacteriol. **161**:1069–1073.

302. **Smith, H. J., and H. B. Moore.** 1988. Isolation of *Mobiluncus* species from clinical specimens by using cold enrichment and selective media. J. Clin. Microbiol. **26**:1134–1137.

303. **Smith, J. W., P. M. Southern, Jr., and J. D. Lehmann.** 1970. Bacteremia in septic abortion: complications and treatment. Obstet. Gynecol. **35**:704–708.

304. **Snydman, D. R., G. J. J. Cuchural, L. McDermott, and M. Gill.** 1992. Correlation of various in vitro testing methods with clinical outcomes in patients with *Bacteroides fragilis* group infections treated with cefoxitin: a retrospective analysis. Antimicrob. Agents Chemother. **36**:540–544.

305. **Snydman, D. R., N. V. Jacobus, L. A. McDermott, S. Supran, G. J. Cuchural, Jr., S. Finegold, L. Harrell, D. W. Hecht, P. Iannini, S. Jenkins, C. Pierson, J. Rihs, and S. L. Gorbach.** 1999. Multicenter study of in vitro susceptibility of the *Bacteroides fragilis* group, 1995 to 1996, with comparison of resistance trends from 1990 to 1996 [In Process Citation]. Antimicrob. Agents Chemother. **43**:2417–2422.

306. **Sobel, J. D.** 1997. Current concepts: vaginitis. N. Engl. J. Med. **337**:1896–1903.

307. **Socransky, S. S., S. Smith, L. Martin, B. J. Paster, F. E. Dewhirst, and A. E. Levin.** 1994. "Checkerboard" DNA-DNA hybridization. Biol. Techniques **17**:788–793.

308. **Solokim, J. S., H. H. Reinhart, E. P. Dellinger, J. M. Bohnen, O. D. Rotstein, S. B. Vogel, et al.** 1996. Results of a randomized trial comparing sequential intravenous/oral treatment with ciprofloxacin plus metronidazole to imipenem/cilastin to intra-abdominal infections. as **223**:303–315.

309. **Sondag, J. E., M. Ali, and P. R. Murray.** 1979. Relative recovery of anaerobes on different isolation media. J. Clin. Microbiol. **10**:756–757.

310. **Soper, D. E., N. J. Brockwell, and H. P. Dalton.** 1992. The importance of wound infection in antibiotic failures in the therapy of postpartum endometritia. Surg. Gynecol. Obstet. **174**:265–269.

311. **Spiegel, C. A., R. Amsel, D. Eschenbach, F. D. Schoenknecht, and K. Holmes.** 1980. Anaerobic bacteria in non-specific vaginitis. N. Engl. J. Med. **303**:601–607.

312. **Spiegel, C. A., and K. J. Krueger.** 1986. A selective agar for isolation of *Mobiluncus* from vaginal fluid. Abstr. Annu. Meet. Am. Soc. Microbiol. **C-334**:383–383.

313. **Steyaert, S., R. Peleman, M. Vaneechoutte, T. De Baere, G. Claeys, and G. Verschraegen.** 1999. Septicemia in neutropenic patients infected with *Clostridium tertium* resistant to cefepime and other expanded-spectrum cephalosporins. J. Clin. Microbiol. **37**:3778–3779.

314. **Stubbs, S. L. J., M. Rupnik, M. Gilbert, J. Brazier, and B. I. Duerden.** Production of actin-specific ADP-ribosyltransferases (binary toxin) by strains of *Clostridium difficile*. VIII-P10, p. 176. Anaerobe 2000. Int. Congr. Confed. Anaerobe Soc.

315. **Suau, A., R. Bonnet, M. Sutren, J. J. Godon, G. R. Gibson, M. D. Rollins, and J. Dore.** 1999. Direct analysis of genes encoding 16S rRNA from complex communities reveals many novel molecular species within the human gut. Appl. Environ. Microbiol. **65**:4799–4807.

316. **Suen, J. C., C. L. Hatheway, A. G. Steigerwalt, and D. J. Brenner.** 1988. *Clostridium argentinense* sp. nov.: a genetically homologous group composed of all strains of *Clostridium botulinum* toxin type G and some nontoxigenic strains previously identified as *Clostridium subterminale* or *Clostridium hastiforme*. Int. J. Syst. Bacteriol. **38**:375–381.

317. **Summanen, P.** 2000. Comparison of effects of medium composition and atmospheric conditions on detection of *Bilophila wadsworthia* ß-lactamase by cefinase and cefinase plus methods. J. Clin. Microbiol. **38**:733–736.

318. **Summanen, P., E. J. Baron, D. M. Citron, C. Strong, H. M. Wexler, and S. M. Finegold.** 1993. Wadsworth anaerobic bacteriology manual. 5th ed. Star Publishing Company, Belmont, Ca.

319. **Summanen, P., and H. Jousimies-Somer.** 1988. Comparative evaluation of RapID ANA and API-20A for identification of anaerobic bacteria. Eur. J. Clin. Microbiol. Infect. Dis. **7**:771–775.

320. **Summanen, P., H. M. Wexler, and S. M. Finegold.** 1992. Antimicrobial susceptibility testing of *Bilophila wadsworthia* by using triphenyltetrazolium chloride to facilitate endpoint determination. Antimicrob. Agents Chemother. **36**:1658–1664.

321. **Summanen, P. H., M. McTeague, M.-L. Vaisanen, C. A. Strong, and S. M. Finegold.** 1999. Comparison of recovery of anaerobic bacteria using the Anoxomat, anaerobic chamber, and the GasPak jar systems. Anaerobe **5**:5–9.

322. **Summanen, P. H., D. A. Talan, C. Strong, M. McTeague, R. Bennion, J. E. Thompson, Jr., M.-L. Väisänen, G. Moran, M. Winer, and S. M. Finegold.** 1995. Bacteriology of skin and soft tissue infections: comparison of infections in intravenous drug users and individuals with no history of intravenous drug use. Clin. Infect. Dis. **20**:S279–S282.

323. **Sussman, M., S. P. Borriello, and D. J. Taylor.** 1998. Gas gangrene and other clostridial infections, p. 669–697. *In* W. J. Hausler, M. Sussman, and E. Arnold (eds.), Topley and Wilson's Microbiology and Microbial Infections, 9th ed.; Bacterial Infections (vol. 3), London.

324. **Sutter, V. L.** 1984. Anaerobes as normal oral flora. Rev. Infect. Dis. **6**:S62–S66.

325. **Sutter, V. L., and S. M. Finegold.** 1971. Antibiotic disc susceptibility tests for rapid presumptive identification of gram-negative anaerobic bacilli. Appl. Microbiol. **21**:13–20.

326. **Swartz, M. N.** 1989. Central nervous system infections, p. 156–212. *In* S. M. Finegold and W. L. George (eds.), Anaerobic infections in humans. Academic Press, San Diego, Ca.

327. **Sweet, R. L.** 1995. Role of bacterial vaginosis in pelvic inflammatory disease. Clin. Infect. Dis. **20**:S271–S275.

328. **Sweet, R. L., S. Roy, S. Faro, W. F. O'Brien, J. S. Sanfilippo, M. Seidlin, et al.** 1994. Piperacillin and tazobactam versus clindamycin and gentamycin in the treatment of hospitalized women with pelvic infection. Obstet. Gynecol. **83**:280–286.

329. **Swenson, R. M., T. C. Michaelson, M. J. Daly, and E. H. Spaulding.** 1973. Anaerobic bacterial infections of the female genital tract. Obstet. Gynecol. **42**:538–541.

330. **Talan, D. A., D. M. Citron, F. M. Abrahamian, G. J. Moran, and E. J. Goldstein.** 1999. Bacteriologic analysis of infected dog and cat bites. Emergency medicine animal bite study group. N. Engl. J. Med. **340**:85–92.

331. **Talan, D. A., P. H. Summanen, and S. M. Finegold.** 2000. Ampicillin/sulbactam and cefoxitin in the treatment of cutaneous and other soft-tissue abscesses in patients with and without a history of drug abuse. Clin. Infect. Dis., 31: 464–471.

332. **Talan, D. A., and G. J. Moran.** 1998. Update on emerging infections: news from the Centers for Disease Control and Prevention. Tetanus among injecting-drug users—California. Ann. Emerg. Med. **32**:385–386.

333. **Tanner, A., M. F. J. Maiden, K. Lee, L. B. Schulman, and H. P. Weber.** 1997. Dental implant infections. Clin. Infect. Dis. **25**:S213–S217.

334. **Tanner, A., M. F. J. Maiden, B. J. Paster, and F. E. Dewhirst.** 1994. The impact of 16S ribosomal RNA-based phylogeny on the taxonomy of oral bacteria. Periodontology 2000 **5**:26–51.

335. **Tanner, A., and N. Stillman.** 1993. Oral and dental infections with anaerobic bacteria: clinical features, predominant pathogens, and treatment. Clin. Infect. Dis. **16**:S304–S309.

336. **Tanner, A. C. R., S. Badger, C.-H. Lai, M. A. Listgarten, R. A. Visconti, and S. S. Socransky.** 1981. *Wolinella* gen. nov., *Wolinella succinogenes* (*Vibrio succinogenes* Wolin *et al.*) comb. nov., and description of *Bacteroides gracilis* sp. nov., *Wolinella recta* sp. nov., *Campylobacter concisus* sp. nov., and *Eikenella corrodens* from humans with periodontal disease. Int. J. Syst. Bacteriol. **31**:432–445.

337. **Tannock, G.** 1999. Medical importance of the normal microflora. Kluwer Academic Publishers, Dordrecht, the Netherlands.

338. **Terpenning, M., W. Bretz, D. Lopatin, S. Langmore, B. Dominiquez, and W. Loesche.** 1993. Bacterial colonization of saliva and plaque in the elderly. Clin. Infect. Dis. **16**:S314–S316.

339. **Thadepalli, H., S. L. Gorbach, and L. Keith.** 1973. Anaerobic infections of the female genital tract: Bacteriologic and therapeutic aspects. Am. J. Obstet. Gynecol. **117**:1034–1040.

340. **Thorpe, T. C., M. L. Wilson, J. E. Turner, J. L. DiGuiseppi, M. Willert, S. Mirrett, and L. B. Reller.** 1990. BacT/Alert: an automated colorimetric microbial detection system. J. Clin. Microbiol. **28**:1608–1612.

341. **Trinh, S., A. Haggoud, G. Reysset, and M. Sebald.** 1995. Plasmids pIP419 and pIP421 from *Bacteroides:* 5-nitroimidazole resistance genes and their upstream insertion sequence elements. Microbiology **141(4)**:927–935.

342. **Trinh, S., and G. Reysset.** 1996. Detection by PCR of the nim genes encoding 5-nitroimidazole resistance in *Bacteroides* spp. J. Clin. Microbiol. **34**:2078–2084.

343. **Trinh, S., and G. Reysset.** 1997. Identification and DNA sequence of the mobilization region of the 5-nitroimidazole resistance plasmid pIP421 from *Bacteroides fragilis.* J. Bacteriol. **179**:4071–4074.

344. **Turner, P., R. Edwards, V. Weston, A. Gazis, P. Ispahani, and D. Greenwood.** 1995. Simultaneous resistance to metronidazole, co-amoxiclav, and imipenem in clinical isolate of *Bacteroides fragilis* [see comments]. Lancet **345**:1275–1277.

345. **Turton, L. J., D. B. Drucker, V. F. Hillier, and L. A. Ganguli.** 1983. Effect of eight growth media upon fermentation profiles of ten anaerobic bacteria. J. Appl. Bacteriol. **54**:295–304.

346. **Uematsu, H., F. Nakazawa, T. Ikeda, and E. Hoshino.** 1993. *Eubacterium saphenus* sp. nov., isolated from human periodontal pockets. Int. J. Syst. Bacteriol. **43**:302–304.

347. **Vandamme, P., M. I. Daneshvar, F. E. Dewhirst, B. J. Paster, K. Kersters, H. Goossens, and C. W. Moss.** 1995. Chemotaxonomic analyses of *Bacteroides gracilis* and *Bacteroides ureolyticus* and reclassification of *B. gracilis* as *Campylobacter gracilis* comb. nov. Int. J. Syst. Bacteriol. **45**:145–152.

348. **Väisänen, M.-L., M. Kiviranta, P. Summanen, S. M. Finegold, and H. R. Jousimies-Somer.** 1997. *Porphyromonas endodontalis*-like organisms isolated from extraoral sources. Clin. Infect. Dis. **23**:S191–S193.

349. **von Konow, L., P. A. Köndell, C. E. Nord, and A. Heimdahl.** 1992. Clindamycin versus phenoxmethylpenicillin in the treatment of acute orofacial infections. Eur. J. Clin. Microbiol. **11**:1129–1136.

350. **Wade, W. G., J. Downes, D. Dymock, S. J. Hiom, A. J. Weightman, F. E. Dewhirst, B. J. Paster, N. Tzellas, and B. Coleman.** 1999. The family *Coriobacteriaceae:* reclassification of *Eubacterium exiguum* (Poco et al. 1996) and *Peptostreptococcus heliotrinreducens* (Lanigan 1976) as *Slackia exigua* gen. nov. and *Slackia heliotrinreducens* gen. nov., comb. nov., and *Eubacterium lentum* (Prevot 1938) as *Eggerthella lenta* gen. nov., comb. nov. Int. J. Syst. Bacteriol. **49**:595–600.

351. **Wade, W. G., J. Downes, M. A. Munson, and A. J. Weightman.** 1999. *Eubacterium minutum* is an earlier synonym of *Eubacterium tardum* and has priority. Int. J. Syst. Bacteriol. **49**:1939–1941.

352. **Wade, W. G., D. A. Spratt, D. Dymock, and A. J. Weightman.** 1997. Molecular detection of novel anaerobic species in dentoalveolar abscesses. Clin. Infect. Dis. **25**:S235–S236.

353. **Wayman, B. E., S. M. Murata, R. J. Almeida, and C. B. Fowler.** 1992. Bacteriological and histological evaluation of 58 periapical lesions. J. Endod. **18**:152–155.

354. **Weinberg, L. G., L. L. Smith, and A. H. McTighe.** 1983. Rapid identification of the *Bacteroides fragilis* group by bile disk and catalase tests. Lab. Med. **14**:785–788.

355. **Weinstein, M. P.** 1996. Current blood culture methods and systems: clinical concepts, technology, and interpretation of results. Clin. Infect. Dis. **23**:40–46.

356. **Weinstein, M. P., M. L. Towns, S. M. Quartey, S. Mirrett, L. G. Reimer, G. Parmigiani, et al.** 1997. The clinical significance of positive blood cultures in the 1990s: a prospective comprehensive evaluation of the microbiology, epidemiology and outcome of bacteremia and fungemia in adults. Clin. Infect. Dis. **24**:584–602.

357. **Wexler, H. M., E. Molitoris, and S. M. Finegold.** 1991. Effect of ß-lactamase inhibitors on the activities of various ß-lactam agents against anaerobic bacteria. Antimicrob. Agents Chemother. **35**:1219–1224.

358. **Wexler, H. M., E. Molitoris, P. R. Murray, J. Washington, R. J. Zabransky, P. H. Edelstein, and S. M. Finegold.** 1996. Comparison of spiral gradient endpoint and agar dilution methods for susceptibility testing of anaerobic bacteria: a multilaboratory collaborative evaluation. J. Clin. Microbiol. **34**:170–174.

359. **Wexler, H. M., D. Reeves, P. H. Summanen, E. Molitoris, M. McTeague, J. Duncan, K. H. Wilson, and S. M. Finegold.** 1996. *Sutterella wadsworthensis* gen. nov., sp. nov., bile-resistant microaerophilic *Campylobacter gracilis*-like clinical isolates. Int. J. Syst. Bacteriol. **46**:252–258.

360. **Whiley, R. A., and D. Beighton.** 1998. Current classification of oral streptococci. Oral Microbiol. Immunol. **13**:195–216.

361. **Whitehead, S. M., R. D. Leach, S. J. Eykyn, and I. Phillips.** 1982. The aetiology of perirectal sepsis. Br. J. Surg. **69**:166–168.

362. **Wideman, P. A., D. M. Citronbaum, and V. L. Sutter.** 1977. Simple disk test for detection of nitrate reduction by anaerobic bacteria. J. Clin. Microbiol. **5**:315–319.

363. **Wideman, P. A., V. L. Vargo, D. Citronbaum, and S. M. Finegold.** 1976. Evaluation of the sodium polyanethol sulfonate disk test for the identification of *Peptostreptococcus anaerobius.* J. Clin. Microbiol. **4**:330–333.

364. **Wilks, M., R. N. Thin, and S. Tabaqchali.** 1984. Quantitative bacteriology of the vaginal flora in genital disease. J. Med. Microbiol. **18**:217–231.

365. **Willems, A., M. Amat-Marco, and M. D. Collins.** 1996. Phylogenetic analysis of *Butyrivibrio* strains reveals three distinct groups of species within the *Clostridium* subphylum of the gram-positive bacteria. Int. J. Syst. Bacteriol. **46**:195–199.

366. **Willems, A., and M. D. Collins.** 1995. 16S rRNA sequences indicate that *Hallella seregens* (Moore and Moore) and *Mitsuokella dentalis* (Haapasalo et al.) are genotypically highly related and are members of the genus *Prevotella:* Emended description of the genus *Prevotella* (Shah and Collins) and description of *Prevotella dentalis* comb. nov. Int. J. Syst. Bacteriol. **45**:832–836.

367. **Willems, A., and M. D. Collins.** 1995. Evidence for the placement of the gram-negative *Catonella morbi* (Moore and Moore) and *Johnsonella ignava* (Moore and Moore) within the *Clostridium* subphylum of the gram-positive bacteria on the basis of 16S rRNA sequences. Int. J. Syst. Bacteriol. **45**:855–857.

368. **Willems, A., and M. D. Collins.** 1995. Phylogenetic analysis of *Ruminococcus flavefaciens,* the type species of the genus *Ruminococcus,* does not support the reclassification of *Streptococcus hansenii* and *Peptostreptococcus productus* as ruminococci. Int. J. Syst. Bacteriol. **45**:572–575.

369. **Willems, A., and M. D. Collins.** 1995. Phylogenetic placement of *Dialister pneumosintes* (formerly *Bacteroides pneumosintes*) within the *Sporomusa* subbranch of the *Clostridium* subphylum of the gram-positive bacteria. Int. J. Syst. Bacteriol. **45**:403–405.

370. **Willems, A., and M. D. Collins.** 1995. Reclassification of *Oribaculum catoniae* (Moore and Moore 1994) as *Porphyromonas catoniae* comb. nov. and emendation of the genus *Porphyromonas.* Int. J. Syst. Bacteriol. **41**:578–581.

371. **Willems, A., and M. D. Collins.** 1996. Phylogenetic relationships of the genera *Acetobacterium* and *Eubacterium* sensu stricto and reclassification of *Eubacterium alactolyticum* as *Pseudoramibacter alactolyticus* gen. nov., comb. nov. Int. J. Syst. Bacteriol. **46**:1083–1087.

372. **Willems, A., W. E. Moore, N. Weiss, and M. D. Collins.** 1997. Phenotypic and phylogenetic characterization of some *Eubacterium*-like isolates containing a novel type B wall murein from human feces: description of *Holdemania filiformis* gen. nov., sp. nov. Int. J. Syst. Bacteriol. **47**:1201–1204.

373. **Williams, B. L., G. F. McCann, and F. D. Schoenknecht.** 1983. Bacteriology of dental abscesses of endodontic origin. J. Clin. Microbiol. **18**:770–774.

374. **Wilson, K. H., J. S. Ikeda, and B. Rhonda.** 1997. Phylogenetic placement of community members of human colonic biota. Clin. Infect. Dis. **25**:S114–S116.

375. **Wilson, M. L., M. P. Weinstein, L. G. Reimer, S. Mirrett, and L. B. Reller.** 1991. Controlled comparison of BacT/Alert and BACTEC NR 660/730 blood culture systems. Abstr. Annu. Meet. Am. Soc. Microbiol. **C-365**:403.

376. **Wong, S. S., P. C. Woo, W. K. Luk, and K. Y. Yuen.** 1999. Susceptibility testing of *Clostridium difficile* against metronidazole and vancomycin by disk diffusion and Etest. Diagn. Microbiol. Infect. Dis. **34**:1–6.

377. **Wüst, J., S. Stubbs, N. Weiss, G. Funke, and M. D. Collins.** 1995. Assignment of *Actinomyces pyogenes*-like (CDC coryneform group E) bacteria to the genus *Actinomyces* as *Actinomyces radingae* sp. nov. and *Actinomyces turicensis* sp. nov. Lett. Appl. Microbiol. **20**:76–81.

378. **Wyss, C.** 1989. Dependence of proliferation of *Bacteroides forsythus* on exogenous *N*-acetylmuramic acid. Infect. Immun. **57**:1757–1759.

379. **Yamamoto, T., S. Kajiura, and T. Watanabe.** 1994. *Capnocytophaga haemolytica* sp. nov. and *Capnocytophaga granulosa* sp. nov. from human dental plaque. Int. J. Syst. Bacteriol. **44**:324–329.

380. **Yamazoe, K., N. Kato, H. Kato, K. Tanaka, Y. Katagiri, and K. Watanabe.** 1999. Distribution of the cfiA gene among *Bacteroides fragilis* strains in Japan and relatedness of cfiA to imipenem resistance [In Process Citation]. Antimicrob. Agents Chemother. **43**:2808–2810.

381. **Ziegler, R., I. Johnscher, P. Martus, D. Lenhardt, and H.-M. Just.** 1998. Controlled clinical laboratory comparison of two supplemented aerobic and anaerobic media used in automated blood culture systems to detect bloodstream infections. J. Clin. Microbiol. **36**:657–661.

Index

Note: *t* = tables/boxes; *f* = figures

B

Index

Index

C

Index

Index

F

G

Index

N

Index

Index

Q

R

S

Index

T

U

V

Index

W

Z